A Primer in Photoemission: Concepts and Applications

Antonio Tejeda

Daniel Malterre

This book has been edited with the support of Laboratoire de Physique des Solides, Institut Jean Lamour, SOLEIL synchrotron and Labex Palm.

Cover illustration: from left to right, graphene bands, schematics of a photoemission process and Fermi surface of bismuth.

ISBN(print): 978-2-7598-2065-8 – ISBN(ebook): 978-2-7598-2391-8

To our recently deceased fathers
Antonio and Daniel

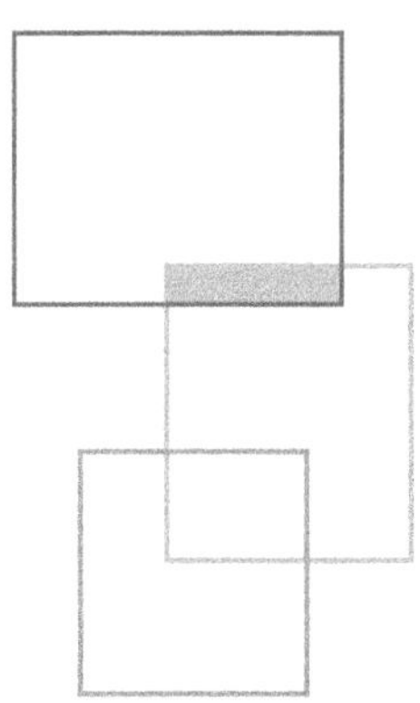

Table of Contents

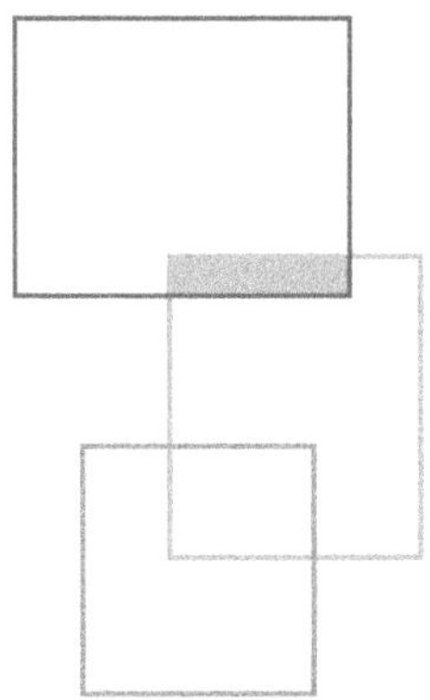

Preface

Photoemission plays a key role in the study of the properties of new materials and therefore, the work proposed by Antonio Tejeda and Daniel Malterre is particularly timely, given the highly dynamic nature of the field. The book will allow the reader to have a comprehensive and in-depth view of the techniques involved, both conceptually and experimentally.

The dynamism of the field is obvious if we mention, for example, the contribution of photoemission to our understanding of the structure and properties of promising materials such as graphene (or its close cousins, silicene and germanene) or topological insulators, just to name only those giving rise to the greatest number of publications nowadays.

The development of the technique has been further strengthened with the development of new sources (high-brightness synchrotrons, ultra-short wavelength lasers), electron detectors for very large angular apertures or operating under conditions around the sample approaching the "real world" (*in situ* studies, *in operando* conditions). For example, at the SOLEIL Synchrotron, our development strategy led us to proposing a large number of experimental photoemission setups to observe time-dependent phenomena (from the second to the femto-second), to analyse samples under ultra-high vacuum conditions up to nearly ambient pressure, to probe the electronic structure of the samples down to the nanoscale in order to ultimately correlate the fundamental properties at the atomic scale with the macroscopic properties.

In the first part of the book, the authors retrace the history of photoemission, recalling the basic concepts, with a detailed analysis of the experimental aspects. They then take us, with judiciously chosen examples, to a deepening of the most contemporary concepts, which will make this book a foremost reference for both student and confirmed research scientists.

Paul Morin
Scientific Director
SOLEIL Synchrotron, Orsay

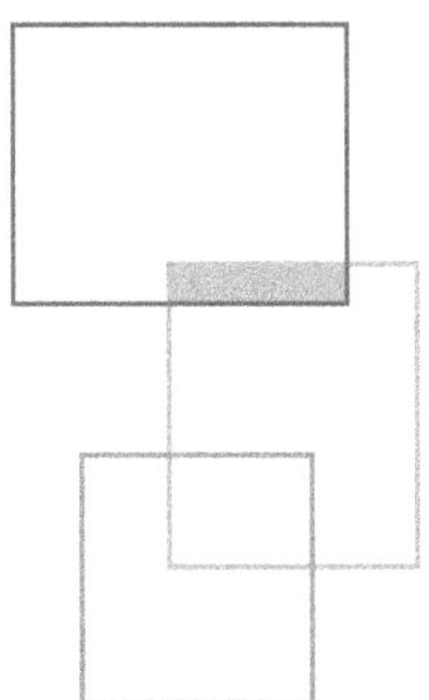

Acknowledgments

We are extremely grateful to all our colleagues and friends that have kindly read this book and have allowed us to improve it before sending it to production. Patrick Le Fèvre has reviewed the chapter on the elementary approach of photoemission. Marco Grioni has focused on the chapter on the basic concepts. The chapter on technical aspects called for input from different readers depending on their speciality: Yannick Fagot-Révurat for conventional photon sources, Marie-Agnès Tordeux for synchrotron radiation, Evangelos Papalazarou, Marino Marsi and Luca Perfetti for time-resolved photoemission, Fausto Sirotti for aspects related to beam lines and the carbon contamination of mirrors, Catalin Miron for electron detectors and François Bertran for spin detection. Amina Taleb-Ibrahimi and Enrique García Michel have concentrated on the chapter on transitions from localized states, Jose-Ángel Martín Gago on that covering photoelectron diffraction and Véronique Brouet on the one of the dispersion relations. Véronique Brouet has also reviewed the technical aspects on the use of bidimensional detectors. The appendices on the quality of the fitting of the core levels and on the determination of the Fermi level have been reviewed by Enrique García Michel. We extend our warm thanks to Olivier Marcouillé and Fabien Briquez for the figures and the calculations of the insertion device brilliance and the flux of the bending magnet. We are also grateful to Guillaume Vasseur and Simon Moser for the figures on the matrix element effects extracted from their PhD manuscript.

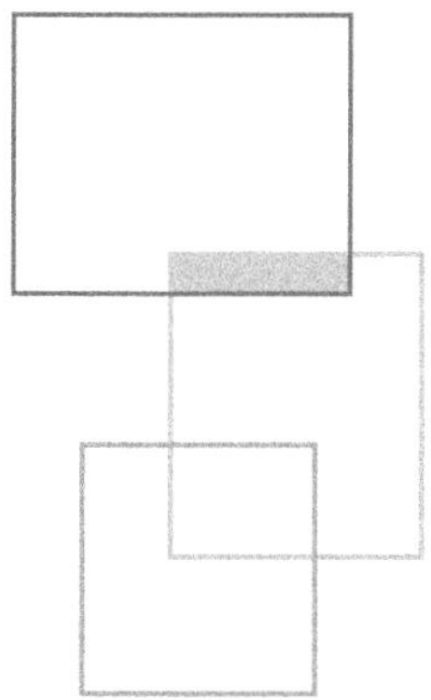

Introduction

Photoemission is the main technique used to study the different subtleties of the electronic structure in materials. From a fundamental point of view, photoemission allows the experimental determination of the band structure. It also provides unique information on many body effects, as has been demonstrated in the recent decades in a host of studies on correlated systems, high temperature superconductors or Mott insulators. Photoemission also gives useful information for applications, as it reports on the physico-chemical properties of surfaces, such as the composition, the degree of oxidation of the different elements, etc. Technological objects such as Schottky barriers in metal-semiconductor interfaces have thus been studied by photoemission. This technique is an extremely valuable spectroscopy for surface physics and nanosciences.

Two Nobel prizes have been awarded for researches related to photoemission. The first was in 1921, when Albert Einstein finally explained the photoelectric effect. The second was awarded on 1981 to Kai Siegbahn for the use of photoelectrons in the chemical analysis of materials. In the last years, photoemission has become a well-stablished technique. Several technical issues have contributed to its spectacular development: the intensity of the new synchrotron sources, the improvement of the beam line optics and the electron detectors. Photoemission can nowadays observe subtle effects associated to very low energy scales. Moreover, new research avenues have been opened to photoemission spectroscopy in recent years. There are now spin electron detectors for gaining insight in the electronic states with spin resolution, and pulsed lasers for time-resolved photoemission. There has also been significant progress on the spatial resolution, which has allowed to develop the photoemission microscopy, combining both structural and spectroscopic information.

This book reviews the basic photoemission concepts while introducing most of the recent developments. The first chapter is based on the photoemission history. It describes how the puzzle of understanding the photoelectric effect was finally solved by the development of the quantum physics. Once the phenomenon was understood, photoemission spectroscopy could be developed when the technical issues, concerning mainly the electron detection, were solved. The second chapter is dedicated to the elementary description of the basic concepts of photoemission without the mathematical formalism. The formalism is developed later in the third chapter, that describes the photoemission concepts, the many-body description of the process, and the detailed analysis of both valence band and core level states. After this chapter, we concentrate on the experimental aspects. Here we describe the tools needed to perform a photoemission experiment, such as the photon sources or the electron detectors. This chapter on the experimental techniques concludes the first part on the general issues on photoemission, followed by a second part setting out examples of photoemission analysis. The first of these chapters is dedicated to core levels. Electrons in these states are very localized. Their wave functions and their energies are very similar to those of an isolated atom. This property is the basis of the core level spectroscopies, as the binding energies of the core levels are a signature of the chemical nature of the emitter. The following chapter presents examples of the photoelectron diffraction, related to the wave-particle duality of photoelectrons diffracting on the atomic lattice. In this way, the anisotropy in the photoemission intensity allows to obtain structural information of the emitter environment. The next chapter shows illustrations on the experimental determination of the band structure, as well as on more subtle information concerning the electronic structure, as those related to the interacting mechanisms between electrons and the collective excitations in a crystal (phonons, magnons, *etc.*). Some technical appendices follow to this last chapter.

Part I

Concepts

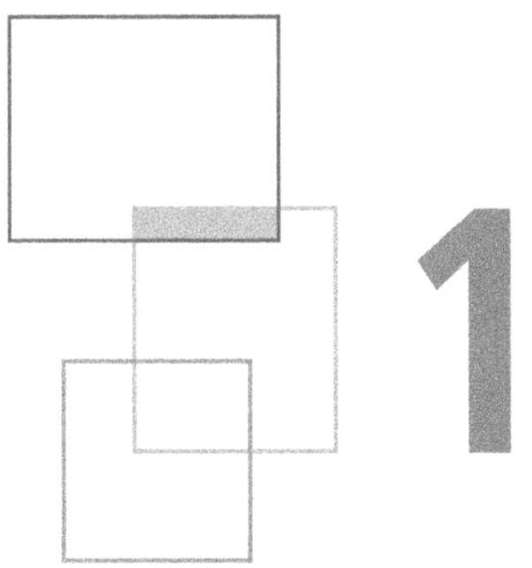

1 The history of photoemission

The twentieth century was an important period for the technological development of humanity that has experienced several fundamental revolutions. Physics is one of the causes of this progress. In particular, electronic devices have transformed our society. Such an advance was made possible thanks to solid state physics and to the appropriate tailoring of material properties, but also through the use of the most modern concepts. For instance, transistors, the building blocks of electronics, cannot be understood without applying quantum mechanics to solids.

Photoemission is a technique allowing us to study the electronic states in solids. It allows us to determine the band structure, which is essential to the understanding of the electronic properties of matter and therefore the electron transport in devices (diodes, transistors). It also provides access to the electronic core levels, characteristic of each atomic element and allows us to determine the chemical composition of a material. The aim of this chapter is to present the history of photoemission and in particular the fundamental principle at the origin of this spectroscopy. Photoemission is a technique derived from the photoelectric effect, which has been a riddle of physics before it was explained by Albert Einstein in 1905. The explanation is based on the quantification of the electromagnetic field and has contributed significantly to the development of quantum physics.

1.1 Origin of photoemission: the photoelectric effect

The discovery of the photoelectric effect by Heinrich Hertz in 1887 was a matter of chance. Hertz had designed an experiment to demonstrate the propagation of the electromagnetic waves predicted by Maxwell theory. He had built an oscillating circuit acting as an emitter and also a receptor consisting of two hollow metallic spheres (Fig. 1.1). Upon reception of an electromagnetic wave, a spark appeared between the spheres. During the experiment, Hertz observed an unexpected effect: the intensity of the spark varied with the light shining on the device. We know today that ultraviolet radiation, that favors electron emission, stimulates the discharge and therefore leads to a more intense spark. Hertz also positioned different materials acting as filters between the source and the detector. In this way he demonstrated that the effect was due to ultraviolet radiation waves.

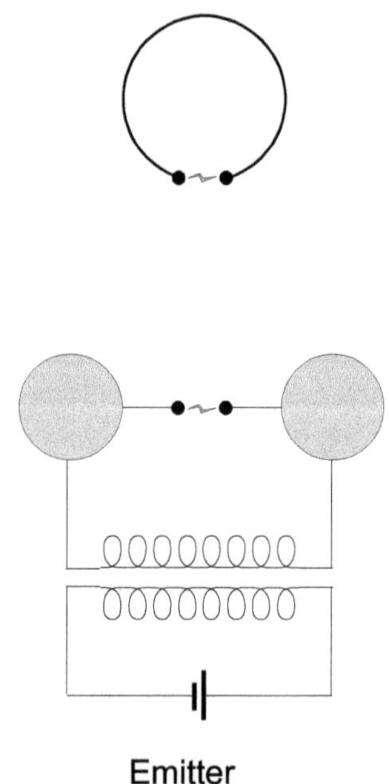

Figure 1.1 *Experimental device used by Hertz to make the propagation of electromagnetic waves explicit.*

A year later, Wilhelm Hallwachs studied how a zinc plate, initially charged, was discharged when submitted to illumination [1]. By connecting it to an electroscope, he studied the discharge kinetics in the presence of light (Fig. 1.2). He was able to show that a negatively charged plate is discharged more quickly when illuminated with ultraviolet light. On the contrary, if the plate is positively charged, no quick discharge was observed.

Later, Philipp Lenard showed that the negatively charged materials emit electrons [2]. He illuminated a metallic cathode and measured the number of electrons reaching the anode depending on the voltage V between the cathode and the anode (Fig. 1.3). For positive V, all electrons emitted from the cathode by photoelectric effect are attracted by the anode and a current is measured. However, for a negative voltage, only electrons with enough kinetic energy to overcome the anode repulsion contribute to the current. The electric potential that cancels the detected current because it repels all the emitted electrons towards the cathode is called the stopping potential. The surprising result is that the photoelectric effect depends on the frequency of the light and not on its intensity. Indeed, the intensity affects only the number of photoemitted electrons and not their kinetic energy. This behavior, *i.e.*, the existence of a threshold frequency, was an enigma under Maxwell's electromagnetism framework that describes waves by electromagnetic fields, *i.e.*, by continuous quantities. Within Maxwell's theory, the energy depends on the square of the field amplitude and varies continuously so it is impossible to understand the existence of a threshold frequency.

It was Albert Einstein who finally explained the photoelectric effect. He applied Planck's ideas to explain the black body radiation (1900)[1]. The spectral distribution of the black body, that is the energy density emitted as a function of the frequency or the wavelength, is a universal curve characterized by a maximum which depends on the temperature. This distribution is used in all the fields of physics. In cosmology, it led, in a Big Bang scenario, to the prediction of a fossil radiation (radiation at 3K) resulting from the cooling of the universe due to its expansion. This 3K radiation was observed by Penzias and Wilson in 1964. Classical physics is unable to describe this black body radiation since thermodynamic arguments lead to the Rayleigh-Jeans law:

$$I(\lambda, T) = 8\pi kT\lambda^{-4}. \quad (1.1)$$

If the agreement with experiments for large wavelengths (small frequencies) is very satisfactory, this law indicates a non-physical divergence for small wavelengths (high frequencies). Experiments on the contrary show that the intensity tends towards zero as is expected in order that the radiated energy remains finite. To explain the experimental behavior, Max Planck offered the postulate that the energy exchanges between the radiation and the black body were quantified, the quantum of energy being proportional to the frequency ν of the radiation :

$$E = h\nu. \quad (1.2)$$

The constant $h = 6.62 \times 10^{-34} Js$ is known since then as the Planck's constant. Albert Einstein pursued with Planck's idea and went much further in proposing that the electromagnetic field itself was quantified. The quantum $h\nu$ become the

[1] A black body is a system that absorbs all the light it receives and emits an electromagnetic spectrum that depends only on the temperature.

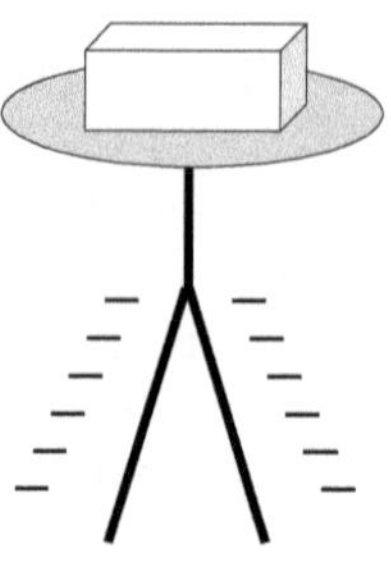

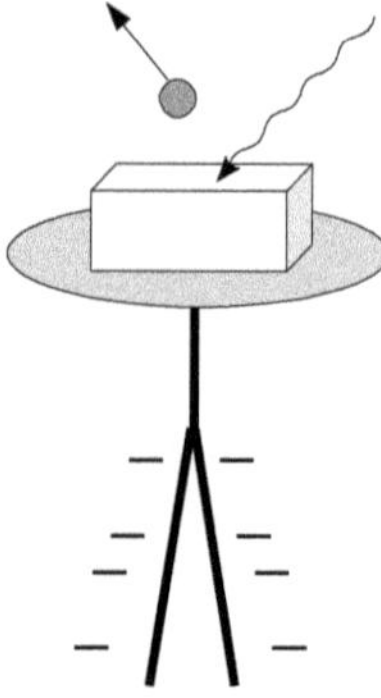

Figure 1.2 *Principle of Hallwachs experiment with an electroscope. The angle between the foil leaves of the electroscope depends on the charge accumulated there. When the top zinc plate is illuminated with ultraviolet radiation, the electroscope discharges, proving that electrons are emitted.*

"grain of light" later called photon. Einstein then showed that the hypothesis of the grain of light allows a simple and elegant description of the photoelectric effect. This explanation is based on the energy conservation. In order for an electron to exit a solid, it must acquire an energy Φ called the work function. This energy is provided by the photon and the excess energy appears as the kinetic energy of the photoemitted electron:

$$E_c = h\nu - \Phi. \quad (1.3)$$

This equation is called Einstein's equation[3]. It makes explicit that it is the frequency and not the intensity which is relevant in the photoelectric effect.

The model developed by Einstein has given rise to various experimental studies including those of Millikan[4, 5]. Millikan carefully determined the repelling voltage as a function of the frequency. The repelling voltage has a linear dependency that does not pass through the origin (figure 1.3). This reflects the fact that electrons the energy of which is lower than the work function cannot be extracted from the metal. From this experience, it was also possible to calculate the h/e ratio. The photoelectric

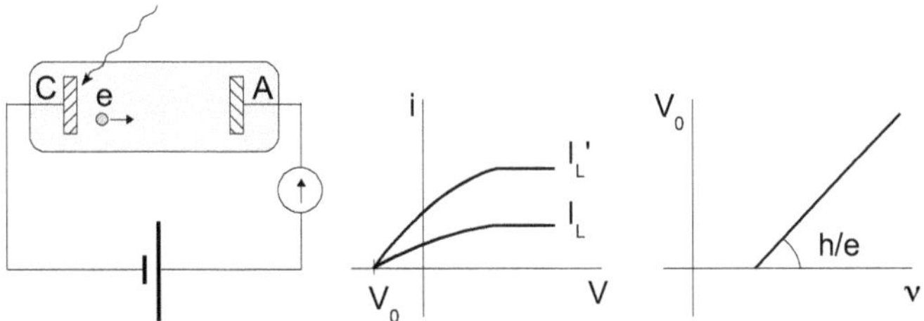

Figure 1.3 *(Left) Lenard's experience: a cathode is illuminated by UV radiation and the current between the cathode and the anode is measured. (Centre) Current i can be compensated by a repelling voltage V_0. In addition, the increase in light intensity $I'_L > I_L$ leads to the increase of current i. (Right) The repelling voltage V_0 as a function of the light frequency is a straight line whose slope gives the h/e ratio.*

effect is therefore a manifestation of the quantum nature of the electromagnetic field, and in particular of the concept of the grain of light (photon).

1.2 Core level spectroscopy

Quantification, firstly introduced for the electromagnetic field, was extended later to the atoms by Niels Bohr (1913). Indeed, classical physics cannot describe the stability of atoms since accelerated charges, like electrons in an atom, must radiate. The electrons should therefore lose energy very quickly and fall back to the nucleus. Moreover, experiments show that the emission spectrum of atoms exhibits discrete lines while in a classical approach a continuous spectrum is expected. This led Niels Bohr to propose the existence of stable orbits in atoms and therefore a discrete emission spectrum corresponding to the transitions between these stable orbits. The emitted frequency is associated to the energy difference between electronic states $(E_i - E_f)$:

$$\nu = \frac{E_i - E_f}{h}. \tag{1.4}$$

The stability condition of the orbits has been obtained from the quantification of the kinetic moment of the electron. The Bohr model of the atom therefore allowed us to explain the spectral lines of the hydrogen atom.

For atoms with several electrons, this simple model does not describe the emission spectra any more. However, quantum mechanics, developed in the 1920s, highlighted the existence of shells and subshells in the atomic levels which explained finally the emission spectra of complex atoms. It should also be noted that the spectral lines are characteristic of each atom, allowing us to exploit them in a spectroscopic technique for the analysis of materials. This sensitivity to the chemical nature of the elements is the basis of Photoelectron Spectroscopy (PES), also known as ESCA

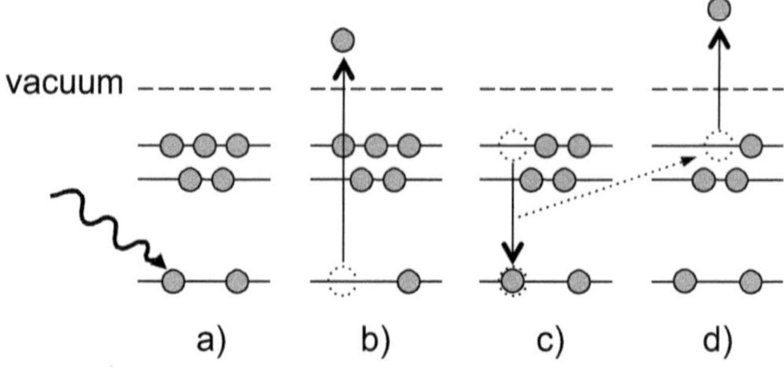

Figure 1.4 *Principle of Auger spectroscopy. A photon is absorbed by a core level electron (a) which is photo-emitted (b). An electron of a higher electronic level fills the hole left by the electron emission (c) in order to reduce the system energy. The energy difference between the two levels is given to a third electron that escapes from the atom (Auger electron).*

(Electron Spectroscopy for Chemical Analysis), since a photoemission spectrum, *i.e.*, the intensity diagram of the photoemitted electrons as a function of the binding energy, allows a chemical analysis of the studied system. In the early years of the photoemission, core levels were studied in detail, in what is called X-ray Photoemission Spectroscopy (XPS). In the spectra, there are also some unexplained lines that do not correspond to any binding energy from the core levels. These lines have been explained after Pierre Auger's experiments on the study of cosmic radiation in a cloud chamber [6, 7]. These experiments highlighted the electron emission resulting from electronic transitions (now called Auger transitions) that are independent of the energy of the cosmic radiation. The Auger transitions result from the relaxation of the hole induced by the photoemission process. Indeed, the emission of a photoelectron creates a hole in an electronic state of the atom. This situation corresponds to an excited state of the atom that it is very unstable. An electron of an electronic layer of higher energy can fill the initial electronic hole to decrease the energy of the atom. The energy earned by the atom is transferred to another electron that is ejected from the atom and that can be detected (Fig. 1.4). Auger transitions are also characteristic of the emitter atom and are at the base of a specific spectroscopy called Auger spectroscopy. The energy of the Auger electron does not depend on the energy of the primary exciting photon but only on the energies of the different energy levels involved in the process.

In the very early days of photoemission, the quality of the spectra was not good enough to make it a powerful spectroscopic technique. Later, photoemission underwent a major development thanks to Kai Siegbahn and his collaborators in Uppsala, Sweden, who were building detectors to analyze the electrons emitted during the decay of radioactive elements. In the 1950s they applied these detectors to the photoelectron analysis [8]. The resolution of these detectors allowed the scientists to accurately determine the kinetic energy of electrons and measure small energy differences due to the chemical environment of the emitter. Indeed, the core level energy depends on the oxidation degree of the photo-emitter atom. This sensitivity allows

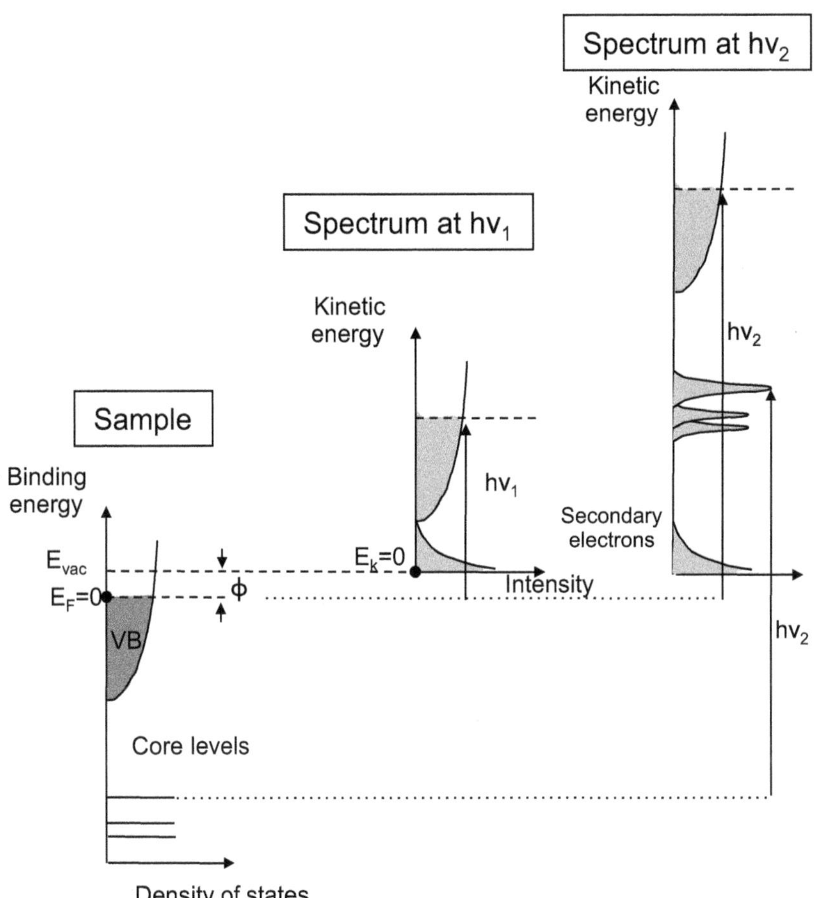

Figure 1.5 *In a solid, there are two types of electrons. There are localized core electrons of atomic nature and also valence band electrons that participate to the chemical bond. The binding energies of the core levels are high while those of the valence states are of the order of just few eV. The photoemission allows to probe these two types of levels by tuning the photon energy. High photon energies ($h\nu_2$) allow to probe both valence and core states while low photon energies ($h\nu_1$) can only probe valence states. In addition, some electrons lose part of their energy (secondary electrons) and contribute to the low kinetic energy spectrum.*

this spectroscopy to study chemical reactions, hence the name Electron Spectroscopy for Chemical Analysis (ESCA) given by Siegbahn.

In one of these experiments, Siegbahn *et al.* observed on NaCl(001) a considerable increase in intensity along the high symmetry directions [9, 10]. They explained the phenomenon by the focusing of electrons along atomic rows. This interpretation was based on a mechanism already known in electron microscopy where the transmitted intensity increases along the dense crystallographic directions (Kikuchi bands).

Later research indicated that these electrons came from the surface [11, 12] and also determined the details of their angular dispersion, leading to a new interpretation. The intensity increase observed in some directions is indeed due to photoelectron diffraction by the atoms surrounding the emitter. The theoretical description of the photoelectron diffraction (PED or PhD) dates from the 70's [13] and was soon used to determine the adsorption sites and the orientation of molecules [14, 15].

Siegbahn's contribution to high resolution electron spectroscopies was therefore extremely valuable. Photoemission spectroscopy of core levels and photoelectron diffraction are still widely used techniques today. Siegbahn was awarded the Nobel Prize in Physics in 1981.

1.3 Band structure

Photoemission is not limited to core level studies. It can also analyze the electrons involved in the chemical bond, *i.e.* the valence electrons (Fig. 1.5). The wave functions of the valence electrons combine to form delocalized states in periodic crystals, that are called Bloch states. A Bloch state corresponds to the coherent propagation of an electronic wave characterized by a wave vector. These states form a continuum (a band) that determine the electronic properties of the system, for example, the electronic transport. We will see below that the wave vector is determined by the emission angle of the photoelectrons and that it is particularly important to measure it accurately.

The experimental determination of the band structure by photoemission was obtained for the first time on copper by Spicer's team [16]. Such a determination is possible thanks to a remarkable property: the conservation of the wave vector parallel to the surface, which is the equivalent, for electrons, to the Descartes refraction law in optics. Therefore we have the conservation laws of the energy and the parallel wave vector which allow us to obtain the band structure in angle-resolved photoemission experiments [17, 18]. Indeed, from the measurement of the energy and the angle of the photoelectron, it is possible to determine the energy and the wave vector of the initial state of the electron and therefore to obtain the dispersion relationships (the band structure). We will see later that ultraviolet photons are often more suitable for this type of study than X-rays and the expression coined was Ultraviolet Photoemission Spectroscopy (UPS). When the angular resolution is used, the technique is called ARUPS or ARPES (Angle Resolved PES).

Finally, it should be noted that the average free path of electrons (average distance between two inelastic collisions) is small in solids. With the photon energies in the UV or in the soft X-ray range, the photoelectrons come from the atomic planes in the close vicinity to the surface. Photoemission is therefore a surface-sensitive technique and it is necessary to master the preparation and the cleaning of the surfaces, which requires the ultra-high vacuum technology ($<10^{-9}$ Torr).

The theory of photoemission was developed in the 1970s. The rigorous description is a theory describing the quantum process in a single-step [19–24]. However, a more intuitive model has been developed. It decomposes the photo-emission process into three independent steps (photon absorption, photoelectron transport to the surface, escape of the electron into the vacuum) [25–29]. This three-step model is still used today, now that photoemission has become a mature spectroscopy for the fine study of the electronic properties of materials.

1.4 Photoemission: a standard technique for the study of electronic properties

We recalled that photoemission spectroscopy requires ultra-high vacuum technology, radiation sources and high-performance detectors. Photoemission benefitted certainly from the development of synchrotrons in the early 1980s as photon sources. Synchrotron radiation, generated by relativistic charges stored in rings, is characterized by high brightness, high-beam focusing, high polarization and the possibility of adjusting the photon energy. More recently it has also allowed scientists to perform time-resolved experiments.

Since the discovery of the photoelectric effect at the end of the 19th century, photoemission has become a widespread technique, with commercial sources and sensors, of very high resolution. Photoemission offers many different possibilities. It can be used for analyzing core levels, band structure and even for obtaining structural information by photoelectron diffraction. XPS on core levels provides access to chemical surface analysis and ARPES is used to determine the band structure. Photoemission not only gives access to the energy, to the wave vector and the spin but also to certain symmetries, such as the parity of the Bloch states. Recently, the photoemission has further enriched its palette of possibilities with the appearance of photoemission microscopy, the use of new sources, such as lasers, or photoemission at near atmospheric pressures.

Photoemission has become a rich and mastered technique that provides essential information for a number of topics. It has strongly contributed to the development of materials science through its chemical sensitivity allowing us to determine the composition and it continues to play a key role in the surface chemistry and catalysis. It has also contributed to a better understanding of the fundamental properties in solid state physics by the experimental determination of the band structure. It allows to determine not only the dispersion relations but also the topology of the Fermi surface, which determines many physical properties such as the electronic transport. It also played a central role in the discovery and in the study of the surface states. Its surface sensitivity also makes it a technique of choice for studying interfaces and also low dimensional systems (quasi-unidimensional and quasi-bidimensional). The latter have complex and exotic electronic properties that can be probed by photoemission (charge density waves, Luttinger liquids, etc.). Finally, it is a spectroscopy particularly well adapted to the study of systems with strong electronic correlations. Indeed, its high energy resolution allows to study in detail low energy excitations such as the opening of band gaps in superconductors or in Peierls insulators and allows to probe directly the characteristic energies associated to electron-electron interactions (Hubbard sub-bands or charge excitations in Kondo systems). Given that it is impossible to draw up an exhaustive panorama of all the contributions of the photoemission in this introduction, so we refer readers interested here to the various chapters of this book and to other review books [30–37] that illustrate how much this technique has contributed to the progress of condensed matter physics.

Bibliography

[1] W. Hallwachs, Ann. Physik und Chemie **269**, 301 (1888).
[2] P. Lenard, Ann. Physik **8**, 147 (1902).
[3] A. Einstein, Ann. Physik **17**, 132 (1905).
[4] R. Millikan, Phys. Rev. **7**, 355 (1916).
[5] R. Millikan, Phys. Rev. **7**, 18 (1916).
[6] P. Auger, J. Phys. Radium **6**, 205 (1925).
[7] P. Auger, Ann. Phys. **6**, 183 (1926).
[8] C. Nordling, E. Sokolowski, and K. Siegbahn, Phys. Rev. **105**, 1676 (1957).
[9] K. Siegbahn, U. Gelius, H. Siegbahn, and E. Olson, Phys. Scr. **1**, 272 (1970).
[10] K. Siegbahn, U. Gelius, H. Siegbahn, and E. Olson, Phys. Lett. **32A**, 221 (1970).
[11] W. Egelhoff, Phys. Rev. B **30**, 1052 (1984).
[12] W. Egelhoff, Phys. Rev. Lett. **59**, 559 (1997).
[13] A. Liebsch, Phys. Rev. Lett. **32**, 1203 (1974).
[14] L.-G. Petersson, S. Kono, N. Hall, and C. Fadley, Phys. Rev. Lett. **42**, 1545 (1979).
[15] C. Fadley et al., Surf. Sci. **89**, 52 (1979).
[16] W. Spicer and C. Berglund, Phys. Rev. Lett. **12**, 9 (1964).
[17] G. Gobeli, F. Allen, and E. Kane, Phys. Rev. Lett. **12**, 94 (1964).
[18] E. Kane, Phys. Rev. Lett. **12**, 97 (1964).
[19] G. Mahan, Phys. Rev. Lett. **24**, 1068 (1970).
[20] G. Mahan, Phys. Rev. B **2**, 4334 (1970).
[21] I. Adawi, Phys. Rev. A **134**, 788 (1964).
[22] W. Schaich and N. Ashcroft, Phys. Rev. B **3**, 2452 (1971).
[23] C. Caroli, D. Lederer-Rozenblatt, B. Roulet, and D. Saint-James, Phys. Rev. B **8**, 4552 (1973).
[24] J. Endriz, Phys. Rev. B **7**, 3464 (1973).
[25] B. Feuerbacher and R. Willis, J. Phys. C **9**, 169 (1976).
[26] H. Puff, Phys. Status Solidi **1**, 636 (1961).
[27] C. Berglund and W. Spicer, Phys. Rev. **136**, A1030 (1964).
[28] C. Berglund and W. Spicer, Phys. Rev. **136**, A1044 (1964).
[29] P. Feibelman and D. Eastman, Phys. Rev. B **10**, 4932 (1974).

[30] S. Hüfner, *Photoelectron spectroscopy* (Springer, Heidelberg, 1995).
[31] F. Himpsel, Adv. Phys. **32**, 1 (1983).
[32] B. Feuerbacher, B. Fitton, and R. Willis, *Photoemission and the electronic properties of surfaces* (Wiley, Chichester, 1978).
[33] *Photoemission in solids*, edited by M. Cardona and L. Ley (Springer, Berlin, 1978), Vol. 1.
[34] *Angle resolved photoemission - Theory and current application*, edited by S. D. Kevan (Elsevier, Amsterdam, 1992).
[35] *Solid-State Photoemission and Related Methods: Theory and Experiment*, edited by W. Schattke and M. V. Hove (Wiley-VCH, Weinheim, 2003).
[36] F. Reinert and S. Hüfner, New J. Phys. **7**, 97 (2005).
[37] in *Electron Spectroscopy: Theory, Techniques and Applications II*, edited by C. Brundle and A. Baker (Academic, New York, 1978).

Elementary approach to photoemission

In the previous chapter we saw how the photoelectric effect was explained once the quantification of the electromagnetic field was understood. Today the photoelectric effect has become the basis of a spectroscopic technique: the photoemission spectroscopy. This chapter is devoted to the basic concepts of this spectroscopy without using the mathematical formalism.

2.1 The photoelectron emission process

We have seen that the absorption of light (or more exactly the radiation[1]) by the electrons, whether the material is a solid, a molecule or an atom, can lead to their extraction. This is possible when the radiation provides the energy needed to overcome the energy barrier that binds the electron to the material. The study of photo-emitted electrons allows us to obtain information about the spectrum of the occupied states.

Let us recall that we distinguish two types of electrons: the core electrons localized on an atom and have therefore a strong atomic character, and the valence electrons delocalized in states associated with the bonds and that can extend throughout the system. The valence electrons are therefore the least bound to the solid and form a continuum of energy levels (a band of energy). This band is filled by satisfying Pauli's exclusion principle (only one electron per state) and it is called the valence

[1] The term "light" refers generally to visible radiation. The photoemission uses instead the ultraviolet part of the electromagnetic spectrum as well as the X-rays.

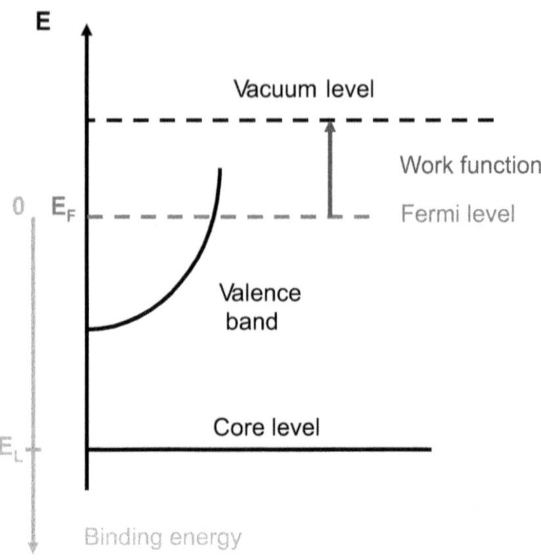

Figure 2.1 *Schematic diagram showing the valence band and a core level state. The diagram defines the different energies: Fermi level, binding energy, work function and the vacuum energy.*

band. The energy of the last occupied state in a metal defines the Fermi level. Very often the binding energy is defined there. The binding energy of a state is therefore the energy required to bring the electron up to the Fermi level. The work function, on the other hand, corresponds to the energy required to extract a electron of the Fermi level and send it to infinity (vacuum level), with zero kinetic energy. The work function value is some eV in solids, so that valence electrons can be photoemitted with UV photons. The core level electrons are more strongly bound to the atoms. The binding energy ranges between few tens of eV to a few tens of keV for the 1s states of heavy atoms. It is therefore necessary to use higher energy photons (X-rays) to extract them. The different quantities defined above are illustrated schematically in figure 2.1.

The binding energy of the core levels is characteristic of every atom. These levels can therefore be used to determine the chemical species that constitute the material by X-ray Photoemission Spectroscopy (XPS). In this way it is possible to determine the composition of a steel or the presence of contaminants on a material. An example is shown on Fig. 2.2, which presents a series of core level spectra. Due to the presence of the 4f lines or the Auger transitions of gold, it can be concluded the presence of gold on the sapphire samples studied there. In some cases it is even possible to highlight differences in the chemical environment of the same type of atom. This is the case for the ethyl trifluoroacetate molecule in which the 1s level of carbon exhibits four structures associated with the four chemically different carbon atoms of the molecule. The energy separation of the different components is called chemical shift.

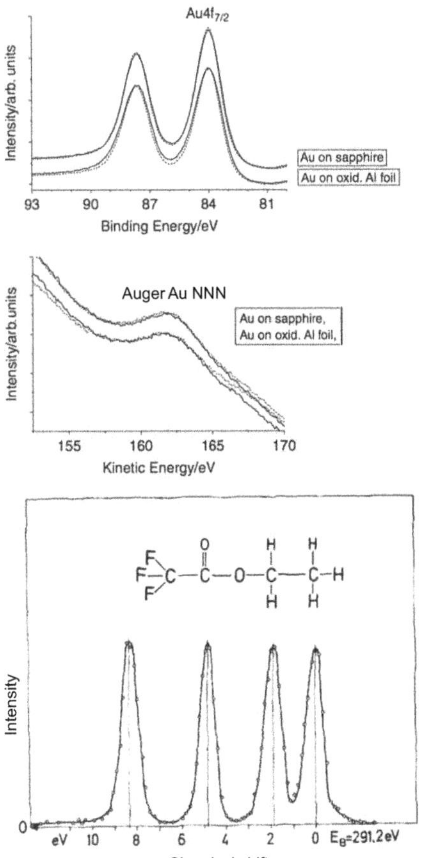

Figure 2.2 *Examples of spectroscopy from localized states. (a) Au 4f core level and (b) Auger NNN transition of a Au film deposited on sapphire and Al oxide on Al [1]. ©2003 John Wiley & Sons, Ltd. (c) C 1s core level in ethyl trifluoroacetate [2]. Reprinted from J. Electron. Spectrosc. Rel. Phenom. Vol. 2, U. Gelius, E. Basilier, S. Svensson, T. Bergmark, K. Siegbahn, "A high resolution ESCA instrument with X-ray monochromator for gases and solids", page 405, ©1974, with permission from Elsevier.*

If the photoemission can be used to study core levels, it also provides information on valence states. Ultraviolet light is here energetic enough to extract electrons. We then speak of Ultraviolet Photoemission Spectroscopy (UPS) which allows to study the valence band states of a solid. The energy of these electrons is often dominated by the kinetic energy contribution $E_{kin} = p^2/2m$. In quantum mechanics, the wave-corpuscle duality associates the momentum of a free particle (corpuscular magnitude) to the wave vector (so-called "undulatory" magnitude):

$$\vec{p} = \hbar\vec{k}.$$

For a free particle, the relationship between energy and wave vector, called dispersion relation, is the quadratic and isotropic function $E(\vec{k}) = \hbar^2 k^2/2m$. In a solid, the interaction with atoms makes this relationship more complex and the knowledge of the dispersion relation $E(\vec{k})$ is essential for understanding the electronic properties of solids. In a photoemission experiment of the valence states, the energy and the wave vector of photoelectrons are determined, the latter from the emission angle of the photoelectron. In this way it is possible to figure out dispersion relations experimentally. This type of measurement, based on the angular resolution, is called Angle Resolved Photoemission Spectroscopy (ARPES) or Angle Resolved Ultraviolet Photoemission Spectroscopy (ARUPS). This technique we will describe in detail in later chapter is undoubtedly one of the best ones to determine the electronic structure of a solid. We illustrate the angle-resolved photoemission on figure 2.3 by showing the dispersion measured on a surface state of Cu(111). This dispersion exhibits an almost free electron behavior, *i.e.*, with a quasi-parabolic dispersion, which is explicit by a dispersion forming a paraboloid of revolution whose constant energy sections form circles.

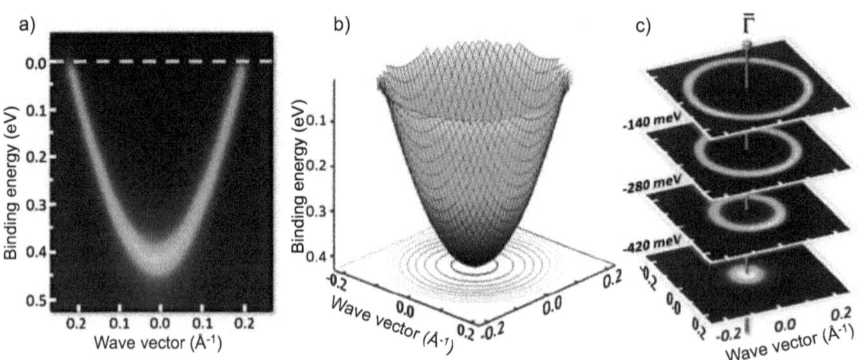

Figure 2.3 *a) Parabolic dispersion ($E(k_x)$) of the Cu(111) surface state along the k_x direction of the reciprocal space. b) Three-dimensional representation of the surface state $E(k_x, k_y)$. c) Constant energy cuts of the electronic structure showing that wave vectors of the same energy form circles.*

We will now continue this brief introduction with technical aspects related to photoemission measurements, followed by a qualitative description of the fundamental mechanisms of the photoemission process. These different aspects will be discussed in more detail in Chapters 3 and 4. Afterwards we will present in a basic way the core level spectroscopy and the angle-resolved photoemission spectroscopy. These two techniques will be described in depth in Chapters 5 and 7, respectively.

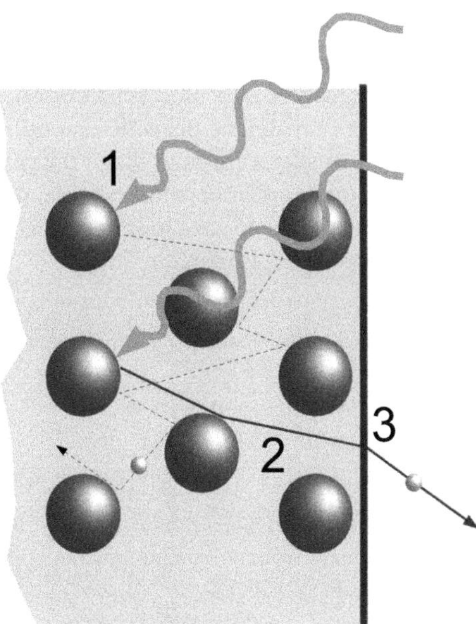

Figure 2.4 *The photoemission models must describe (1) the excitation of the electron by a photon, (2) its propagation to the surface and (3) the escape of the electron to the vacuum when crossing the surface.*

2.2 Technical aspects of a photoemission experiment

The essential elements to carry out photoemission measurements are a radiation source and an electron detector. Many different radiation sources exist. They range from X-ray tubes to rare gas discharge lamps, up to lasers and synchrotrons. The detectors must be capable of accurately determining the kinetic energy of the photoemitted electrons and possibly their emission angle, which determines the wave vector.

How do these different photon sources work? The operation principle is very different depending on the source. X-ray tubes work as follows: high energy electrons are directed against an anode consisting of a metallic element. Incident electrons excite a core electron from the atoms in the anode. When these electrons relax, they emit X-ray radiation characteristic of the excited core level. In a discharge lamp, a discharge in a rare gas ionizes the atoms and creates a plasma. When the atoms relax, ultraviolet photons are emitted. Lasers use stimulated emission to provide radiation. However, currently their energy is in the infrared or in the visible range of the electromagnetic spectrum. This is an energy range that does not allow to extract electrons from a solid, so frequency doubling devices must be used to create, for example, photons of 2ν frequency from two photons of ν frequency. Finally, there are also synchrotrons, which are particle accelerators dedicated to the production of electromagnetic radiation. Synchrotron sources provide photons in a very wide frequency domain ranging from infrared to very hard X-rays. In a synchrotron, relativistic electrons travel in rings under the effect of magnetic fields. Radiation theory shows that an accelerated charge radiates in a smaller solid angle when the particle speed is close to the speed of light. Unlike the other radiation sources presented above, the radiated spectrum is very wide and it is necessary to use a monochromator to select the desired photon energy for a photoemission measurement.

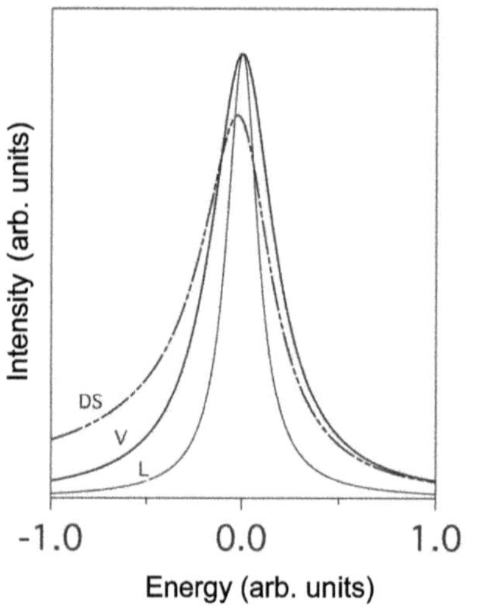

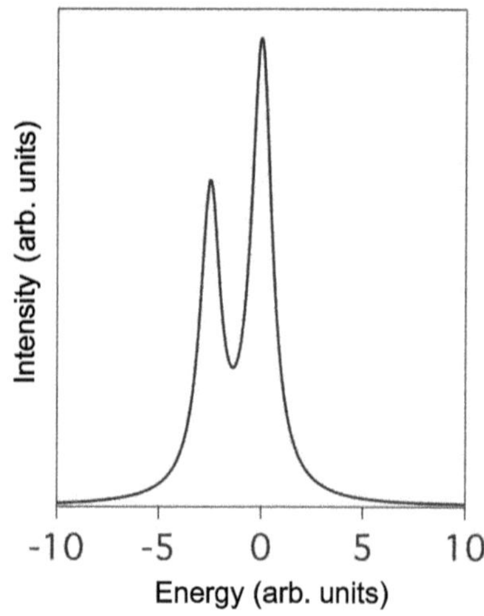

Figure 2.5 *Core level line shape. (Left) Core level line shape of an s level. The line shape of an optical transition is a lorentzian function (L), but many gaussian factors broaden it and transform the line shape into a more complex shape: a Voigt (V) function for non-metallic systems, or a Doniach-Sunjic (DS) function for metals. (Right) Core level splitting due to spin-orbit coupling.*

As explained previously, the absorption of a photon of sufficiently high energy can lead to the extraction of an electron. In a photoemission experiment it is necessary to measure the energy E of these photoelectrons but also their emission angle relative the surface normal. Most modern detectors discriminate electrons with electric fields. In these detectors, generally of hemispherical shape, the photoelectron beam is deflected by electrostatic fields. The trajectory of electrons depends on their kinetic energy, which therefore allows you to select the desired photoelectron energy. In addition, the detectors also allow to determine the emission angle of the photoelectrons and thus to obtain the wave vector $\vec{k}$ as a function of the emission angle. The simultaneous knowledge of E and $\vec{k}$ magnitudes is required to obtain the initial state of the electrons, as we will see in the next section.

2.3 Model to describing the photoemission process

In order to determine the initial state of the electrons in a solid, it is necessary to understand the mechanisms taking place as of photon absorption to the detection of the emitted electron. The simplest model that describes all these phenomena is the three-step model (Fig. 2.4), which decomposes, in an artificial but effective way, the processes that electrons undergo before being detected. The first step is the optical excitation and corresponds to the absorption of a photon of energy $h\nu$ by an electron of energy E_i. The final state is an excited state of energy $E_i + h\nu$. The probability of this electronic transition depends on the photon energy, the light polarization, the initial and final states, the experimental geometry... The photoelectron then propagates into the solid from its location at the time of the photon absorption: this is the second step. *A priori*, the photoelectron can propagate in all directions but only electrons propagating to the surface will be able to exit the solid and reach the detector. In solids, the high energy electrons interact strongly with the atoms, leading to inelastic collisions. When losing part of their energy, photoelectrons lose also their information on the initial state. These electrons are called secondary electrons. They only contribute to the unstructured photoemitted signal (background). The inelastic diffusion mechanisms limit the depth probed by photoemission so that only the electrons emitted in the vicinity of the surface will be able to exit the material without inelastic diffusion. The last step is the surface crossing. The surface constitutes a potential barrier for electrons. This barrier absorbs part of the kinetic energy of the photoelectrons and decreases the component of the wave vector perpendicular to the surface. In contrast, the component of the wave vector parallel to the surface is preserved. Consequently, the crossing of the surface corresponds to a refraction similar to the Snell-Descartes refraction in optics. The electrons are finally detected in the vacuum by a detector which measures their energy and also their emission direction. These measurements are used to determine the energy and the wave vector (at least its component parallel to the surface) when the photon energy and the work function are known.

2.4 The core levels

As explained previously, the core level electrons are not involved in the chemical bond because of their atomic nature. The energies of these levels are therefore characteristic of each element. But the core level energy contains more information than the atomic composition alone. The precise energy of a level is sensitive to the atomic environment. It is therefore possible to identify the different carbon atoms of an organic molecule, the oxidation states of an element or the components associated with surface or bulk atoms. That is what led Siegbahn to introduce (for this technique) the term ESCA (Electron Spectroscopy for Chemical Analysis). The physical principle of ESCA is that each local environment creates an electrostatic potential that displaces the core level energy of the emitter. For example, the binding energy

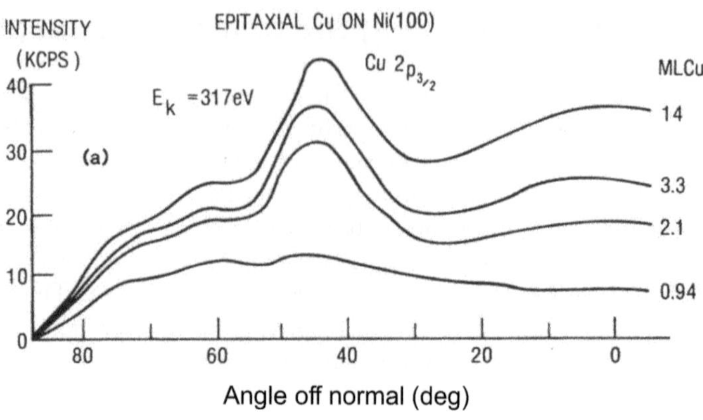

Figure 2.6 *Photoemitted intensity of Cu 2p core level as a function of the emission angle for a system composed of a copper thin film deposited on a nickel substrate. When at least two layers are present, there are peaks due to constructive interference in preferential directions [3]. Reprinted with permission from W. F. Egelhoff, Jr., Phys. Rev. B 30, 1052 (1984). ©1984 by the American Physical Society.*

depends on the degree of oxidation, so that Fe^{2+} level is displaced when compared to that of Fe^{3+} (chemical shift).

The photoemission lines of a core level can sometimes have a complex shape with several components of special shapes. Fine analysis requires there modelling the shape of the core level spectra. The spectral shape of a core level corresponding to a single orbital (s state) is a priori a peak of Lorentzian shape, whose width is characteristic of the lifetime of the excited state (*i.e.*, that of the hole left in the material by the excitation of the electron). However, many factors broaden this contribution (experimental resolution, lattice vibrations, defects...) that can be described by a Gaussian broadening. The convolution of these two functions is called a Voigt function (Fig. 2.5) which usually describes the line shape for non-metallic systems. For metals, a specific screening mechanism of the hole by electron-hole pairs appears, which leads to asymmetric shapes described by the Doniach-Sunjic function (Fig. 2.5).

Let us consider the core levels of initial states that are degenerated in energy ($\ell \neq 0$). The spectrum exhibits generally two peaks whose energy separation is due to the spin-orbit coupling. This interaction of relativistic origin depends on the coupling between the spin moment of the hole $s = 1/2$ and its orbital kinetic moment ℓ, $\ell = 1$ for a p state, $\ell = 2$ for a d state etc. The two peaks correspond to the two possible final states that are characterized by the total kinetic moment $j = \ell + 1/2$ and $j = \ell - 1/2$. The respective intensities of the two structures depend on the degeneracy of the states, *i.e.* $2j+1$.

Core levels can also provide structural information. Indeed, the photoelectron can be considered as an electronic wave originated in the emitter atom and diffracting in surrounding atoms. The wave originated in the emitter interferes with the waves scattered by the neighboring atoms. The interference depends on the atomic structure and in the particular directions between the emitter and the neighbors as well as on the interatomic distances. This is the photoelectron diffraction. Figure 2.6 shows the photoemitted intensity as a function of the angle for several thicknesses of copper films on nickel. For a single atomic layer, the emission from the core level of copper is isotropic. On the contrary, when there are several layers, there is an interference phenomenon between the directly emitted wave and the waves scattered by the atoms on the layers above the emitter, leading to photoemitted intensity maxima and minima depending on the electron emission angle.

Diffusion can be more or less complex, depending on the kinetic energy of outgoing electrons. High kinetic energy electrons are mainly diffused towards the front, *i.e.*, in the emitter-scatterer direction. This regime is called *forward scatttering*. On the other hand, when photoelectrons have low kinetic energies, the diffusion takes place in almost any direction, especially to the rear. This regime is called *backward scattering*. Each regime has its own specificities. The high kinetic energy regime is not very sensitive to the surface and the scattering is simpler to describe. It directly provides the directions between the emitter and the neighboring atoms. On the contrary, the low kinetic energy regime is more complex because of the different scattering directions and because of a greater number of scattering events to be considered. In this regime, the interpretation of the data is far from intuitive and simulations are required.

These mechanisms allow us to understand the principle of photoelectron diffraction. The main characteristic of the technique is the chemical specificity. Since photoelectron diffraction is based on the excitation of core levels, it allows to study the atomic environment of an element or even those of inequivalent atoms of the same element. Another feature of photoelectron diffraction is the short-range sensitivity because of the mean free path of electrons in the solid. As electrons explore only short distances from the emitter, the long-range order is not necessary, which differentiates photoelectron diffraction from X-ray or low-energy electron diffraction. Last but not least, it provides structural information with exactly the same experimental setup as the one necessary to obtain the electronic structure.

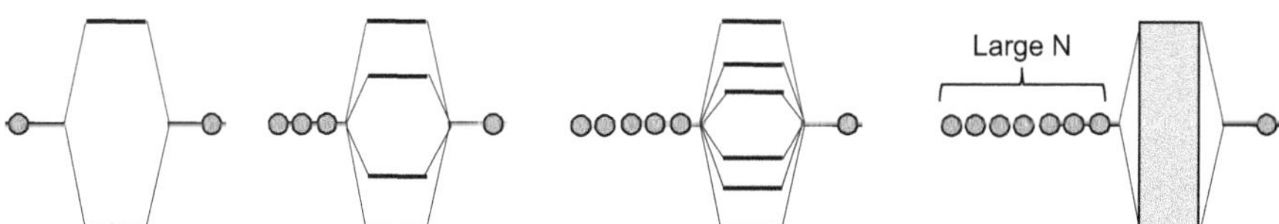

Figure 2.7 *Band formation from atomic orbitals. In a diatomic molecule, a bonding and an anti-bonding bonds are formed. When the number of atoms is increased, the additional bonds have an energy between the bonding and the anti-bonding states which leads in a solid to a continuum of available energies: an electronic band.*

2.5 The valence band

The states involved in the chemical bond are the orbitals in molecules or the band states in solids. In a molecule, the combination of atomic orbitals gives rise to molecular orbitals whose energies are different from those of the initial atomic orbitals. In a diatomic molecule, for instance, a bonding orbital and an anti-bonding orbital are obtained. When increasing the number of atoms, the additional bonds have energies between the energies of the bonding and the anti-bonding orbitals. In a solid, this leads to a continuum of available energies called an energy band (Fig. 2.7).

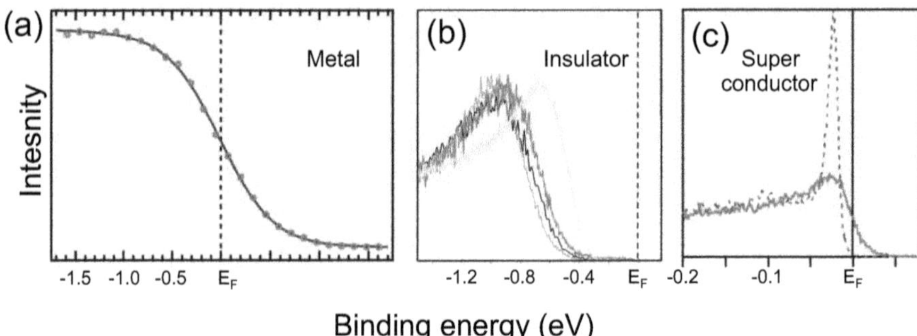

Figure 2.8 *(a) Photoemission spectrum of gold. The intensity at the Fermi level is indicative of a metallic character [4]. Reprinted with permission from T. Kiss, F. Kanetaka, T. Yokoya, T. Shimojima, K. Kanai, S. Shin, Y. Onuki, T. Togashi, C. Zhang, C. T. Chen, and S. Watanabe, Phys. Rev. Lett. 94, 057001 (2005). ©2005 by the American Physical Society. (b) $Sr_2CuO_{3+\delta}$ photoemission spectrum for different preparation conditions. The absence of intensity at the Fermi level indicates the insulating character of the compound [5]. Reprinted with permission from T. E. Kidd, T. Valla, P. D. Johnson, K. W. Kim, G. D. Gu, and C. C. Homes, Phys. Rev. B 77, 054503 (2008). ©2008 by the American Physical Society. (c) Photoemission spectra of the high critical temperature superconductor Bi2212 in normal (solid line) and superconductor states (dashed line) [6]. Reprinted with permission from J. D. Koralek, J. F. Douglas, N. C. Plumb, Z. Sun, A. V. Fedorov, M. M. Murnane, H. C. Kapteyn, S. T. Cundiff, Y. Aiura, K. Oka, H. Eisaki, and D. S. Dessau, Phys. Rev. Lett. 96, 017005 (2006). ©2006 by the American Physical Society.*

Important information can be extracted from the analysis of the electronic bands. In particular, it is possible to determine whether a material is metallic, insulating or semiconductor (Fig. 2.8). That depends on the electronic filling of the last occupied band. If the number of electrons from the solid leads to a partial filling of the band, a metal is obtained and the energy of the last occupied state defines the Fermi level. On the other hand, if the band is completely occupied, there is a finite energy separation between the last occupied state and the first empty state. The material is then an insulator, a semiconductor or a semi-metal, depending on the amplitude of the band gap (up to 3 eV for large gap semiconductors and 0 eV for semi-metals). There is also a particular state of matter: superconductivity. In conventional superconductors the superconducting state is characterized by conduction through electron pairs (Cooper pairs). The energy spectrum presents a band gap close to the Fermi level, matching the energy of the Cooper pairs. We will see in more detail the interest of photoemission for the study of electronic properties in Chapter 7.

Bibliography

[1] S. Oswald *et al.*, Surf. Interf. Anal. **35**, 991 (2003).
[2] U. Gelius *et al.*, J. Electron Spectrosc. Rel. Phenom. **2**, 405 (1974).
[3] W. Egelhoff, Phys. Rev. B **30**, 1052 (1984).
[4] T. Kiss *et al.*, Phys. Rev. Lett. **94**, 057001 (2005).
[5] T. Kidd *et al.*, Phys. Rev. B 77, 054503 (2008).
[6] J. Koralek *et al.*, Phys. Rev. Lett. **96**, 017005 (2006).

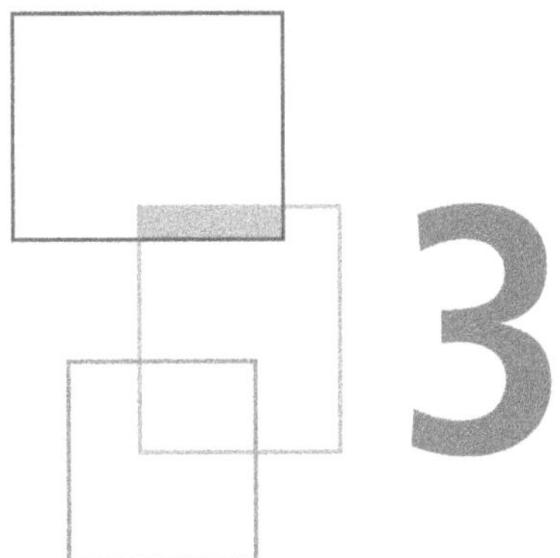

Basic concepts

3.1 Photoemission modelling

Photoemission on a solid is a complex process. The description of the absorption of a photon and the emission of an electron in the vacuum requires considering a finite crystal and the presence of a surface (solid-vacuum interface). In a perturbation treatment of the interaction with the electromagnetic field, one would have to consider the eigenstates of the system composed by the finite crystal and the vacuum, and the absorption of the photon inducing transitions between these states. It is a problem of rare complexity because of the presence of the surface (descriptions of crystals generally use translation symmetry and consider infinite crystals). Moreover one has to take into account the interactions between electrons and between electrons and phonons. However, as we shall see, it is possible to simplify the problem by making approximations which are found to be legitimate in a large number of cases. A detailed description of the photoemission theory can be found in several synthesis papers [1–5]. We will introduce the most relevant concepts in this chapter.

3.1.1 *Hamiltonian of the electron-photon interaction and transition probability*

The problem of electronic interactions is one of the most difficult problems of solid-state physics. Fortunately, one-electron approximations, where particle interactions are only considered on average, have proved very satisfactory for most systems. They are the basis of the calculations of band structures, especially in the functional density approaches. We will first describe this framework in which the electrons are considered as independent and described by monoelectronic functions. This electron wave function is characterized by quantum numbers associated with the symmetry of the problem. Whereas, as in the case of an atom these are the quantum numbers corresponding to the angular momentum and for a molecule to those corresponding to the molecular orbital describing the electron, they correspond (for a solid) to the wave vector of the electron in the first Brillouin zone (other quantum numbers associated with the symmetry of the space group of the solid also characterize the wave function). Within the independent electrons framework, the energy of the system is simply the sum of the energies of each electron:

$$E_{tot} = E_1 + E_2 + ... + E_n = \sum_{i=1}^{n} E_i \tag{3.1}$$

and the state of a n electron system is the antisymmetrized product of monoelectronic states $|\phi_i\rangle$ compliant with the Pauli principle:

$$|\Psi\rangle = A(|\phi_1\rangle|\phi_2\rangle...|\phi_n\rangle) \tag{3.2}$$

where A is the antisymmetrisation operator with respect to the permutation of electrons. In position representation, the wave function of a n independent electron system can be written as a Slater determinant of monoelectronic functions ($\phi_i(\vec{r}, s)$ where s is the spin component):

$$\Psi(\vec{r}_1, s_1, \vec{r}_2, s_2, \ldots, \vec{r}_n, s_n) = \frac{1}{\sqrt{n!}} \begin{vmatrix} \phi_1(\vec{r}_1, s_1) & \phi_1(\vec{r}_2, s_2) & \ldots & \phi_1(\vec{r}_n, s_n) \\ \phi_2(\vec{r}_1, s_1) & \phi_2(\vec{r}_2, s_2) & \ldots & \phi_2(\vec{r}_n, s_n) \\ \vdots & & & \vdots \\ \phi_n(\vec{r}_1, s_1) & \phi_n(\vec{r}_2, s_2) & \ldots & \phi_n(\vec{r}_n, s_n) \end{vmatrix}. \tag{3.3}$$

In the photoemission process, a photon is absorbed by an electron and transmits its energy and momentum to it. To describe the process in quantum mechanics and to take into account the interaction of the system with the electromagnetic field, the electron momentum $\vec{p}$ in the Hamiltonian has to be substitute by by $\vec{p} - q\vec{A}$ (where $\vec{A}$ is the vector potential of the field) [6]:

$$H = \frac{[\vec{p} - q\vec{A}(\vec{r}, t)]^2}{2m} + q\varphi(\vec{r}, t) + V \tag{3.4}$$

V is the potential energy due to all other particles of the system and $\varphi(\vec{r}, t)$ is the scalar potential of the field. This Hamiltonian can be written as the sum of two terms, the Hamiltonian H_0 which describes the whole electronic system and H_{int} which is associated with the coupling with the radiation field:

$$H = H_0 + H_{int} \tag{3.5}$$

$$H_0 = \frac{p^2}{2m} + V \tag{3.6}$$

$$H_{int} = -\frac{q}{2m}(\vec{A} \cdot \vec{p} + \vec{p} \cdot \vec{A}) + \frac{q^2}{2m}|\vec{A}|^2 + q\varphi \tag{3.7}$$

The quadratic term $\frac{q^2}{2m}|\vec{A}|^2$ is small with usual photon sources. As $[\vec{p}, \vec{A}] = -i\hbar\vec{\nabla} \cdot \vec{A}$, H_{int} can be written as:

$$H_{int} = \frac{1}{2m}\left[-2q\vec{A} \cdot \vec{p} + \frac{q\hbar}{i}(\vec{\nabla} \cdot \vec{A}) + q^2|\vec{A}|^2\right] + q\varphi\,. \tag{3.8}$$

Each of the terms of 3.8 gives rise to contributions specific to the transition probability. This expression can be modified according to the choice of the gauge, which does not modify the observable electric and magnetic fields but changes the scalar and vector potentials according to:

$$\begin{aligned} \vec{A}(\vec{r}, t) &\rightarrow \vec{A}'(\vec{r}, t) = \vec{A}(\vec{r}, t) + \vec{\nabla} f(\vec{r}, t) \\ \varphi(\vec{r}, t) &\rightarrow \varphi'(\vec{r}, t) = \varphi(\vec{r}, t) - \tfrac{1}{c}\tfrac{\partial}{\partial t} f(\vec{r}, t) \end{aligned} \tag{3.9}$$

The gauge generally used in low-energy physics is the Coulomb gauge with $\vec{\nabla} \cdot \vec{A} = 0$ and $\varphi = 0$ in such a way that the interaction Hamiltonian can be written as:

$$H_{int} = \frac{1}{2m}\left[-2q\vec{A} \cdot \vec{p} + q^2|\vec{A}|^2\right] \tag{3.10}$$

It should be mentioned, however, that the discontinuity of $\vec{A}$ due to the surface can lead to a surface photoemission process associated with the $\vec{\nabla} \cdot \vec{A}$ term in the Hamiltonian. We will not discuss the processes due to this singular term.

To simplify the Hamiltonian, it is possible to use the dipolar approximation. Recall that the vector potential is always:

$$\vec{A}(\vec{r}, t) = A_0(\omega)\; e^{i(\vec{k} \cdot \vec{r})}\; \vec{\epsilon} \tag{3.11}$$

where $A_0(\omega) = e^{-i\omega t}$. The spatial dependence can be developed in Taylor series:

$$e^{i(\vec{k} \cdot \vec{r})} = 1 + i\vec{k} \cdot \vec{r} - \frac{(\vec{k} \cdot \vec{r})^2}{2} + \ldots \tag{3.12}$$

the dipolar approximation consisting of keeping only the constant term $e^{i(\vec{k} \cdot \vec{r})} \sim 1$, in such a way that:

$$\vec{A}(\vec{r}, t) = \vec{A}_0 = A_0(\omega)\, \vec{\epsilon}\,. \tag{3.13}$$

In the dipolar approximation we neglect the spatial variation of $\vec{A}$. This approximation is legitimate when the wavelength of the radiation is greater than the distances characteristic of the electronic states. The square of matrix element M_{fi} is then:

$$|M_{fi}|^2 = \frac{q^2}{m^2}|\langle\phi_f|\vec{p}\cdot\vec{A}_0|\phi_i\rangle|^2 = \frac{q^2}{m^2}|\langle\phi_f|\vec{p}|\phi_i\rangle\cdot\vec{A}_0|^2 = \frac{q^2\hbar^2}{m^2}|\langle\phi_f|\vec{\nabla}|\phi_i\rangle\cdot\vec{A}_0|^2 \quad (3.14)$$

In a perturbative approach, the electron-photon interaction Hamiltonian induces transitions between eigenstates of the non-perturbed Hamiltonian H_0. The probability of transition between a monoelectronic initial state $|\phi_i\rangle$ with energy E_i and the final states $|\phi_f\rangle$ with energy E_f is given by the Fermi golden rule. The transition probability per unit of time w_{fi} between the two states is given by:

$$w_{fi} = \frac{2\pi}{\hbar}|\langle\phi_f|H_{int}|\phi_i\rangle|^2\delta(E_f - E_i - \hbar\omega) \quad (3.15)$$

where $\hbar\omega$ is the photon energy. The matrix element $\langle\phi_f|H_{int}|\phi_i\rangle \equiv M_{fi}$ contains most of the physics of the process and the Dirac function reflects the conservation of energy. In this formula, the spin of the electron is not included. It could have been added, but since $\vec{A}$ only interacts with orbital degrees of freedom, the spin does not change in the photoemission process described by this Hamiltonian. It is thus possible to know the spin in the initial state by detecting that of the final state at least if the spin-orbit coupling is negligible.

Conservation of momentum must be considered since the photon transfers its momentum to the electron. As the electron that absorbs the photon is in a potential (atom, molecule or solid) its momentum is not a constant of motion. However, in the periodic potential of a crystal, the translation invariance of the crystal lattice allows to define the crystal momentum or quasi-momentum $\vec{p} = \hbar\vec{k}$ where $\vec{k}$ is the wave vector of the Bloch function. This quantity, which must be distinguished from the electron momentum, is defined modulo a reciprocal lattice vector ($\vec{k} \equiv \vec{k} + \vec{G}$). This means that momentum transfers to the lattice can only take discrete values equal to $\hbar\vec{G}$. Then it is possible to obtain a lessened form of momentum conservation (conservation modulo $\hbar\vec{G}$) in a periodic potential. Moreover, in the ultraviolet domain, the pulse of the photon is negligible compared to that of the electron. Indeed, the momentum of a photon of 20 eV is 0.01 Å and therefore is very low in comparison with quasi-momenta in a typical Brillouin zone. Consequently, during an optical absorption, the allowed transitions are vertical in a reduced Brillouin zone scheme, i.e. the wave vector of the initial $\psi_{\vec{k}}$ and final $\phi_{\vec{k}'}$ states are the same. This can be verified directly by computing the matrix element $\langle\phi_{\vec{k}'}|\vec{\nabla}|\psi_{\vec{k}}\rangle$ which is zero if $\vec{k} \neq \vec{k}'$ modulo a vector of the reciprocal lattice. However, it is necessary to distinguish the direction perpendicular to the surface of the two parallel directions. Indeed, the translation symmetry is broken perpendicularly to the surface so that the perpendicular component of $\vec{k}$ is not conserved. Physically this reflects the fact that when the electron crosses the potential associated with the surface, it can give it a momentum (the surface potential is abrupt and its Fourier spectrum is very large). We will study this point in detail in the following sections.

3.1.2 Qualitative approach: nearly-free electrons

We have seen that the eigenstates of a periodic crystal are characterized by their wave vectors in the first Brillouin zone. The periodicity of the energy allows to represent the band structure in the first zone by folding back the bands of increasing energy (reduced zone picture). Although this representation is useful, the optical absorption and photoemission processes are more easily understood in the extended zone picture. Consider a system of nearly free electrons. In such a system, the periodic potential is very low and it weakly distorts the dispersion with respect to the free electron behavior but allows momentum transfers. In this case, absorption of a photon can allow the transition from an electron state of energy E_i and wave vector $\vec{k}_i$ ($E_i = \hbar^2 k_i^2/2m$) to a final state of energy $E_f = E_i + \hbar\omega$ and wave vector $\vec{k}_f$ ($E_f = \hbar^2 k_f^2/2m$), the transition being permissible if and only if the crystal provides the required momentum to the electron $\vec{k}_f = \vec{k}_i + \vec{G}$ (figure 3.1).

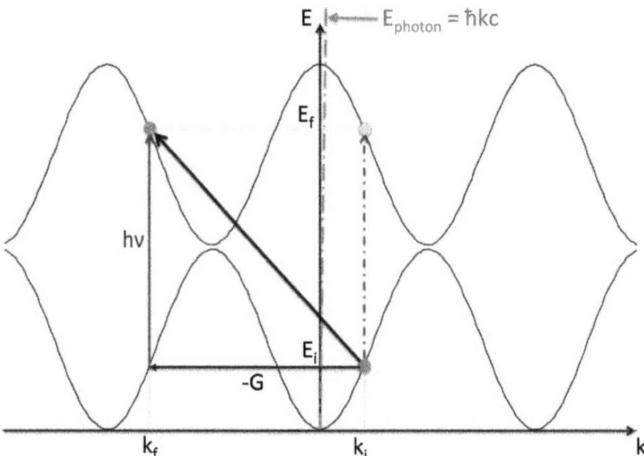

Figure 3.1 *Band structure in the extended zone scheme. In a photon absorption process, conservation of energy suggests that the final state is characterized by E_f et $\vec{k}_f$. Since the photon momentum is negligible, the wave vector is conserved modulo a reciprocal lattice vector $\vec{k}_f = \vec{k}_i + \vec{G}$.*

In this extended zone scheme, the energy of the electron being greater in the final state, its momentum must also be greater. The problem in photoemission is thus to determine the vectors of the reciprocal lattice involved in the optical absorption process (they depend of course on the energy of the photon) and likewise to know the distribution of the electrons of given energy within and outside the solid. We will show below that in the case of nearly free electrons, the distribution of the final energy states E_f forms a cone centered along the reciprocal lattice vector $\vec{G}$. This vector is necessary for energy and momentum conservations. This distribution is called the first Mahan cone. To know the distribution outside the solid, one has to take into account the modifications induced by the potential of the surface and to project the different cones through the crystal surface.

A finite potential modifies this simplified picture of band structure. Indeed, a Bloch state can be written as a linear combination of plane waves whose wave vectors are related by a reciprocal lattice vector. Each plane wave component will contribute to the distribution of electrons outside the solid. The main component will give the primary distribution (first Mahan cone) while the other components will give distributions of lower intensity (secondary cones). These distributions can also be explained as follows: the solutions outside the solid (essentially plane waves) must be continuously connected with the solutions inside the solid (Bloch waves). Consequently, it is necessary to have many escaping plane waves and therefore many emission directions, to achieve this connection.

Let us explicitly calculate the angle distribution of the final states at a given energy. Mahan proposed a solution in the limit of nearly free electrons. Indeed, in a periodic potential, a Bloch state can be decomposed on plane waves:

$$|\phi_{\vec{k}}\rangle = \left[1 + \sum_{\vec{G}'\neq 0} u^2_{\vec{k},\vec{G}'}\right]^{-1/2} \left[|\vec{k}\rangle + \sum_{\vec{G}'\neq 0} u_{\vec{k},\vec{G}'}|\vec{k}+\vec{G}'\rangle\right] \qquad (3.16)$$

where by definition $\vec{k}$ belongs to the first Brillouin zone. When the potential is very low, a contribution (a plane wave) dominates. Thus the first band corresponds to the plane wave $|\vec{k}\rangle$, the second band to the plane wave $|\vec{k}+\vec{G}_1\rangle$ etc. We have seen that an optical transition is vertical in the first Brillouin zone, but in an extended zone diagram, involves a reciprocal lattice vector to account for the increase in energy and momentum transfer. Let us call $\vec{K} = \vec{k_f}$ the wave vector of the final state of the photoelectron. The vector $\vec{K}$ is related to the energy by:

$$E_f = \frac{\hbar^2 K^2}{2m}$$

In order for the transition to be possible, there must exist a reciprocal lattice vector $\vec{G}$ to obey the conservation of the momentum. The wave vector of the initial state, which we assume to be in the first Brillouin zone, is therefore $(\vec{K}-\vec{G})$, and conservation of energy allows us to write:

$$\frac{\hbar^2 K^2}{2m} = \frac{\hbar^2(\vec{K}-\vec{G})^2}{2m} + \hbar\omega$$

which leads to:

$$\frac{\hbar^2}{m} KG\cos\theta_0 = E_G + \hbar\omega \equiv \Lambda_G$$

where θ_0 is the angle between $\vec{K}$ and $\vec{G}$. Then we have:

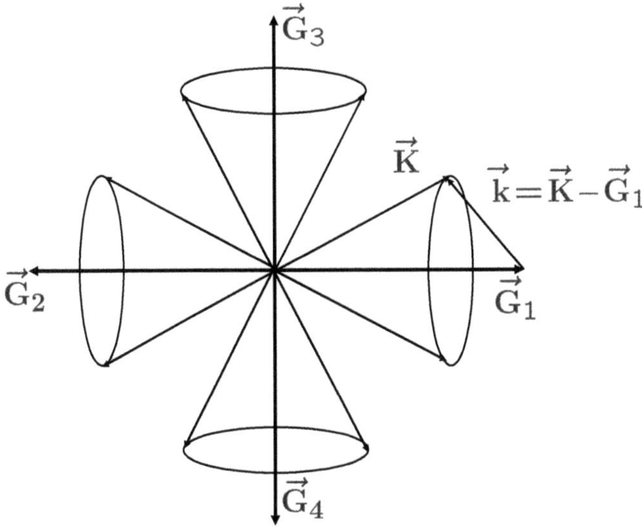

Figure 3.2 *Distribution of wave vectors of the final states with the same energy: they form a cone around the vector of the reciprocal lattice, necessary for the conservation of energy and momentum. For each vector $\vec{K}$ of the cone, the wave vector of the initial state can be obtained by $\vec{k} = \vec{K} - \vec{G}$. More precisely, because of the equivalence of vectors of the reciprocal lattice due to the crystal symmetry, we have several cones corresponding to the same energy as illustrated in figure 3.2 for a square lattice.*

$$E_f = \frac{\hbar^2 K^2}{2m} = \frac{\Lambda_G^2}{4E_G \cos^2 \theta_0} \tag{3.17}$$

This equation is easily interpreted. An electron whose final state wave vector makes an angle θ_0 with the reciprocal lattice vector needed for the momentum conservation has an energy given by the last equation. Conversely, all electrons of the same final energy are characterized by the angle their wave vectors make with $\vec{G}$. Their distribution forms a cone oriented in a direction of reciprocal space (the first Mahan cone, see figure 3.2). Moreover, because of the symmetry of the crystal, several vectors of the reciprocal lattice are equivalent so that we have a a series of cones around the equivalent vectors. Physically, this means that the photoelectrons of the same energy can propagate in the crystal in directions which make the same angle with a family of equivalent reciprocal lattice vectors.

The higher the θ_0 angle, the greater the energy of the electrons. It should be noted, however, that the maximum energy of the electrons is $E_{max} = E_F + \hbar\omega$ where E_F is the energy of Fermi. The maximum angle is therefore:

$$\theta_0^{max} = \cos^{-1}\left[\Lambda_G/2(E_F + \hbar\omega)^{1/2} E_G^{1/2}\right]. \tag{3.18}$$

Table 3.1 *Values of E_Z and $E_{\|}$ associated with the main reciprocal lattice vectors for (100), (011) and (111) surfaces of Na.*

Orientation	$\mathbf{G}(a/2\pi)$	$E_{\|}/E_G$	E_z/E_G
(001)	(101) $(\bar{1}01)$ (011) $(0\bar{1}1)$	1/2	1/2
(011)	(011)	0	1
	(101) $(\bar{1}01)$ (110) $(\bar{1}10)$	3/4	1/4
(111)	(101) (011) (110)	1/3	2/3

Given that $-1 \leq \cos\theta_0 \leq 1$, the optical absorption is only allowed for:

$$E_G - 2(E_G E_F)^{1/2} \leq \hbar\omega \leq E_G + 2(E_G E_F)^{1/2} \tag{3.19}$$

For example, for Na and the different reciprocal vectors (110) and a photon energy $\hbar\omega = 0.6\,E_G$, one finds $\theta_0^{max} = 26°$. Since the different vectors (110) make an angle of 60° to each other, the cones centered on these vectors do not overlap.

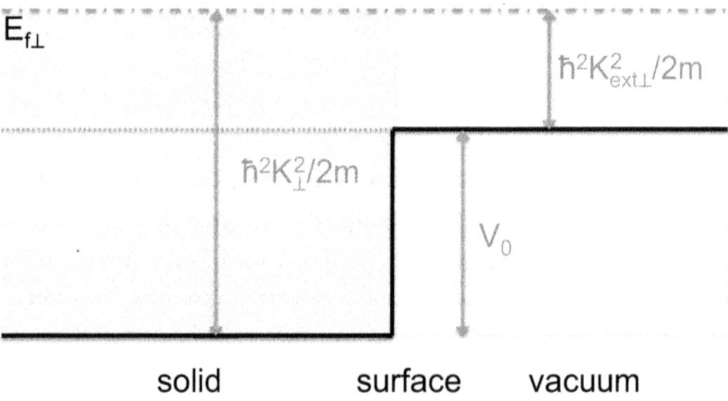

Figure 3.3 *Potential diagram in the direction perpendicular to the surface (the potential parallel to the surface is flat).*

It is clear that with photoelectrons propagating in different directions, only some of them will be able to leave the solid and cross the surface (for example, those that propagate towards the interior of the solid will not leave the solid!). To escape from the crystal the kinetic energy of the photoelectron in the direction perpendicular to the surface must exceed the surface potential (figure 3.3), viz.,

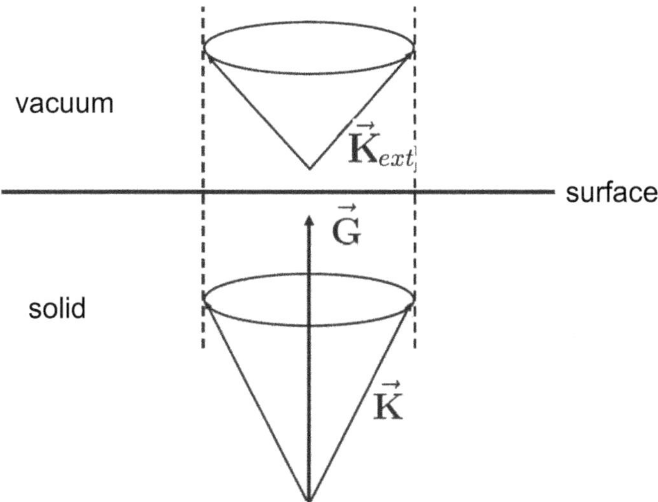

Figure 3.4 *Modification of the emission cone in and outside the solid for a $\vec{G}$ vector perpendicular to the surface.*

$$\frac{\hbar^2 K_\perp}{2m} > V_0$$

where V_0, called the inner potential of the crystal, corresponds in this simple approach to the difference between the vacuum level and the energy of the bottom of the band. V_0 is therefore the sum of the energy of Fermi (E_F) and the work function (Φ) ($V_0 = E_F + \Phi$). The kinetic energy in the perpendicular direction denoted by $E_{f\perp}$ is therefore:

$$E_{f\perp} = \frac{\hbar^2 K_\perp}{2m} = \frac{\hbar^2 K_{ext\perp}}{2m} + V_0$$

where $K_\perp$ and $K_{ext\perp}$, respectively, are the perpendicular components of the photoelectron wave vectors in the solid and in the vacuum. In order for the photoelectron to cross the surface, the wave vector transfer due to the crystal lattice $\hbar\vec{G}$ must therefore have a sufficiently high contribution in the direction perpendicular to the surface. Consider the most favorable case where $\vec{G}$ is perpendicular to the surface. In this case illustrated in the figure 3.2, all the electrons of the cone have the same perpendicular component in the solid but also outside the solid where their distribution is also conical. It is no longer the case if the vector $\vec{G}$ is arbitrary. Consider the example of a $\vec{G}$ vector in the plane of the surface. The perpendicular component of the photoelectron wave vector in the solid is then identical to that of the initial state, and the electron does not have enough kinetic energy in the perpendicular direction to escape from the crystal.

For a $\vec{G}$ reciprocal lattice vector making an angle with the normal to the surface, the perpendicular components of the different wave vectors of a cone around $\vec{G}$ are different. Only the electrons whose perpendicular component is greater than $\sqrt{2mV_0/\hbar^2}$ (in bold in figure 3.5) have enough energy $E_{f\perp}$ to escape the surface. The distribution outside the solid is no longer conical.

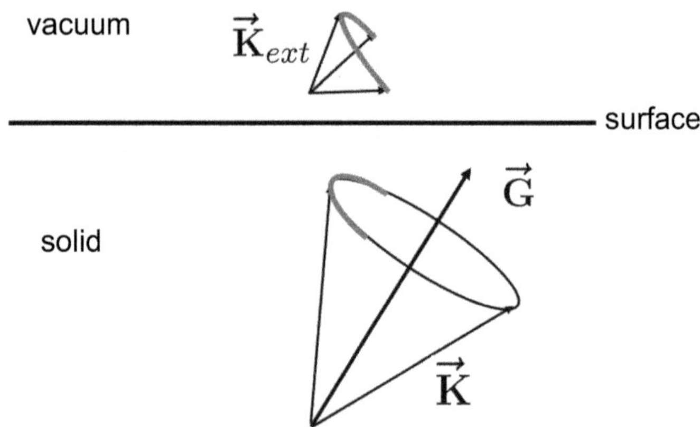

Figure 3.5 *Case of a $\vec{G}$ vector making an arbitrary angle with the normal to the surface. The perpendicular components of the wave vector in the associated cone are no longer constant and then not all photoelectrons can escape the surface. The momentum distribution of the escaping photoelectron (in bold in the figure) no longer exhibits a conical shape in the vacuum.*

To quantitatively determine the momentum distribution outside the crystal, we decompose the reciprocal vectors, in fact each reciprocal vector of the family of equivalent vectors, into a component parallel to the surface $\vec{G}_{\|}$ and a perpendicular component G_z. Outside the crystal, the energy of the photoelectron with respect to the vacuum level is simply:

$$E = \frac{p^2}{2m} = \frac{p_{\|}^2 + p_z^2}{2m}$$

with the conservation of the parallel component of the momentum:

$$\vec{p}_{\|} = \hbar \vec{K}_{\|} \, . \tag{3.20}$$

Inside the crystal, the energy of the final state with respect to the vacuum level can be written:

$$E = \frac{\hbar^2 K^2}{2m} - V_0$$

where V_0 is the inner potential. We obtain:

$$\hbar^2 K_z^2 = p_z^2 + 2mV_0. \tag{3.21}$$

The final state energy can also be written from the initial state one and the photon energy:

$$E = \frac{\hbar^2(\vec{K} - \vec{G})^2}{2m} - V_0 + \hbar\omega = \frac{\hbar^2 K^2}{2m} - V_0 + \Lambda_G - \frac{\hbar^2(K_z G_z + \vec{K}_{\parallel} \cdot \vec{G}_{\parallel})}{m}.$$

If we call φ the angle between $\vec{K}_{\parallel}$ and $\vec{G}_{\parallel}$ et ϑ, and ϑ the angle between $\vec{p}$ and the normal to the surface, we can write:

$$\frac{\Lambda_G}{2} = \frac{\hbar^2(K_z G_z + \vec{K}_{\parallel} \cdot \vec{G}_{\parallel})}{2m} = E_z^2[E\cos^2\vartheta + V_0]^{1/2} + (EE_{\parallel})^{1/2}\sin\vartheta\cos\varphi \tag{3.22}$$

with $E_z = \hbar^2 G_z^2/2m$ and $E_{\parallel} = \hbar^2 \vec{G}_{\parallel}^2/2m$. For a given energy E, the angles φ and ϑ are related by:

$$\sin\vartheta = \frac{\Lambda_G}{2dE^{1/2}}\left[E_{\parallel}^{1/2}\cos\varphi \pm E_z^{1/2}\left(\frac{4d(E+V_0)}{\Lambda_G} - 1\right)^{1/2}\right] \tag{3.23}$$

where $d = E_{\parallel}\cos^2\varphi + E_z$ and we can draw the lines of constant energies. Some such lines for sodium with photon energy of $\hbar\omega = 5$ eV are shown in Figure 3.6. The Na crystallizes in the body-centered cubic structure so that its reciprocal lattice has a face-centered cubic structure. The reciprocal vectors closest to the Γ point are those of the $\vec{G} = 2\pi/a(110)$ type. We can describe Na by a model of nearly free electrons with the following parameters: $E_F = 3.16$ eV, $V_0 = 5.41$ eV and $E_G = 16.4$ eV. For photon energy $\hbar\omega = 5.0$ eV, we find $E_{max} = 2.75$ eV. For each kind of surface, for example (001), (011) or (111), it is necessary to determine among the 12 $\vec{G} = 2\pi/a(110)$ reciprocal vectors, those which lead to photoelectron emission outside the solid. Indeed, to have a photoelectron emitted ($p_z > 0$), the final state kinetic energy perpendicular to the surface must be greater than V_0 (equation 3.21). This is achieved by the G_z contribution and the associated energy E_z. We must know how the E_G energy is decomposed in the parallel and perpendicular contributions. The table 3.1.2 summarizes the orientations of the different $\vec{G}$ vectors (110) leading to the emission of photoelectrons with respect to the three main surfaces.

In figure 3.6, we can see the energy lines (in steps of 0.25 eV from $E_{max} = 2.75$ eV) associated with one of the $\vec{G} = 2\pi/a(110)$ reciprocal vectors for three surfaces of Na. It is found that the conical shape is generally distorted outside the crystal by refraction at the surface. For the surface (110) the reciprocal vector being perpendicular to the surface, the cone is not deformed and the surfaces of the same energies remain as circles.

We have just seen that in the nearly free electron approach in which the final state is dominated by a single plane wave, the distribution of electrons after optical absorption forms a cone centered on the reciprocal vector $\vec{G}$ necessary for the energy and

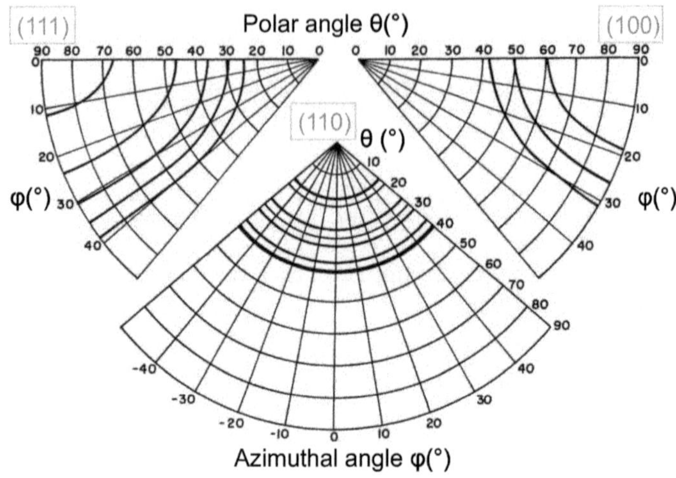

Figure 3.6 *Angular distributions of the primary Mahan cones for Na: (a) (111) surface* $E_Z = \frac{2}{3}E_G$ *and* $E_{\|} = \frac{1}{3}E_G$*; (b) (100)* $E_Z = E_{\|} = \frac{1}{2}E_G$*; (c) (110) surface* $E_Z = E_G$ *and* $E_{\|} = 0$ *(according to [7]. Figure reprinted with permission from G. D. Mahan, Phys. Rev. B 2, 4334 (1970). ©1970 by the American Physical Society.*

momentum conservations. An effect of the band structure is to lead to the emission of electrons in other directions. These distributions are called secondary cones, even if they are not conical, and have lower intensities. Indeed, the Bloch function contains not only the $|\vec{K}\rangle$ state but other contributions $|\vec{K}+\vec{G}'\rangle$, with smaller weight. The Bloch function can be written:

$$\Psi_{\vec{K}}(\vec{r}) = \exp\left(i\vec{K}\cdot\vec{r}\right)\sum_{\vec{G}'} u_{\vec{K},\vec{G}'} \exp\left(i\vec{G}'\cdot\vec{r}\right) \tag{3.24}$$

which means that the electron after the optical absorption process has an amplitude of probability of propagating in the $\vec{K}+\vec{G}'$ directions. Thus, the electron in the $|\Psi_{\vec{K}}\rangle$ state propagates not only with the $\vec{K}$ wave vector but also with the other $\vec{K}+\vec{G}'$ wave vectors. Consequently, the photoelectrons outside the solid propagate in different directions characterized by the conservation of the parallel component:

$$\vec{p}_{\|} = \hbar(\vec{K}_{\|} + \vec{G}'_{\|})$$

but it is noteworthy that the energy in the final state does not depend on $\vec{G}'$ and is close to $\hbar^2K^2/2m$. Outside the solid, the energy with respect to the vacuum level is equal to:

$$E = \frac{p^2}{2m} = \frac{\hbar^2K^2}{2m} - V_0$$

and then:

$$\hbar^2 K_z^2 = p^2 + 2mV_0 - (\vec{p}_{\|} - \hbar\vec{G}'_{\|})^2 = p_z^2 + 2mV_0 - \hbar^2 \vec{G}'^2_{\|} + 2\hbar p_{\|} G'_{\|} \cos\varphi' \quad (3.25)$$

Evidently, K_z and p_z have to be both real. This condition limits the possible values of the $\vec{G}'_{\|}$ vectors. The secondary cones exist only for some $\vec{G}'_{\|}$ values that depend on the value of the $\vec{G}_{\|}$ vector involved in the optical transition. Indeed, we have seen above that the energy of the initial state, characterized by the $\vec{K}_{\|} - \vec{G}_{\|}$ wave vector is connected to the final energy by:

$$E = E_i + \hbar\omega = \frac{\hbar^2(\vec{K} - \vec{G})^2}{2m} + \hbar\omega - V_0$$

that leads, using the equations giving K_z and $\vec{G}'_{\|}$ as well as the equation 3.22:

$$\begin{aligned}\frac{\Lambda_G}{2} &= E_z^{1/2}[E\cos^2\vartheta + V_0 - E'_{\|} + 2\sin\vartheta\cos\varphi'(EE'_{\|})^{1/2}]^{1/2} \\ &+ (E_{\|})^{1/2}[\sin\vartheta\cos\varphi\, E^{1/2} - \cos(\varphi - \varphi')(E'_{\|})^{1/2}] \qquad (3.26)\end{aligned}$$

where $E'_{\|} = \hbar^2\vec{G}'^2_{\|}/2m$. We can solve this equation and determine the energy according to the photoelectron emission angles:

$$E(\vartheta,\varphi) = \left[B\sin\vartheta - \left[B^2\sin^2\vartheta + D(\lambda^2 - E_z V_0 + E_z E'_{\|})\right]^{1/2}\right]^2 D^{-2} \quad (3.27)$$

with

$$\begin{aligned}\lambda &= \Lambda_G/2 + \cos(\varphi - \varphi')(E_{\|}E'_{\|})^{1/2} \\ B &= \lambda(E_{\|})^{1/2}\cos\varphi + E_z(E'_{\|})^{1/2}\cos\varphi.\end{aligned}$$

The lines of constant energy $E = E_{max}$ on the two distributions of the figure 3.7 represent the primary cone corresponding to the $\vec{G} = 2\pi/a(101)$ vector and the secondary cones of the Na (100) surface for $\hbar\omega = 5$ eV and 10 eV. The values of $\vec{G}'_{\|} = (G'_x, G'_y)$ are given in parentheses. We can see that a single cone is present for $\hbar\omega = 5$ eV while many cones appear for $\hbar\omega = 10$ eV. This increase in the number of cones is easy to understand. Indeed, the increase in photon energy leads to final states of higher energy and hence to larger p_z. This allows, according to the equation 3.25, more components $\vec{G}'_{\|}$ to be compatible with the real character of K_z.

This approach using nearly-free electron picture captures the physics of photoemission and can easily be extended to any band structures. In the following section, we develop the one-electron formalism of photoemission and then show in the sub-section (3.2.2) how we can account for the interactions between electrons by developing a many body description of photoemission.

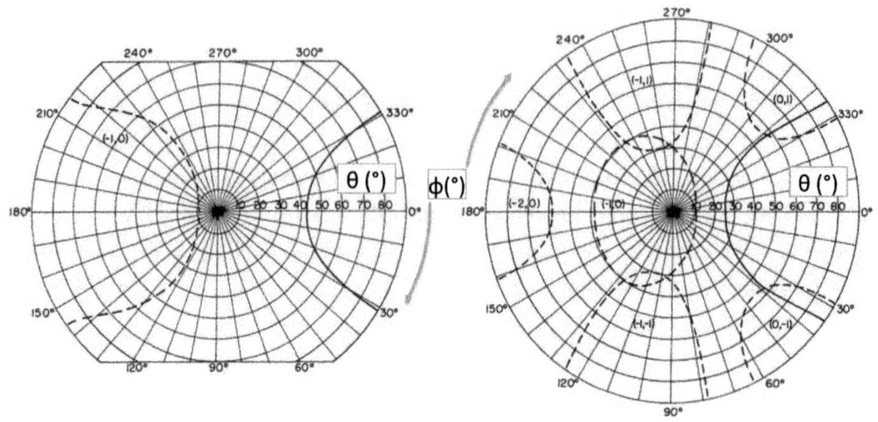

Figure 3.7 *Angle distributions corresponding to E_{max} outside the solid for $\hbar\omega = 5$ eV (left) and $\hbar\omega = 10$ eV(right). The primary cone is in solid line, and the secondary cones in dashed lines (from [7]. Figure reprinted with permission from G. D. Mahan, Phys. Rev. B 2, 4334 (1970). ©1970 by the American Physical Society.*

3.1.3 *Qualitative approach: core levels*

It is possible to photo-excite core electrons provided that photon energy is sufficiently high (X-rays). The photon energies used can vary from the soft X-ray range (a few hundred eV) to hard X-rays (several keV) in X-ray photoemission spectroscopy (XPS). As the energy of the core levels is a quasi-atomic signature, this technique can be used for a chemical analysis of the surface of the sample. That is the reason why this technique was initially called electron spectroscopy for chemical analysis (ESCA).

The energies of the spectral structures associated with core levels are characteristic of each element (see Fig. 3.8 or compilations [9, 11]). Since core levels are localized, the energy of the spectral structures are independent of the emission angle. Nevertheless, the intensity may vary with $h\nu$ because of the diffraction in the crystalline lattice.

XPS can also be used to study the valence band with significant differences with UPS. The energy dependence of the mean free path shows that in the XPS regime, photoemission is much less sensitive to the surface than in the UPS regime. In addition, the use of high photon energy leads to the loss of the vertical character of the transitions in the Brillouin since the momentum of the photon is no longer negligible in this range of energy. Another consequence is the wave vector integration of a significant part of the Brillouin zone due to the experimental angular resolution. Indeed, for $h\nu \sim 1000$ eV, an angular acceptance of $\Delta\theta = 1°$ allows you to explore a significant part of the Brillouin zone: from equation 3.37, $\Delta k_{||} = 0.28$ Å^{-1}. Moreover, at high energy many reciprocal lattice vectors are available to fold up the bands of the photoelectron in the first Brillouin zone so that the final states form

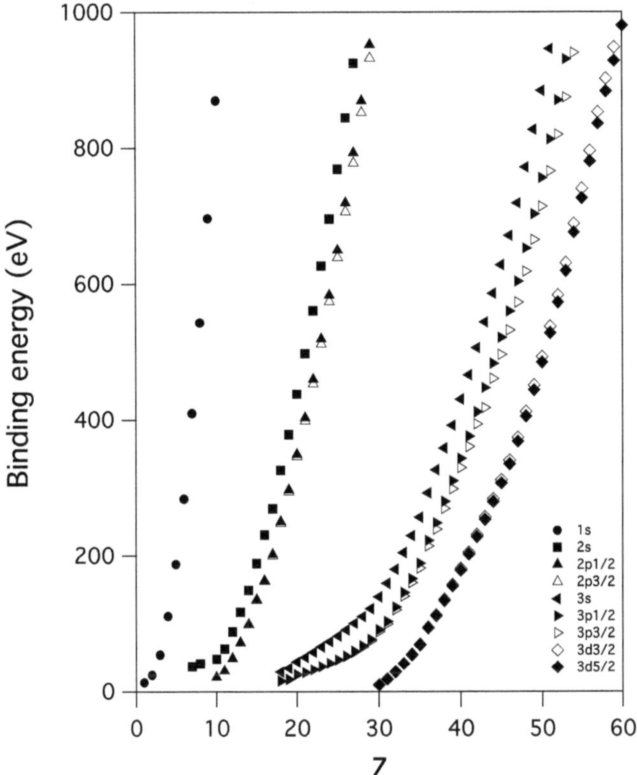

Figure 3.8 *Variation of the electron binding energy as a function of the atomic number. These values were compiled by G. Williams (http://www.jlab.org/g̃wyn / ebindene.html) from [8–10].*

a continuum with a nearly constant density leading to the softening of the matrix element effects. Consequently, because of the large integration in wave vector and a quasi-constant final state density, an XPS spectrum of the valence band recorded with photon energy of several keV mainly reflects the density of the initial states (DOS density of states).

Shape of the core level spectra

A first application of XPS is the determination of the chemical composition at the sample surface (Fig. 3.8, table 3.2). However, it is possible to obtain more information. The binding energy of an element changes slightly depending on its chemical state or its surrounding. It is therefore possible to obtain structural information about the inequivalent atoms from the shape of the core level spectra. These studies require the adjustment of the core level spectra.

We have seen in the previous chapter that the core level spectra usually exhibit a doublet shape due to the spin-orbit interaction. This interaction of relativistic origin

Table 3.2 *Binding energies of the different elements.*

Z		$1s$	$2s$	$2p_{1/2}$	$2p_{3/2}$	$3s$	$3p_{1/2}$	$3p_{3/2}$	$3d_{3/2}$	$3d_{5/2}$	$4s$
1	H	13.6									
2	He	24.6									
3	Li	54.7									
4	Be	111.5									
5	B	188									
6	C	284.2									
7	N	409.9	37.3								
8	O	543.1	41.6								
9	F	696.7									
10	Ne	870.2	48.5	21.7	21.6						
11	Na	1070.8	63.5	30.65	30.81						
12	Mg	1303	88.7	49.78	49.5						
13	Al	1559.6	117.8	72.95	72.55						
14	Si	1839	149.7	99.82	99.42						
15	P	2145.5	189	136	135						
16	S	2472	230.9	163.6	162.5						
17	Cl	2822.4	270	202	200						
18	Ar	3205.9	326.3	250.6	248.4	29.3	15.9	15.7			
19	K	3608.4	378.6	297.3	294.6	34.8	18.3	18.3			
20	Ca	4038.5	438.4	349.7	346.2	44.3	25.4	25.4			
21	Sc	4492	498	403.6	398.7	51.1	28.3	28.3			
22	Ti	4966	560.9	460.2	453.8	58.7	32.6	32.6			
23	V	5465	626.7	519.8	512.1	66.3	37.2	37.2			
24	Cr	5989	696	583.8	574.1	74.1	42.2	42.2			
25	Mn	6539	769.1	649.9	638.7	82.3	47.2	47.2			
26	Fe	7112	844.6	719.9	706.8	91.3	52.7	52.7			
27	Co	7709	925.1	793.2	778.1	101	58.9	59.9			
28	Ni	8333	1008.6	870	852.7	110.8	68	66.2			
29	Cu	8979	1096.7	952.3	932.7	122.5	77.3	75.1			
30	Zn	9659	1196.2	1044.9	1021.8	139.8	91.4	88.6	10.2	10.1	
31	Ga	10367	1299	1143.2	1116.4	159.5	103.5	100	18.7	18.7	
32	Ge	11103	1414.6	1248.1	1217	180.1	124.9	120.8	29.8	29.2	
33	As	11867	1527	1359.1	1323.6	204.7	146.2	141.2	41.7	41.7	
34	Se	12658	1652	1474.3	1433.9	229.6	166.5	160.7	55.5	54.6	
35	Br	13474	1782	1596	1550	257	189	182	70	69	
36	Kr	14326	1921	1730.9	1678.4	292.8	222.2	214.4	95	93.8	27.5
37	Rb	15200	2065	1864	1804	326.7	248.7	239.1	113	112	30.5
38	Sr	16105	2216	2007	1940	358.7	280.3	270	136	134.2	38.9
39	Y	17038	2373	2156	2080	392	310.6	298.8	157.7	155.8	43.8
40	Zr	17998	2532	2307	2223	430.3	343.5	329.8	181.1	178.8	50.6
41	Nb	18986	2698	2465	2371	466.6	376.1	360.6	205	202.3	56.4
42	Mo	20000	2866	2625	2520	506.3	411.6	394	231.1	227.9	63.2
43	Tc	21044	3043	2793	2677	544	447.6	417.7	257.6	253.9	69.5
44	Ru	22117	3224	2967	2838	586.1	483.5	461.4	284.2	280	75
45	Rh	23220	3412	3146	3004	628.1	521.3	496.5	311.9	307.2	81.4
46	Pd	24350	3604	3330	3173	671.6	559.9	532.3	340.5	335.2	87.1
47	Ag	25514	3806	3524	3351	719	603.8	573	374	368.3	97
48	Cd	26711	4018	3727	3538	772	652.6	618.4	411.9	405.2	109.8
49	In	27940	4238	3938	3730	827.2	703.2	665.3	451.4	443.9	122.9
50	Sn	29200	4465	4156	3929	884.7	756.5	714.6	493.2	484.9	137.1
51	Sb	30491	4698	4380	4132	946	812.7	766.4	537.5	528.2	153.2
52	Te	31814	4939	4612	4341	1006	870.8	820	583.4	573	169.4
53	I	33169	5188	4852	4557	1072	931	875	630.8	619.3	186
54	Xe	34561	5453	5107	4786	1148.7	1002.1	940.6	689	676.4	213.2
55	Cs	35985	5714	5359	5012	1211	1071	1003	740.5	726.6	232.3
56	Ba	37441	5989	5624	5247	1293	1137	1063	795.7	780.5	253.5
57	La	38925	6266	5891	5483	1362	1209	1128	853	836	274.7
58	Ce	40443	6549	6164	5723	1436	1274	1187	902.4	883.8	291
59	Pr	41991	6835	6440	5964	1511	1337	1242	948.3	928.8	304.5
60	Nd	43569	7126	6722	6208	1575	1403	1297	1003.3	980.4	319.2
61	Pm	45184	7428	7013	6459		1471	1357	1052	1027	
62	Sm	46834	7737	7312	6716	1723	1541	1420	1110.9	1083.4	347.2
63	Eu	48519	8052	7617	6977	1800	1614	1481	1158.6	1127.5	360
64	Gd	50239	8376	7930	7243	1881	1688	1544	1221.9	1189.6	378.6
65	Tb	51996	8708	8252	7514	1968	1768	1611	1276.9	1241.1	396
66	Dy	53789	9046	8581	7790	2047	1842	1676	1333	1292.6	414.2
67	Ho	55618	9394	8918	8071	2128	1923	1741	1392	1351	432.4
68	Er	57486	9751	9264	8358	2207	2006	1812	1453	1409	449.8
69	Tm	59390	10116	9617	8648	2307	2090	1885	1515	1468	470.9
70	Yb	61332	10486	9978	8944	2398	2173	1950	1576	1528	480.5
71	Lu	63314	10870	10349	9244	2491	2264	2024	1639	1589	506.8
72	Hf	65351	11271	10739	9561	2601	2365	2108	1716	1662	538
73	Ta	67416	11682	11136	9881	2708	2469	2194	1793	1735	563.4
74	W	69525	12100	11544	10207	2820	2575	2281	1872	1809	594.1
75	Re	71676	12527	11959	10535	2932	2682	2367	1949	1883	625.4
76	Os	73871	12968	12385	10871	3049	2792	2457	2031	1960	658.2
77	Ir	76111	13419	12824	11215	3174	2909	2551	2116	2040	691.1
78	Pt	78395	13880	13273	11564	3296	3027	2645	2202	2122	725.4
79	Au	80725	14353	13734	11919	3425	3148	2743	2291	2206	762.1
80	Hg	83102	14839	14209	12284	3562	3279	2847	2385	2295	802.2
81	Tl	85530	15347	14698	12658	3704	3416	2957	2485	2389	846.2
82	Pb	88005	15861	15200	13035	3851	3554	3066	2586	2484	891.8
83	Bi	90526	16388	15711	13419	3999	3696	3177	2688	2580	939
84	Po	93105	16939	16244	13814	4149	3854	3302	2798	2683	995
85	At	95730	17493	16785	14214	4317	4008	3426	2909	2787	1042

Table 3.2 *continued.*

$4p_{1/2}$	$4p_{3/2}$	$4d_{3/2}$	$4d_{5/2}$	$4f_{5/2}$	$4f_{7/2}$	$5s$	$5p_{1/2}$	$5p_{3/2}$	$5d_{3/2}$	$5d_{5/2}$
14.1	14.1									
16.3	15.3									
21.3	20.1									
24.4	23.1									
28.5	27.1									
32.6	30.8									
37.6	35.5									
42.3	39.9									
46.3	43.2									
50.5	47.3									
55.7	50.9									
63.7	58.3									
63.9	63.9	11.7	10.7							
73.5	73.5	17.7	16.9							
83.6	83.6	24.9	23.9							
95.6	95.6	33.3	32.1							
103.3	103.3	41.9	40.4							
123	123	50.6	48.9							
146.7	145.5	69.5	67.5			23.3	13.4	12.1		
172.4	161.3	79.8	77.5			22.7	14.2	12.1		
192	178.6	92.6	89.9			30.3	17	14.8		
205.8	196	105.3	102.5			34.3	19.3	16.8		
223.2	206.5	109		0.1	0.1	37.8	19.8	17		
236.3	217.6	115.1	115.1	2	2	37.4	22.3	22.3		
243.3	224.6	120.5	120.5	1.5	1.5	37.5	21.1	21.1		
242	242	120	120							
265.6	247.4	129	129	5.2	5.2	37.4	21.3	21.3		
284	257	133	127.7	0	0	32	22	22		
286	271		142.6	8.6	8.6	36	28	21		
322.4	284.1	150.5	150.5	7.7	2.4	45.6	28.7	22.6		
333.5	293.2	153.6	153.6	8	4.3	49.9	26.3	26.3		
343.5	308.2	160	160	8.6	5.2	49.3	30.8	24.1		
366.2	320.2	167.6	167.6		4.7	50.6	31.4	24.7		
385.9	332.6	175.5	175.5		4.6	54.7	31.8	25		
388.7	339.7	191.2	182.4	2.5	1.3	52	30.3	24.1		
412.4	359.2	206.1	196.3	0.9	7.5	57.3	33.6	26.7		
438.2	380.7	220	211.5	15.9	14.2	64.2	38	29.9		
463.4	400.9	237.9	226.4	23.5	21.6	69.7	42.2	32.7		
490.4	423.6	255.9	243.5	33.6	31.4	75.6	45.3	36.8		
518.7	446.8	273.9	260.5	42.9	40.5	83	45.6	34.6		
549.1	470.7	293.1	278.5	53.4	50.7	84	58	44.5		
577.8	495.8	311.9	296.3	63.8	60.8	95.2	63	48		
609.1	519.4	331.6	314.6	74.5	71.2	101.7	65.3	51.7		
642.7	546.3	353.2	335.1	87.6	84	107.2	74.2	57.2		
680.2	576.6	378.2	358.8	104	99.9	127	83.1	64.5	9.6	7.8
720.5	609.5	405.7	385	122.2	117.8	136	94.6	73.5	14.7	12.5
761.9	643.5	434.3	412.2	141.7	136.9	147	106.4	83.3	20.7	18.1
805.2	678.8	464	440.1	162.3	157	159.3	119	92.6	26.9	23.8
851	705	500	473	184	184	177	132	104	31	31
886	740	533	507	210	210	195	148	115	40	40

(its order of magnitude is given by v^2/c^2 where v is the velocity of the electron) is described for the hydrogen atom in quantum mechanics textbooks. Let us recall the essential points. The electron has an intrinsic angular momentum (spin 1/2) which is associated with magnetic moment:

$$\vec{M}_s = \frac{q}{m}\vec{s}.$$

This magnetic moment interacts with the magnetic field created by the relative motion of the proton with respect to the electron by a Zeeman term:

$$H_{int} = -\vec{M}_s \cdot \vec{B}$$

The magnetic field can be calculated in the electron reference frame. In this frame, the proton rotates around the electron and the associated current loop generates a magnetic field which can be written from the laws of magnetostatics:

$$\vec{B} = \frac{1}{qmc^2}\frac{1}{r}\frac{dV(r)}{dr}\vec{p} \times \vec{r} = -\frac{1}{qmc^2}\frac{1}{r}\frac{dV(r)}{dr}\vec{\ell}$$

showing that the field is proportional to the electron's orbital angular momentum and to the potential energy gradient $V(r)$. The spin-orbit interaction term is written taking into account a factor 1/2 due to the Thomas precession[1]:

$$H_{S.O} = -\frac{1}{2}\vec{M}_s \cdot \vec{B} = \frac{1}{2m^2c^2}\frac{1}{r}\frac{dV}{dr}\,\vec{\ell} \cdot \vec{s}. \qquad (3.28)$$

This spin-orbit term also occurs for multi-electron atoms and must be considered to account for the final state of a photoemission process on a core state. Indeed in the final state, we have a hole in an electronic sublayer, characterized by an orbital angular momentum ℓ ($\ell = 1$ for a p level, 2 for a d level, etc.) and spin s ($s = 1/2$). The eigenstates of the spin-orbit are eigenstates of the total angular momentum. Rigorously, they are eigenstates of the j^2 operator arising from the sum the orbital and the spin angular momenta of the electron:

$$\vec{j} = \vec{\ell} + \vec{s} \quad \text{so} \quad \vec{\ell} \cdot \vec{s} = j^2 - \ell^2 - s^2.$$

The rules of combination of angular momenta lead at the total angular momentum equal to $j = \ell \pm 1/2$. We thus have two spin-orbit levels associated with the two values of j with $2j + 1$ degeneracy. The notation of these levels is $n\ell_j$ (*c.f.* table 3.3).

The spin-orbit separation depends on the gradient of V and the electron velocity. For the same atomic potential, it decreases with decreasing the principal quantum number. It can be noted that the electron velocity increases with decreasing the

[1] The reference frame of the electron is accelerated, which results in an additional rotation of the spin with respect to the laboratory reference frame.

Table 3.3 *Notation of core level structures*

n	ℓ	j	Level
1	0	1/2	$1s_{1/2}$
2	0	1/2	$2s_{1/2}$
2	1	1/2	$2p_{1/2}$
2	1	3/2	$2p_{3/2}$
3	0	1/2	$3s_{1/2}$
3	1	1/2	$3p_{1/2}$
3	1	3/2	$3p_{3/2}$
3	2	3/2	$3d_{3/2}$
3	2	5/2	$3d_{5/2}$
4	0	1/2	$4s_{1/2}$
4	1	1/2	$4p_{1/2}$
4	1	3/2	$4p_{3/2}$
4	2	3/2	$4d_{3/2}$
4	2	5/2	$4d_{5/2}$
4	3	5/2	$4f_{5/2}$
4	3	7/2	$4f_{7/2}$

electron-nucleus average distance and then with increasing the binding energy. Moreover, in a series of the periodic table, for the same sublayer $n\ell$, the spin-orbit coupling increases with increasing the nuclear charge of the atom because of the increase in the electron velocity.

A core level $(n\ell)$ thus manifests itself in a photoemission spectrum by two structures associated with the two values of j $\ell \pm s)$ whose spectral weight is proportional to $2j + 1$. Each peak exhibits a lorentzian shape (in insulator without multiplet effect):

$$A_h = \frac{\Gamma_h/\pi}{(E_f - E_i - \hbar\omega)^2 + \Gamma_h^2} \tag{3.29}$$

where the lorentzian width (Γ_h) is inversely proportional to the lifetime and depends on the hole desexcitation mechanisms.

3.1.4 *The one-step and the three-step models*

Let us return to the probability of transition. In the one-step model (Fig. 3.9b), introduced by Pendry *et al.* [12–14], initial $|\phi_i\rangle$ and final $|\phi_f\rangle$ states are eigenstates of a semi-infinite problem (crystal with its surface and vacuum). Due to the translation symmetry breaking at the surface, these states are Bloch states only in the directions parallel to the surface. The precise description of these states proves somewhat difficult. However, a qualitative description can be given. The initial state, far from the surface, must resemble a Bloch state but exhibit an evanescent form in the vacuum near the surface. On the other hand, the final state is, in the vacuum away from the surface, a plane wave and attenuates within the crystal because of the finite mean free path in matter. Technically, the description of the attenuated final state is obtained with a complex potential whose imaginary part is responsible for the losses and therefore for the attenuation; the surface potential remains real. The one-step model naturally considers several waves that arrive at the surface, which spontaneously describes the photoelectron diffraction behavior. This model leads to a better description of the phenomenon of the photoemission but at the price of a high degree of complexity.

A simpler and more intuitive model has been developed [15–17] that allows a simple interpretation of the photoemission spectra. We show that this approach can be deduced from the more general one-step model when the mean free path of the electrons is large (weak attenuation)[18]. It breaks down the photoemission process into three successive stages (Fig. 3.9a):

- Optical excitation of the electron between an initial and final state, inside the material.
- Propagation of the electron toward the surface.
- Crossing the surface: from the solid to the vacuum.

This model proves very useful and allows to obtain a very intuitive picture of the photoemission process. Beyond the decomposition into three stages, it also supposes several assumptions that may not be valid in certain situations. For example, the model considers that photoelectrons are distributed isotropically in space [15], without taking into account that the final state should be the interference between photoelectron waves that escapes directly from the solid and those that are dispersed elastically or inelastically by the atoms of the lattice. It is possible, however, to include multiple scattering in the three-step model [12, 19, 20].

Remark : surface photoemission

The singularity of the vector potential at the surface leads to a non-zero contribution of the term in $\vec{\nabla}\cdot\vec{A}$ [21, 22]. This contribution is negligible if the photon frequency is very large compared to the plasmon frequency (of the order of 10–20 eV). The surface can be considered in a first approximation as a step of $\varepsilon(\omega)$ (for a more elaborate surface model, see [23]). Maxwell's equations at the vacuum-surface interface show that the component of the radiation perpendicular to the surface is not constant.

3.1.5 *The three-step model*

Step 1 : optical excitation The first step describes the optical transition between two states of an infinite crystal, i.e., between two Bloch states. As we have seen

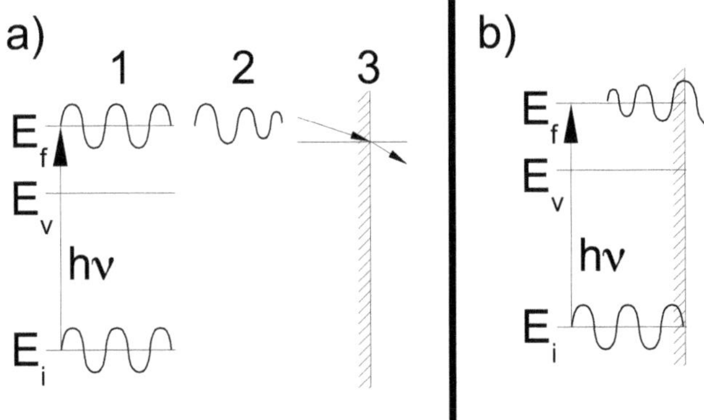

Figure 3.9 *Models for photoemission (a) in the three-step and (b) in the one-step approach. In the three-step model, there is an artificial division between the optical excitation process, the surface propagation and the surface-to-vacuum crossing.*

above, such a transition satisfies the conservation of the wave vector, more precisely of the reduced wave vector in the first Brillouin zone when the photon momentum can be neglected. The initial and final states are therefore characterized by the same wave vector $\vec{k}$, which we define inside the first Brillouin zone. We will now write them $|\phi^{(i)}_{\vec{k}}\rangle$ and $|\phi^{(f)}_{\vec{k}}\rangle$, so that the transition probability is written according to the equation 3.15:

$$w_{fi} = \frac{2\pi}{\hbar}|\langle\phi^{(f)}_{\vec{k}}|H_{int}|\phi^{(i)}_{\vec{k}}\rangle|^2\delta(E_f(\vec{k}) - E_i(\vec{k}) - \hbar\omega) \tag{3.30}$$

In this formula, it is considered that the interaction Hamiltonian, or in other words the vector potential of the electromagnetic wave, is the same as in vacuum. In reality, the electromagnetic wave is attenuated in the medium but the penetration length of the photons is generally very large compared to the mean free path of the electrons which limits the thickness probed by the photoemission, so this electromagnetic wave attenuation can be neglected.

The final state energy $E_f(\vec{k})$ depends on the photon energy. With UV photons (usually a few tens of eV), the final state belongs to a high energy band for which the kinetic energy is very large compared to the potential energy. We will see below the importance of this remark for the modelling of this excited final state. One can legitimately ask if final states exist for all $\vec{k}$ values and a given photon energy. We will discuss this point in detail below but note now that for high energy photons, this condition is always satisfied. It is noted that this first step depends only on the band structure of the crystal and we already understand the interest of the photoemission for the experimental determination of the band structure.

Step 2 : Propagation toward the surface

The first step leads to a transition inside the solid between two Bloch states (eigenstates of the infinite crystal with E_i and E_f energies). Before escaping the solid, photoelectrons propagate in the solid and can interact with other electrons. These interactions lead to a change in the energy and momentum (inelastic collisions) of the photoelectrons and to the loss of the information about the initial state. Only electrons that have not suffered a collision are of interest for photoemission, they are called primary electrons. Electrons which have lost energy during collisions are called secondary electrons. Their contribution to the photoemission spectrum is a continuous background with no structure whose intensity increases when the kinetic energy decreases before falling abruptly to $E_{kin} = 0$ as shown in Figure 3.10. This cut-off reflects the work function and depends on the photon energy [15].

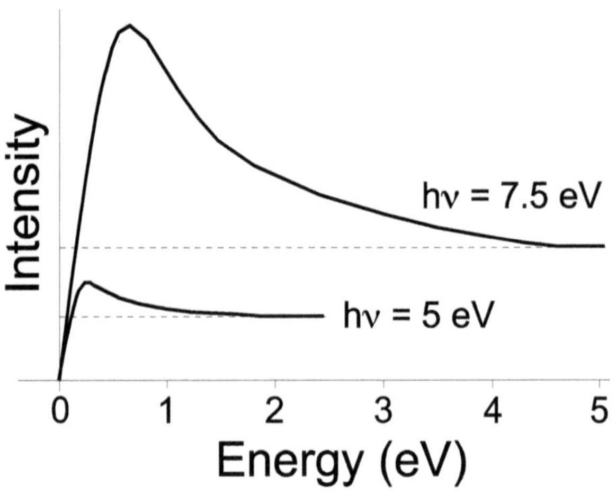

Figure 3.10 *Energy dependence of the contribution of secondary electrons to the photoemission spectra.*

The energy loss mechanisms are very varied: interactions with plasmons, phonons, excitations of electron-hole pairs ... It is often considered that inelastic scattering is isotropic. Inelastic collisions are described phenomenologically by the inelastic mean free path λ (*Inelastic Mean Free Path* - IMFP), which represents the average distance between two inelastic scatterings of an electron in a solid. This inelastic scattering limits the life time of the electronic state of the photoelectron. Indeed, a photoelectron described by a wave packet characterized by the group velocity v_g, remains in the same state during the time τ according to:

$$\lambda(E) = \tau v_g = \frac{\tau}{\hbar}\frac{dE}{dk} \tag{3.31}$$

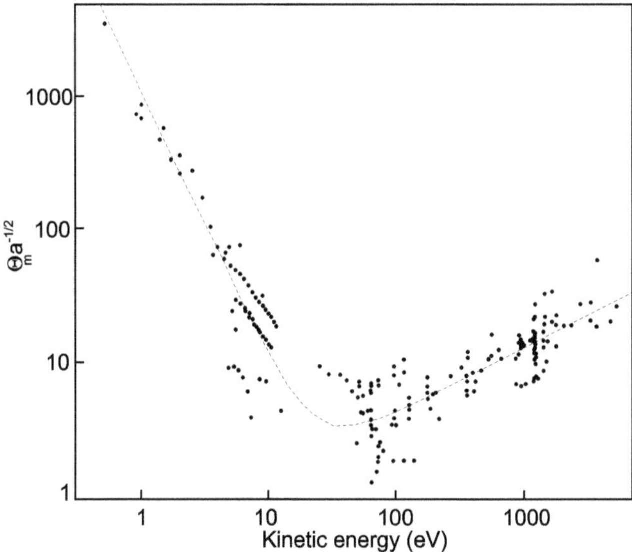

Figure 3.11 *Mean free path dependence versus kinetic energy of the outgoing electron for different elements (from [24]). ©1979 John Wiley & Sons, Ltd.*

Therefore, because of the uncertainty relation $\Delta E \cdot \tau \geq \hbar/2$, the inelastic scattering processes lead to an energy broadening of the photoelectron electronic states.

The inelastic mean free path depends on the kinetic energy of the electrons. This so-called universal energy dependence (because it is observed in all materials Fig. 3.11) exhibits a minimum of the order of a few Å, for photoelectron kinetic energies between 40 and 100 eV. As the probability that a photoelectron created at the distance z from the surface does not undergo inelastic scattering varies in $\exp(-z/\lambda)$, photoemission only probes a thickness in the order of λ below the surface. It is possible to understand qualitatively the energy dependence of the mean free path. On the one hand, at high kinetic energy, the increase of the average free path with energy is a characteristic of the behavior of the scattering cross section with energy. The faster a particle goes, the shorter the time spent in the interaction region, hence the decrease in the probability of scattering. This generic behavior is also valid in classical mechanics: the faster a vehicle crosses a crossroads, the shorter the crossing time and the lower the probability of an accident occurring! On the other hand, the low energy behavior is more subtle. It reflects a decrease in the number of inelastic scattering channels when the kinetic energy of the electrons decreases. Indeed, an electron with kinetic energy E can only interact with energy excitations less than E. For example, a high energy electron can excite a plasmon but when its energy goes below the threshold energy of plasmons this mechanism is forbidden. This universal curve shows that to be less sensitive to the surface, it is necessary to use either high energy photons (of several keV) or photons of very low energy (less than 10 eV). We will return below to the consequences of this property. Finally, it should be noted that the probed thickness introduces an uncertainty on the perpendicular component of the wave vector. Indeed, Heisenberg's uncertainty relation:

$$\Delta k_{\perp} \Delta z \sim 1$$

leads for the photoelectron to $\Delta k_{\perp} \sim 1/\lambda$.

Surface sensitivity of photoemission has important technical consequences as it requires surface preparation (atomic cleanliness) and ultra-vacuum technology to keep surfaces clean for as long as possible.

Step 3: Escaping from the surface

The last step is the crossing of the surface. This process obeys conservation laws associated with the symmetry of the problem. The surface corresponds to a potential barrier that breaks the translational invariance of the crystal in the perpendicular direction. This leads to the non-conservation of the perpendicular component of the wave vector. On the other hand, translational invariance in parallel directions is conserved (the medium outside the solid is invariant by any translation). Using the notations defined in the previous section:

$$\vec{K}_{ext\|} = \vec{K}_{\|} + \vec{G}_{\| } \tag{3.32}$$

where we recall that $\vec{K}_{\|}$ and $\vec{K}_{ext\|}$ are respectively the component parallel to the surface of the wave vector of the photoelectron in the solid and in the vacuum (Fig. 3.12). Note that wave vector conservation is modulo a vector of the reciprocal lattice to take into account the periodicity of the surface. This leads to the emission of photoelectrons according to the different cones of Mahan (primary and secondary).

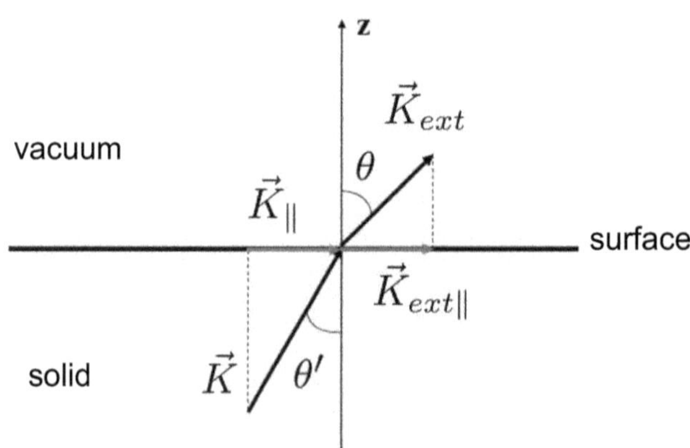

Figure 3.12 *Conservation of the momentum parallel to the surface for the first cone of Mahan $\vec{G}_{\|} = 0$.*

Photoemission spectroscopy is based on the conservation laws of of energy and parallel components of the wave vector. It is possible from the measurement of the energy and emission angles (θ and ϕ) of the photoelectron, to determine the band structure of the solid. In the vacuum, the energy of the photoelectron is dominated by its kinetic energy:

$$E_{kin} = \frac{\hbar K_{ext}^2}{2m} \tag{3.33}$$

If we consider the first Mahan cone ($\vec{G}_{\|} = 0$), as $K_{ext\|} = K_{ext} \sin\theta$, where θ is the emission angle of the photoelectron, we have:

$$K_{ext\|} = K_{\|} = \sqrt{\frac{2m}{\hbar^2}(h\nu - E_i - \Phi)} \sin\theta = \sqrt{\frac{2m}{\hbar^2} E_{kin}} \sin\theta \tag{3.34}$$

which gives numerically:

$$K_{\|} = 0.512\text{Å}^{-1} \sqrt{E_{kin}(\text{eV})} \sin\theta \tag{3.35}$$

The transition across the surface in the first cone of Mahan recalls the law of Snell-Descartes of the refraction (figure 3.12):

$$K_{\|} = \sin\theta \left[\frac{2m}{\hbar^2} E_{kin}\right]^{1/2} = \sin\theta' \left[\frac{2m}{\hbar^2}(E_{kin} + V_0)\right]^{1/2}, \tag{3.36}$$

where θ and θ' represent the propagation angles of the photoelectron inside and outside the crystal. An uncertainty of the polar angle of $\Delta\theta$ leads to an uncertainty on $K_{\|}$:

$$\Delta K_{\|} = 0.512\text{Å}^{-1} \sqrt{E_{kin}(\text{eV})} \cos\theta \; \Delta\theta \tag{3.37}$$

This relation shows that for the same angular resolution $\Delta\theta$, the wave vector resolution is higher ($\Delta K_{\|}$ decreases) with decreasing the photon energy (E_{kin}).

Translational symmetry leads to the conservation of $K_{\|}$ but does not impose a condition for $K_{\perp}$. The relation between $K_{ext\perp}$ and $K_{\perp}$ is a function of the crystal structure and the inner potential V_0. Since photoelectrons have a high kinetic energy, it is often legitimate to neglect the periodic potential of the solid and to consider that the final states in the solid can be described by the approach of free electrons [25–29]. The problem of determining $K_{\perp}$ is then reduced to the determination of the potential V_0 (figure 3.3), which can be obtained either experimentally or from theoretical calculations of band structure. The energy of the photoelectron in the solid $\hbar^2 K^2/2m$ is equal to the sum of the kinetic energy E_{kin} outside the solid, that is, with respect to the vacuum level, and the inner potential V_0:

$$\frac{\hbar^2}{2m}(K_{\|}^2 + K_{\perp}^2) = E_{kin} + V_0 \tag{3.38}$$

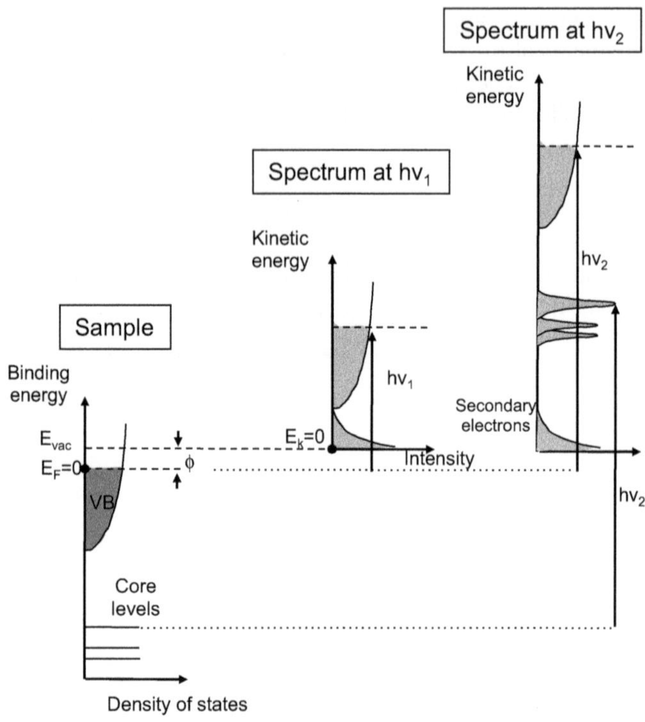

Figure 3.13 *Schematic picture of angle-resolved photoemission and definition of emission θ and ϕ angles.*

which makes it possible to obtain, using the relation 3.34, the perpendicular component of the wave vector in the final state:

$$K_{\perp} = \sqrt{\frac{2m}{\hbar^2}(E_{kin}\cos^2\theta + V_0)} \tag{3.39}$$

To obtain the wave vector of the initial state $\vec{k}_i$, one need only return to the first step and consider that the transition between the initial state is vertical, i.e., for the perpendicular component:

$$k_{i\perp} = K_{\perp} \tag{3.40}$$

modulo of course a reciprocal lattice vector $G_{\perp}$. It should be noted, however, that the finite mean free path of the photoelectron in the solid introduces uncertainty on the wave vector. Indeed, the photoelectron detected in the analyzer was emitted near the surface in a layer thickness of about $\Delta z \approx \lambda$. The Heisenberg uncertainty relations thus lead to the relation:

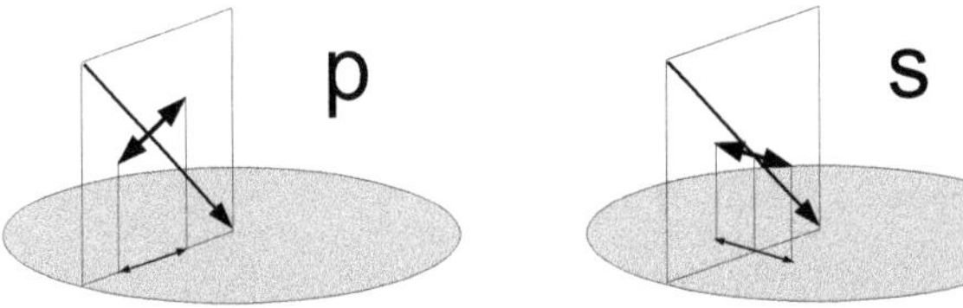

Figure 3.14 *Definition of linear polarizations. The p polarization corresponds to the electric field vector in the plane of incidence whereas for the s polarization, the electric field is perpendicular to the plane of incidence.*

$$\Delta k_{\perp} \sim \frac{1}{\lambda} \tag{3.41}$$

However, it should be noted that in most cases, $\Delta k_{\perp}$ remains small with respect to the size of the Brillouin zone so that $k_{\perp}$ remains fairly well defined.

Summary of the three steps

From the summary of the three steps, it is possible to obtain the transition probability and the photoemission intensity. The kinetic energy E_{kin} and momentum $\vec{p}$ dependence of the photoelectron intensity can be written:

$$I^{int}(E_{kin}, \hbar\omega, \vec{p}, \theta_r) \sim \sum_{f,i} |M_{fi}|^2 \; f(E_i)\delta(E_f - E_i - \hbar\omega)\delta(E_{kin} - E_f + \Phi)$$
$$\delta(\vec{K} - \vec{k}_i - \vec{G})\,\delta(\vec{K}_{\parallel} - \vec{p}_{\parallel}/\hbar - \vec{G}_{\parallel}) \tag{3.42}$$

where $f(E)$ is the Fermi function and Φ the work function of the analyser.

The Fermi functions appearing in this intensity equation reflect the different conservation laws. Indeed, the energy of the final state is the sum of the energy of the photon and the initial state whereas the kinetic energy of the photoelectron outside the solid is reduced by the work function. The conservation of the wave vector is more complicated. Since the transition is vertical in the first step between two Bloch states, the wave vector is conserved, but only the component parallel to the surface is conserved due to the crossing of the surface (third step).

If the matrix element M_{fi} were constant, the intensity would be proportionate to the density of the occupied states. If this hypothesis permits a rough interpretation of the photoemission spectra, the modulation of the matrix elements of M_{fi} with angle, photon polarization or energy can lead to significant changes in the measured intensity (matrix element effects). Indeed according to the equation 3.14, the matrix element involves the potential vector $\vec{A}_0$ of the electromagnetic field:

$$M_{fi} = \frac{-q}{m}\langle\phi_f|\vec{p}|\phi_i\rangle \cdot \vec{A}_0$$

We consequently have two cases according to which the vector potential, and therefore the electric field of the electromagnetic wave, is in the plane (polarization p) or perpendicular to the plane (polarization s) defined by the normal to the surface and the direction of photons (plane of incidence).

3.2 Detailed analysis of valence states: N-body approach

When interactions between electrons are important, that is, when the valence electrons are highly localized, band calculations are no longer appropriate to describe the electronic structure. Likewise, the description of photoemission in a one-electron model is no longer valid and a many-body approach is necessary to take into account the effect of electronic correlations on the excitation spectrum. Before developing the many-body approach of photoemission, we would like to recall some generalities on the electronic correlations in solids.

3.2.1 *Fermi liquid and quasi-particles*

In an interacting particle system, only magnitudes associated with all of the particles have a physical meaning, for example, the energy of a particle is not conserved and only the total energy is defined. To illustrate this behavior, let us imagine that we place, at a given moment in time, an electron in a Bloch state. In a system without interaction, the electron would remain there for an infinite time ($\vec{k}$ is in this case a good quantum number). In contrast, electronic interactions will induce transitions between different Bloch states so the electron will only stay for a finite time in the initial state. Landau introduced a concept, the concept of quasi-particle, to give individual entities a physical sense. Indeed, in systems where interactions can be treated as a perturbation, we can build more complex, almost independent entities: the quasi-particles (QP) which describe the one-particle excitations of a Fermi liquid.

The behavior of these entities is reminiscent of the electrons in an electron gas without interactions. Let us specify the main results of the Landau model of a Fermi liquid. The interaction between quasi-particles moves asymptotically to zero at the Fermi level and increases as their energy increases. We can then describe the low-energy physical properties by the excitations of quasi-particles; at high energy the quasi-particles are no longer defined (overstrong interactions between the QP). While in the electron gas at finite temperature, we have excitations of the electron-hole type, these are quasi-particles ("quasi-electron-quasi-hole" pairs) in the correlated systems. For example, the resistivity in a system of nearly-free electrons is interpreted by the diffusion of electrons by defects and it corresponds to the diffusion of quasi-particles when one has interactions. The properties of quasi-particles can be qualitatively understood by invoking the Heisenberg time-energy uncertainty relation. Because of residual interactions between quasi-particles, they have a finite life time τ (which diverges nevertheless at Fermi level). The quasi-particle energy E is therefore defined with an uncertainty $\Delta E = \hbar/\tau$. A quasi-particle therefore remains meaningful as long as $\Delta E \ll E$, or in other words, as long as its life time remains higher at the characteristic time of the excitation.

The fundamental interest of quasi-particles lies in the fact that the expressions of the thermodynamic quantities deduced from a free electron approach remain valid in the correlated systems as soon as we use renormalized quantities. For example, Pauli's susceptibility to free electron gas is proportional to the density of low energy excitations (density of states $N(E_F)$). In an interacting electron system, we have the same expression:

$$\chi = \mu_0 \mu_B N^{Q.P.}(E_F) \tag{3.43}$$

but where $N^{Q.P.}(E_F)$ is the quasiparticle state density. Most thermodynamic techniques involve renormalized quantities associated with quasi-particles because they only concern low-energy excitations ($E \approx kT$). On the other hand, dynamic techniques, such as photoemission, involve the excitations of an electron. Indeed, in a photoemission experiment an electron is extracted not a quasi-particle! In the vicinity of E_F (low energy excitations), the photoemission spectrum gives information on quasi-particles. But, the further we get from E_F the higher the excitations of high energy that inform directly as to the interactions between electrons.

3.2.2 *Many-body formalism*

In the N-body models, the second quantization formalism is particularly well-adapted. This formalism allows to treat systems with a variable number of particles and uses operators of creation and annihilation of particles. For the following, it is not necessary to know all the mathematical tools of the second quantification. We only need to recall that the creation operator of a Bloch state ($|\vec{k}\rangle$), noted $a^\dagger_{\vec{k}}$, is defined by its action on the vacuum state (state of the system with 0 particle noted $|0\rangle$):

$$a^\dagger_{\vec{k}}|0\rangle = |\vec{k}\rangle.$$

A state of N independent particles (a Slater determinant) can be written:

$$\prod_i^N a^\dagger_{\vec{k}_i}|0\rangle = |\vec{k}_1, \vec{k}_2, \ldots \vec{k}_N\rangle.$$

In the same way we define destruction or annihilation operators denoted $a_{\vec{k}_j}$ the action of which, on a state with N electrons leads to a state with (N$-$1) electrons. for example:

$$a_{\vec{k}_j}|\vec{k}_1,\vec{k}_2,\ldots,\vec{k}_j,\ldots\vec{k}_N\rangle = |\vec{k}_1,\vec{k}_2,\ldots,0,\ldots\vec{k}_N\rangle.$$

Finally, the algebra of these operators is defined by the antisymmetrization equations:

$$\{a^\dagger_{\vec{k}_i}, a^\dagger_{\vec{k}_j}\} = \{a_{\vec{k}_i}, a_{\vec{k}_j}\} = \delta(\vec{k}_i,\vec{k}_j) \text{ and } \{a^\dagger_{\vec{k}_i}, a_{\vec{k}_j}\} = 0 \tag{3.44}$$

to satisfy the fermionic character of electrons[2]. In the second quantization formalism, all the operators can be written from creation and destruction operators.

3.2.2.1 Hamiltonian and eigenstates.

The electromagnetic field-matter interaction induces, during the photoemission process, a transition from a monoelectronic state i (initial) to another monoelectronic state f (final). In the second quantization formalism, this interaction is simply described by the destruction of the initial i state followed by the creation of the f final state. The interaction Hamiltonian can then written from the creation and destruction operators:

$$H_{int} = \sum_{i,f} M_{fi} a^\dagger_f a_i \tag{3.45}$$

where we sum over all possible initial and final states, M_{fi} is the matrix element of the electron-photon interaction between the initial and final monoelectronic states (it is the same term as the one introduced above in the one-electron description of photoemission). To obtain the transition probability, one must take the matrix element of H_{int} between the N-electron states describing the system as a whole. We will call $|N,0\rangle$ the system ground state and $|N,s\rangle$ the different excited states. As we have seen above, these different states, in the approximation of independent electrons, correspond to simple determinants of Slater with N electrons. On the other hand, in the presence of interactions, these states are more complex and can be written as linear combinations of Slater determinants. The transition probability is written for a photon energy $\hbar\Omega$:

$$p(\hbar\Omega) = \frac{2\pi}{\hbar}\sum_s |\langle N,s|H_{int}|N,0\rangle|^2 \delta(E_s^{(N)} - E_0^{(N)} - \hbar\Omega) \tag{3.46}$$

where $E_s^{(N)}$ and $E_0^{(N)}$ are respectively the eigenstate energies of the system in the initial and final state.

We can now re-examine the notion of sudden approximation. For the photoemission on the valence states, the interaction Hamiltonian given by the equation 3.45 can be written:

[2] These are the antisymmetrization fermion relations. Remember that $\{a^\dagger_i a_j\} = a^\dagger_i a_j + a_j a^\dagger_i$.

$$H_{int} = \sum_{\vec{k},\vec{\kappa}} M_{\vec{\kappa},\vec{k}}\, a^{\dagger}_{\vec{\kappa}} a_{\vec{k}} \tag{3.47}$$

where the initial state is a Bloch state labelled by the wave vector $\vec{k}$ whereas the final state is a free electron state in the vacuum characterized by the wave vector $\vec{\kappa}$. In the photoemission process, we can neglect the interaction of the photoelectron with the system, so that the final state with N electrons is written as the antisymmetric product of the monoelectronic state $|\phi_{\vec{\kappa}}\rangle$ of energy $\varepsilon_{\vec{\kappa}}$ and momentum $\hbar\vec{\kappa}$, describing the photoelectron in a vacuum with a state at $(N-1)$ electrons describing the remaining system:

$$|N, s\rangle = A(|\phi_{\vec{\kappa}}\rangle \otimes |N-1, s\rangle) = a^{\dagger}_{\vec{\kappa}}|N-1, s\rangle$$

$|N-1, s\rangle$ is an eigenstate of the remaining system with $(N-1)$ electrons of energy $E_s^{(N-1)}$ and A is the operator which antisymmetrizes the total electronic state.

With this assumption, the probability to detect a photoelectron of energy $\varepsilon_{\vec{\kappa}}$ and wave vector $\vec{\kappa}$ (equation 3.46) is written:

$$p(\varepsilon_{\vec{\kappa}}, \vec{\kappa}) = \frac{2\pi}{\hbar} \sum_{s} |\langle N-1, s| a_{\vec{\kappa}} H_{int} |N, 0\rangle|^2 \delta(E_s^{(N-1)} + \varepsilon_{\vec{\kappa}} - E_0^{(N)} - \hbar\Omega) \tag{3.48}$$

The square of the matrix element can be developed:

$$|\langle N-1, s| a_{\vec{\kappa}} \sum_{\vec{k},\vec{\kappa}'} M_{\vec{\kappa}',\vec{k}}\, a^{\dagger}_{\vec{\kappa}'} a_{\vec{k}} |N, 0\rangle|^2 = \sum_{\vec{k}} |M_{\vec{\kappa},\vec{k}}|^2\, |\langle N-1, s| a_{\vec{k}} |N, 0\rangle|^2$$

We will take into account the wave vector conservation presented above. Only the parallel component is conserved due to the crossing of the surface. We obtain:

$$p(\varepsilon_{\vec{\kappa}}, \vec{\kappa}) = \frac{2\pi}{\hbar} \sum_{s,\vec{k}} |M_{\vec{\kappa}\vec{k}}|^2\, |\langle N-1, s| a_{\vec{k}} |N, 0\rangle|^2 \delta(\varepsilon_{\vec{\kappa}} + E_s^{(N-1)} - E_0^{(N)} - \hbar\Omega)\, \delta(\vec{k}_{\parallel} - \vec{\kappa}_{\parallel} - \vec{G}_{\parallel}) \tag{3.49}$$

The interpretation of this expression is immediate: because of the interactions between electrons, $a_{\vec{k}}|N, 0\rangle$ is not an eigenstate of the system and can be expressed in the base of eigenstate of the system with $(N-1)$ electrons:

$$a_{\vec{k}}|N, 0\rangle = \sum_{s} c_s |N-1, s\rangle = \sum_{s} \langle N-1, s| a_{\vec{k}} |N, 0\rangle\, |N-1, s\rangle \tag{3.50}$$

where c_s is the amplitude of each transition given by the matrix element of $a_{\vec{k}}$ between the ground state $|N, 0\rangle$ and the final state $|N-1, s\rangle$. Photoemission appears as the response of an electronic system to the sudden annihilation of an electron. This condition is not satisfied in other spectroscopies, such as absorption, for which the excited electron remains in the system and continues to interact with it. In the context of the sudden approximation, the transition probability is proportional to a fundamental quantity, the spectral function $A^{-}(\vec{k}, \omega)$, which is a characteristic of the Hamiltonian system and describes the one-particle excitations:

$$p(\varepsilon_{\vec{\kappa}},\vec{\kappa}) = \frac{2\pi}{\hbar}\sum_{\vec{k}} |M_{\vec{\kappa}\vec{k}}|^2 A^-(\vec{k},\omega)\,\delta(\vec{k}_\parallel - \vec{\kappa}_\parallel - \vec{G}_\parallel)\,\delta(\varepsilon_{\vec{\kappa}} + \varepsilon_s^{(N-1)} - \mu - \hbar\Omega) \tag{3.51}$$

with

$$A^-(\vec{k},\omega) = \sum_s |\langle N-1,s|a_{\vec{k}}|N,0\rangle|^2 \delta(\omega + \varepsilon_s^{(N-1)}) \tag{3.52}$$

μ is the chemical potential defined by $\mu = E_0^{(N)} - E_0^{(N-1)}$ and $\varepsilon_s^{(N-1)}$ is the excitation energy of the $N-1$ electron system defined by $\varepsilon_s^{(N-1)} = E_s^{(N-1)} - E_0^{(N-1)}$[3]. In a metal at zero temperature, $\mu = \varepsilon_F$ is the Fermi energy. In the spectral function $A^-(\vec{k},\omega)$, $\omega = 0$ corresponds to the Fermi level. The spectral function is a property of the Hamiltonian of the system and can be shown to express itself as the imaginary part of the one-electron Green function:

$$A^-(\vec{k},\omega) = \frac{1}{\pi} Im\, G^-(\vec{k},\omega) = \int_{-\infty}^{+\infty} dt\; e^{i\omega t}\Big(i\langle N,0|a^\dagger_{\vec{k}}(t)a_{\vec{k}}(0)|N,0\rangle\Big). \tag{3.53}$$

The many-body effects are completely described by the spectral function $A^-(\vec{k},\omega)$. The experimental intensity, proportionate to $p(\varepsilon_{\vec{\kappa}},\vec{\kappa})$, corresponds to the spectral function modulated by the one-electron matrix element $M_{\vec{\kappa}\vec{k}}$. We have seen that this matrix element describes the distribution of transition probability in the different accessible Brillouin zones associated with the different Mahan cones ($\vec{\kappa}_\parallel = \vec{k}_\parallel + \vec{G}_\parallel$). However, in a first approximation, we could consider that the matrix element $M_{\vec{\kappa}\vec{k}}$ in a given Brillouin zone is constant so that photoemission measures directly the spectral function (imaginary part of the one-electron Green function). Thus in each Brillouin zone we observe a replica of $A^-(\vec{k},\omega)$ whose intensity is given by the variation of the one-electron matrix element between the different zones. This approximation is the more satisfactory as the kinetic energy of the photoelectrons (and therefore the photon energy) is larger. However, at low energy and for experimental geometries leading to intensity extinctions due to symmetry, the wave vector variation of the matrix elements can lead to an experimental intensity which is no longer simply proportionate to the spectral function.

This approach provides a better understanding of what a photoemission spectrum represents. It gives the one-particle excitations of the electronic system. It is very important to distinguish these excitations energies of the energies appearing in the dispersion relations obtained from a band structure calculation. The Bloch state energies ($\varepsilon_{\vec{k}}$) correspond to the ground state and cannot be interpreted as excitations of the system. Strictly speaking, these quantities have no physical meaning and only the sum over all states corresponds to a physical quantity, *viz.*, the total energy of the

[3] The sum on the final states becomes an integral and a continuous spectral function in the case of an infinite system.

system. Nevertheless, when the electronic correlations are weak, the $\varepsilon_{\vec{k}}$ constitute satisfactory approximations of the excitations of the system justifying the comparison of the photoemission spectra with band structure calculations.

The fundamental difference with the monoelectronic approach developed in Chapter 3 is that when an electron is emitted, the whole system reacts and the other electrons do not remain, so to speak, as spectators. One can, of course, find the monoelectronic formalism from the formalism to the many body approach. In the limit of electrons without interactions, when an electron is removed, we obtain an eigenstate of the $(N-1)$ electron system so that the equation 3.50 becomes:

$$a_{\vec{k}}|N,0\rangle = |N-1,s_i\rangle$$

and the spectral function is reduced to a delta function:

$$A^{-}(\vec{k},\omega) = \sum_{s} |\langle N-1,s|a_{\vec{k}}|N,0\rangle|^2 \delta(\omega + \varepsilon_s^{(N-1)}) = \delta(\omega + \varepsilon_{s_i}^{(N-1)}),$$

since only one term (the one corresponding to $s = s_i$) is non-zero. On the other hand, $\varepsilon_{s_i}^{(N-1)}$ is the energy with respect to the Fermi level $(\varepsilon_{\vec{k}})$ of the Bloch state probed by the photoemission process.

3.2.2.2 Spectral density and spectral function

We have shown that an angle-resolved photoemission spectrum is proportionate to the spectral function $A^{-}(\vec{k},\omega)$ which is a quantity associated with the Hamiltonian of the system (this is the imaginary part of the one-electron Green function) and therefore it characterizes the interactions between electrons. In this section, we will examine the properties of this spectral function. Consider the simple case of independent electrons. The Hamiltonian can be written as a sum of terms to one electron:

$$H_0 = \sum_{k} \varepsilon_{\vec{k}}^0 a_{\vec{k}}^+ a_{\vec{k}} = \sum_{k} \varepsilon_{\vec{k}}^0 n_{\vec{k}},$$

where $n_{\vec{k}}$ is the occupation number operator of the state with energy $\varepsilon_{\vec{k}}^0$. Applied to the ground state, it gives 1 for occupied states and 0 for unoccupied states[4]. We find that the total energy is the sum of the occupied state energies:

$$E_0^{(N)} = \sum_{k}^{k_F} \varepsilon_{\vec{k}}^0.$$

[4] At non-zero temperature, $n_{\vec{k}}$ for electrons without interactions is the occupation probability of a state, it is therefore the Fermi function.

As we already saw at the end of the previous paragraph, the system being without interactions, $a_{\vec{k}}|N,0\rangle$ is an eigenstate of the N−1 electron system with $E_0^{(N)} - \varepsilon_{\vec{k}}^0$, so that the spectral function is a single Dirac peak. It is customary to place the origin of the one-electron excitation energies at the Fermi level and to have negative ω for occupied levels probed by photoemission. We then write:

$$E_0^{(N)} - \varepsilon_{\vec{k}}^0 = E_0^{(N-1)} + \mu - \varepsilon_{\vec{k}}^0 = E_0^{(N-1)} - \varepsilon_{\vec{k}}$$

where $\varepsilon_{\vec{k}}$ is negative for occupied states ($\varepsilon_{\vec{k}} = \varepsilon_{\vec{k}}^0 - \mu$). With this definition, the spectral function is then:

$$A^-(\vec{k},\omega) = \delta(\omega - \varepsilon_{\vec{k}}) = +\frac{1}{\pi} \lim_{\eta \to 0^+} Im \frac{1}{\omega - \varepsilon_{\vec{k}} - i\eta} \tag{3.54}$$

where we used the Cauchy theorem which states:

$$\lim_{\eta \to 0} \frac{1}{x \pm i\eta} = \mathcal{P}\frac{1}{x} \mp i\pi\delta(x)$$

where $\mathcal{P}$ indicates the Cauchy principal part. One checks that for occupied states $\varepsilon_{\vec{k}}^0 \leqslant \mu$ and therefore $\varepsilon_{\vec{k}}$ negatif ($\varepsilon_{\vec{k}} = 0$ corresponds to the Fermi energy). This formula expresses that the spectral function is the imaginary part of the one-particle Green function ($G^{0-}(\vec{k},\omega)$) which in momentum-energy representation is written for a system without interactions:

$$G^{0-}(\vec{k},\omega) = \lim_{\eta \to 0} \frac{1}{\omega - \varepsilon_{\vec{k}} - i\eta}, \quad \omega \leqslant 0. \tag{3.55}$$

For an interacting electron system, we find a similar expression with a sum on the final states:

$$G^-(\vec{k},\omega) = \lim_{\eta \to 0} \sum_s \frac{|\langle N-1, s|a_{\vec{k}}|N,0\rangle|^2}{\omega + E_s^{(N-1)} - E_0^{(N)} + \mu - i\eta}, \quad \omega \leqslant 0. \tag{3.56}$$

The reader can check that the imaginary part gives the spectral function $A^-(\vec{k},\omega)$ given by the relation 3.52.

$$A^-(\vec{k},\omega) = \sum_s |\langle N-1, s|a_{\vec{k}}|N,0\rangle|^2 \delta(\omega + \varepsilon_s^{(N-1)}) \quad \omega < 0.$$

The Green function corresponding to the positive energies and to the addition of an electron and thus to transitions to states at $N+1$ electrons is given by:

$$G^{+}(\vec{k},\omega)=\lim_{\eta\to 0}\sum_{s}\frac{|\langle N+1,s|a_{\vec{k}}^{\dagger}|N,0\rangle|^{2}}{\omega-E_{s}^{(N+1)}+E_{0}^{(N)}+\mu+i\eta},\qquad \omega\geqslant 0. \tag{3.57}$$

The corresponding spectral function $A^{+}(\vec{k},\omega)$ corresponds to inverse photoemission spectroscopy and gives information about unoccupied states $\omega \geqslant 0$:

$$A^{+}(\vec{k},\omega)=\sum_{s}|\langle N+1,s|a_{\vec{k}}^{\dagger}|N,0\rangle|^{2}\delta(\omega-\varepsilon_{s}^{(N+1)})\quad \omega\geqslant 0.$$

where $\varepsilon_{s}^{(N+1)}$ is the excitation energies of the system with $N+1$ electrons, *i.e.*,

$$\varepsilon_{s}^{(N+1)}=E_{s}^{(N+1)}-E_{0}^{(N+1)}.$$

Thus the total spectral function, describing both the addition and the annihilation of an electron and corresponding to photoemission and inverse photoemission spectroscopies is simply written:

$$A(\vec{k},\omega)=A^{+}(\vec{k},\omega)+A^{-}(\vec{k},\omega), \tag{3.58}$$

it obeys the following sum rule:

$$\int_{-\infty}^{+\infty}A(\vec{k},\omega)\,d\omega=\int_{-\infty}^{+\infty}\left(A^{+}(\vec{k},\omega)+A^{-}(\vec{k},\omega)\right)\,d\omega=1 \tag{3.59}$$

which simply reflects the normalization of a fermionic state. However, since the $\vec{k}$ wave vector is not conserved in an interacting system, the $A(\vec{k},\omega)$ function may have non-zero values for positive and negative ω, i.e., the spectral function for $\vec{k} < \vec{k}_F$ may have a non-zero contribution for $\omega > 0$. If we limit ourselves to the spectral function measured in photoemission, we have the equation (at zero temperature):

$$\int_{-\infty}^{+\infty}A^{-}(\vec{k},\omega)\,d\omega=n(\vec{k}) \tag{3.60}$$

where $n(\vec{k})$ is the occupation function of the state labelled by $\vec{k}$. We will come back below on this function in the study of Fermi liquids.

The spectral function, which is a simple Dirac peak for a non-interacting electron system, is significantly modified by the interactions. It can be put in a more useful form by introducing a quantity called self-energy. Indeed, the Green function ($G = G^{+} + G^{-}$) obeys an iterative equation called the Dyson equation which allows to define a fundamental quantity, the self-energy $\Sigma(\vec{k},\omega)$, which contains all the interaction effects:

$$G(\vec{k},\omega)=G^{0}(\vec{k},\omega)+G^{0}(\vec{k},\omega)\,\Sigma(\vec{k},\omega)\,G(\vec{k},\omega) \tag{3.61}$$

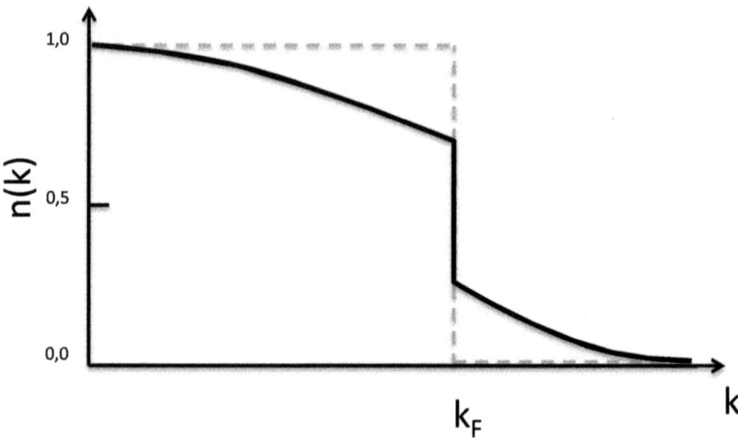

Figure 3.15 *$n(\mathbf{k})$ function for a non-interacting electron system (in dotted lines) and for an interacting electron system (solid line). The Fermi surface corresponds to the discontinuity for $\vec{k} = \vec{k}_F$.*

where $G^0(\vec{k}, \omega)$ is the Green function of the non-interacting electron system. According to our convention, $\omega = 0$ corresponds to the Fermi level which is also the energy origin of the excitations $\varepsilon_{\vec{k}}$. We can put this expression in the form:

$$G(\vec{k}, \omega) = \frac{1}{\omega - \varepsilon_{\vec{k}} - \Sigma(\vec{k}, \omega)} . \tag{3.62}$$

If we call Σ' and Σ'' the real and imaginary parts of the self-energy ($\Sigma(\vec{k}, \omega) = \Sigma'(\vec{k}, \omega) + i\Sigma''(\vec{k}, \omega)$), we obtain easily for the spectral function:

$$A^{\pm}(\vec{k}, \omega) = \mp\frac{1}{\pi} Im \frac{1}{\omega - \varepsilon_{\vec{k}} - \Sigma(\vec{k}, \omega)} = \frac{1}{\pi} \frac{|\Sigma''(\vec{k}, \omega)|}{[\omega - \varepsilon_{\vec{k}} - \Sigma'(\vec{k}, \omega)]^2 + [\Sigma''(\vec{k}, \omega)]^2} . \tag{3.63}$$

This relation shows that all the effects of the interactions on the spectral function are contained in the self-energy. It is not usually possible to determine this function exactly, but methods have been developed to obtain approximate forms. For example, Landau has shown that the asymptotic form of the self-energy in the vicinity of the Fermi surface for Fermi liquids (standard metals) is written:

$$\Sigma_{FL}(\vec{k}, \omega) \simeq \alpha\, \omega + i\beta \left[\omega^2 + (\pi\, k_B T)^2\right], \tag{3.64}$$

which characterizes the low energy excitations $\omega \approx 0$ near the Fermi surface.

We will briefly describe the characteristics of the spectral function and return to the notion of the quasi-particle that we introduced above. For a system without interactions, we have seen that the spectral function is a Dirac function at the energy

$\varepsilon_{\vec{k}}$ of the considered electronic state. The expression of the spectral function in the general case shows that it has a maximum at the energy shifted by the real part of the self-energy $\varepsilon_{\vec{k}} + \Sigma'(\vec{k}, \omega)$. It is therefore interesting to develop around the pole $E_{\vec{k}} = \varepsilon_{\vec{k}} + \Sigma'(\vec{k}, E_{\vec{k}})$:

$$\Sigma'(\vec{k}, \omega) \approx \Sigma'(\vec{k}, E_{\vec{k}}) + (\omega - E_{\vec{k}}) \left.\frac{d\Sigma'}{d\omega}\right|_{\omega = E_{\vec{k}}}$$

leading to:

$$\omega - \varepsilon_{\vec{k}} - \Sigma'(\vec{k}, \omega) \approx \frac{1}{Z_{\vec{k}}}(\omega - E_{\vec{k}})$$

where

$$Z_{\vec{k}} = \left\{1 - \left.\frac{d\Sigma'}{d\omega}\right|_{\omega = E_{\vec{k}}}\right\}^{-1}.$$

With this development, it is possible to decompose the spectral function in a lorenzian component (coherent part) and an incoherent part:

$$A(\vec{k}, \omega) = Z_{\vec{k}} \frac{\Gamma_{\vec{k}}/\pi}{(\omega - E_{\vec{k}})^2 + \Gamma_{\vec{k}}^2} + A_{inc}(\vec{k}, \omega)$$

where $\Gamma_{\vec{k}} = Z_{\vec{k}} |\Sigma''(\vec{k}, E_{\vec{k}})|$. The development of the self-energy in the vicinity of the pole makes it possible to decompose the spectral function in two parts: the coherent part or quasi-particle peak of weight $Z_{\vec{k}}$ which has a lorentzian shape and an incoherent part of weight $1 - Z_{\vec{k}}$ (figure 3.16). The coherent part is reminiscent of the electron gas without interactions for which $\vec{k}$ is conserved and for which the spectral function is a Dirac peak. As we have already discussed above, the interactions between electrons break the invariance of $\vec{k}$ but if the imaginary part of the self-energy remains weak, an excitation of momentum $\vec{k}$ has a fairly well-defined energy ($E_{\vec{k}}$ broadened by $\Gamma_{\vec{k}}$) called quasi-particle. In a Fermi liquid, the imaginary part vanishes at the Fermi level ($\omega = 0$), so that the spectral width vanishes and the spectral function for $\vec{k} = \vec{k}_F$ is a delta function. This singular behavior, which defines the Fermi surface, leads to a discontinuity at $\vec{k}_F$ of the $n(\vec{k})$ function(figure 3.15).

The spectral function highlights the different effects due to interactions and contained in self-energy. They are of 3 orders: i) The spectral function exhibits a narrow peak (quasi-particle peak). Its position ($E_{\vec{k}}$) is shifted relative to the non-interacting case ($\varepsilon_{\vec{k}}$) of a quantity that is the real part of the self-energy $\Sigma(\vec{k}, \omega = E_{\vec{k}})$. This shift reflects the change in energy due to interactions. Whatever the nature and sign of the interaction (repulsive or attractive), $|E_{\vec{k}}| \leqslant |\varepsilon_{\vec{k}}|$ so that the peak of quasi-particle is closer to the Fermi level than would be the Dirac peak of the non-interacting corresponding system. This leads to a quasi-particle bandwidth narrower than the bandwidth of non-interacting states.

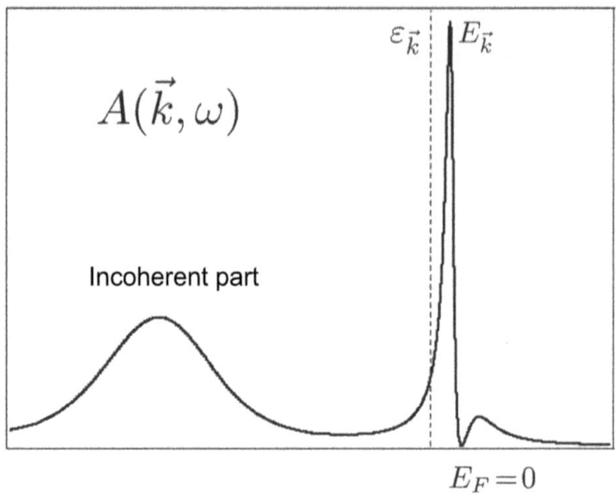

Figure 3.16 *Schematic representation of the spectral function for $\vec{k} < \vec{k}_F$ for a Fermi liquid. Without electron-electron interactions the spectral function is a delta peak at the energy $\varepsilon_{\vec{k}}$. The interactions lead to a broadened peak at $E_{\vec{k}}$ (shifted by Σ_R). Note that the spectral function vanishes at the Fermi energy E_F.*

ii) The second effect is the broadening of the peak. It reflects the finite life time of the excitations (quasi-particles) and the spectral width is proportional to the imaginary part of the self-energy ($\Gamma(\vec{k},\omega) = Z_{\vec{k}}|\Sigma''(\vec{k},\omega)|$). This lifetime reflects the residual interaction between quasi-particles. While the interaction between two "naked" electrons is important, the interaction between two quasi-particles is weak. From the broadening magnitude, it is possible to understand the limit of the quasi-particle approach. Indeed, when $\Gamma(\vec{k},\omega) > E_{\vec{k}}$, there is no more peaked signature in the spectral function and the notion of quasi-particle with sufficiently well-defined energy loses its meaning. In the Landau model of Fermi liquids, the imaginary part tends to 0 on the Fermi surface so that the life time of the quasi-particles tends to infinity. The interactions between quasi-particles vanished at E_F and the spectral signature is a delta peak at E_F of weight $Z_{\vec{k}_F}$:

$$A^-(\vec{k}_F,\omega) = Z_{\vec{k}_F}\delta(\omega) + A^-_{inc}(\vec{k}_F,\omega) \tag{3.65}$$

iii) The third effect is associated with the decrease of the spectral weight of the quasi-particle peak and with the transfer of weight towards a continuum of energy (incoherent part). The physical origin of this effect is more subtle. This transferred spectral weight reflects the virtual excitations associated with a quasi-particle. Indeed, a quasi-particle consists of a naked particle (an electron) surrounded by a cloud of virtual excitations that depend on the interactions and therefore on the Hamiltonian of the system. For example, if the Hamiltonian describes the electron-phonon interaction, a quasi-particle in this case called polaron, can be considered as an electron surrounded by virtual phonons. The virtual term means that the electron emits phonons and reabsorbs them. This description in terms of virtual phonons reflects the distortion of the lattice around an electron. These virtual phonons are not real, i.e. are not observable, but they manifest themselves in the physical properties of the system. One can have high energy virtual excitations since the time-energy uncertainty relation allows to violate the conservation of the energy over a very short time.

Thus in the transport properties, the entity that propagates is the quasi-particle, i.e. the electron accompanied by its cloud of virtual excitations. This cloud of virtual excitations is all the more important as the interactions are strong, and affect the effective mass of the electron. On the other hand, in high-energy spectroscopy techniques, such as photoemission, the situation is completely different. Indeed, a "naked" electron is extracted from the solid without its cloud of virtual excitations. A high energy photon is able to break the quasi-particle by removing the naked electron and revealing the nature of the cloud of virtual excitations. Indeed, a polaronic system in the final state can be left without phonon (this corresponds to the coherent part or the quasi-particle peak) but also with one, two .... real phonons. These different states correspond to eigenstates of the Hamiltonian (of energy $E_s^{(N-1)}$). The photoemission process allows to make virtual excitations real, the energy necessary to create real phonons is given by the photon. This mechanism leads to spectral weight at high energy (incoherent part of the spectral function). The nature of the virtual excitations and the corresponding weight of the incoherent part depend on the Hamiltonian of the system and are completely contained in the self-energy.

In principle, a photoemission spectrum contains all information on one-particle excitations of electronic systems. If the peak in the spectral function gives information on the energy and dispersion of quasi-particles (low-energy excitations responsible for most thermodynamic properties), the analysis of the incoherent part gives information on the interactions between electrons or with the other degrees of freedom of the system. Indeed, in highly correlated systems, Coulomb interactions between electrons in solids have a magnitude of a few eV. These energies are several orders of magnitude higher than the thermodynamic scale kT (1 eV = 11 605 K) and the corresponding states cannot be excited thermally. But, the photon gives the energy to excite them. That is why they contribute to the photoemission spectrum. In the following section, we will illustrate these concepts using some examples.

We have seen that in an interacting system we can introduce entities, quasi-particles, whose mutual interaction vanishes at the level of Fermi. This is why the expressions of the thermodynamic relations established for free electrons apply provided that the density of states at the Fermi level is replaced by the density of the quasi-particle states. It is necessary to distinguish these different densities of state. The spectral density is obtained from the spectral function by integrating on the wave vector. At Fermi level ($\omega = 0$) we obtain:

$$N(E_F) = \int_{-\infty}^{+\infty} A(\vec{k}, \omega = 0)\, d\vec{k} \tag{3.66}$$

On the other hand, the quasi-particle density of states is obtained by integrating only the coherent part (peak of quasi-particle) whose weight is $Z_{\vec{k}}$. We then obtain the equation:

$$N^{Q.P.}(E_F) = Z_{\vec{k}_F}^{-1}\, N(E_F) \tag{3.67}$$

which shows that the quasi-particle density of states is renormalized with respect to the spectral density. This effect is particularly important in strongly correlated systems (heavy fermions) for which $Z_{\vec{k}_F} \ll 1$.

3.2.3 *Illustrations*

We have just seen that the many-body effects lead to important modifications of the spectral function: (i) shift of the spectral lines towards the Fermi level compared to the case of the non-interacting system and thus narrower bands than predicted by the band structure calculations, (ii) broadening of quasi-particle peaks reflecting a finite life time of these quasi-particles, and (iii) spectral weight transfer at high energy (incoherent part). Below we will illustrate these effects by some experimental examples selected from the literature. If experiments frequently show the shift of the spectral structures with respect to the calculations, the quantitative interpretation of the spectral widths by correlation effects is more delicate and finally the demonstration of a spectral weight transfer proves much more problematic. Indeed, the incoherent part is generally weak, broad and not structured so that only systems where the transfers will be particularly important are likely to exhibit this effect.

3.2.3.1 Sodium : a non-interacting system?

Alkalis are often presented as archetypes of nearly free electron systems. The dispersion functions calculated in the framework of the local density approximation (LDA) are parabolic and in quantitative agreement with a very simple model of nearly free electrons. The photoemission measurements confirm this parabolic dispersion. Nevertheless, the experimental bandwidth (occupied part) is narrower by about 20 % than the calculated band (figure 3.17).

This difference between experimental and calculated dispersions reflects the effects of interactions between electrons. We saw in section 4.2 that the shift of the quasi-particle peak is equal to the real part of the self-energy. The difference between the excitation energy measured in photoemission and the band calculation energy is $Re\Sigma(\vec{k}, \omega)$. The particularly simple electronic structure of Na makes it possible to estimate this quantity [31]. When considering self-energy as in a GW approach, it is possible to quantitatively reproduce the experimental dispersion of the quasi-particle structure (figure 3.18).

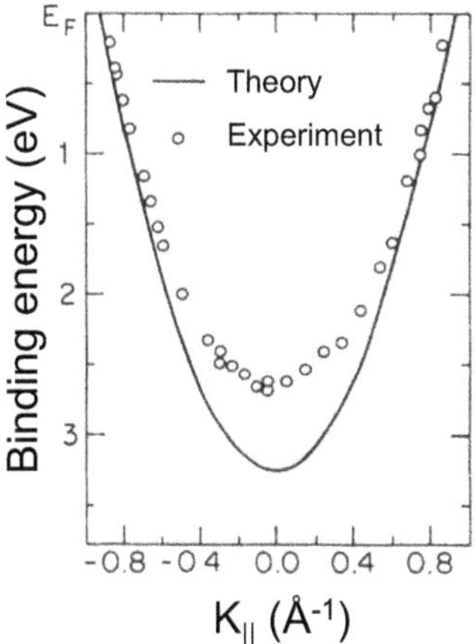

Figure 3.17 *Experimental and calculated dispersions of the Na valence band [30]. Figure reprinted with permission from In-Whan Lyo and E. W. Plummer, Phys. Rev. Lett. 60, 1558 (1998). ©1998 by the American Physical Society. The disagreement can be attributed to the correlation effects.*

This result highlights the essential difference between the dispersion relations of the band structure calculation and the experimental dispersions. The band calculation gives access to the properties of the ground state (total energy and electron density), the monoelectronic energies $\varepsilon_{\vec{k}}$, appear only as intermediate calculations and cannot *a priori* be interpreted as the excitations of the system. A spectroscopy experiment, on the other hand, gives access to the excited states. Thus, even in the case where the ground state is very well-described by the band calculation, as for Na, the spectrum of the excitations may differ significantly from the dispersion relations calculated in a band approach. On the other hand, it is found that the Fermi surface is well-described (the experimental and calculated values of k_F are identical).

3.2.3.2 Strongly correlated systems

In many transition metal compounds and rare earth compounds, the conduction electrons of d or f symmetries are very localized. The intra-atomic interactions are very large because the energy cost for 2 electrons on the same site is U, which may be larger than the bandwidth. The band structure models are then inappropriate to satisfactorily describe such systems characterized by highly localized electrons and an atomic character. Some quite unique electronic properties are often observed. In some transition metal oxides, an insulating behavior can be observed while the calculated band structure predicts a metal. Likewise, certain ytterbium, uranium or cerium compounds exhibit, at low temperature, exotic thermodynamic properties, very high Pauli susceptibility and linear coefficient of specific heat, which suggest a very large density of states at the Fermi level and an anomalous high effective

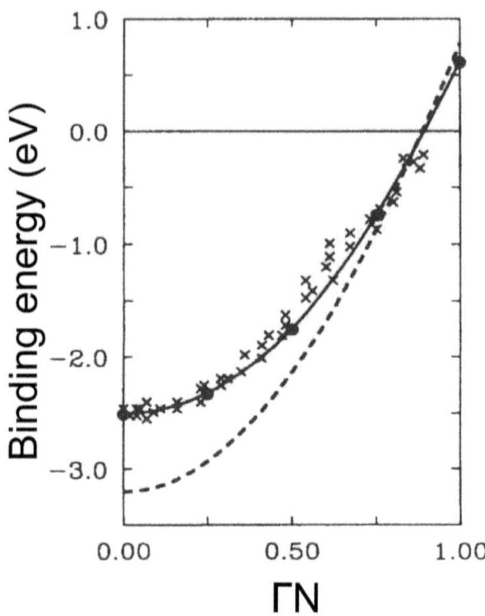

Figure 3.18 *Experimental dispersions (cross) and calculated by LDA (dotted line) and GW (point and solid line) of the Na band. The correction made by the GW calculation corresponds to the real part of the self-energy (from [31]). Repdrinted figure with permission from John E. Northrup, Mark S. Hybertsen, and Steven G. Louie, Phys. Rev. Lett. 59, 819 (1987). ©1987 by the American Physical Society.*

electronic mass: such systems are called heavy fermion materials. These systems are also characterized by a very small energy scale, a few meV to a few tens of meV, which manifests itself in the thermal evolutions of the physical quantities. For example, the compounds are often non-magnetic at low temperature and exhibit, above a characteristic temperature (Kondo temperature T_K), a localized magnetic moment behavior. Like most strongly correlated systems, the low energy excitations that determine the thermodynamic properties (of the order of kT) are spin excitations while the very energetic charge excitations can be revealed by high energy spectroscopies such as photoemission. The low temperature phase is described by heavy quasi-particles, corresponding to a dense cloud of virtual excitations, which disappear at high temperature. To summarize, the very large electronic correlations lead to the appearance of an emergent energy scale of energy often very small (kT_K) which governs most of the thermodynamic properties. From a phenomenological point of view, this energy scale is associated with the width of the quasi-particle band.

The thermodynamic properties of these systems are therefore associated with particularly heavy quasi-particles (100 to 1000 times the mass of a free electron). It is therefore a favorable case to see the weight of the incoherent part in photoemission due to a very small renormalization factor $Z_{\vec{k}_F}$. In the case of cerium compounds, a heavy f quasi-particle consists of a bare f electron whose weight is $Z_{\vec{k}_F} \ll 1$ surrounded by a cloud of virtual excitations of weight $1 - Z_{\vec{k}_F}$. These virtual excitations are essentially charge excitations with characteristic energies ε_f (for the charge excitations $f^1 - f^0$) and U_{ff} (for the charge excitations $f^1 - f^2$).

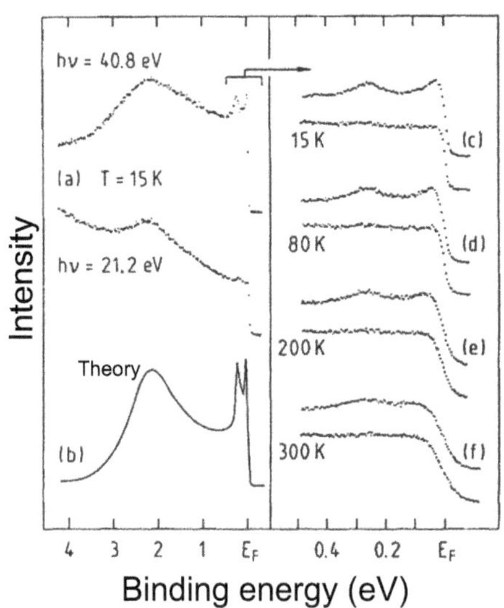

Figure 3.19 *(a) Photoemission spectra of $CeSi_2$ at 15 K with photons of 21.2 eV (HeI) and 40.8 eV (HeII). The 4f cross section is very weak in HeI and increases with the photon energy so that the difference of the HeII and HeI spectra can be interpreted as the signature of the f states. (b) Simulation of the 4f spectral function including many body effects. (c) – (f) HeII and HeI spectra near the Fermi level for different temperatures (from [32]). Figure reprinted with permission from F. Patthey, W.-D. Schneider, Y. Baer, and B. Delley, Phys. Rev. Lett. 58, 2810 (1987). ©1987 by the American Physical Society.*

What is observed in photoemission? We have reported in figure 3.19, the spectrum of a typical heavy fermion compound, $CeSi_2$ [32]. The spectral function is dominated by a broad structure at -2 eV, and 2 narrow peaks appear at the Fermi level and -0.3 eV. The E_F structure represents the quasi-particle peak. Its low intensity reflects the large mass of quasi-particles and its width suggests a very narrow band. The spectral weight is essentially in the incoherent part, and specific models allow to identify the nature of the virtual excitations constituting the quasi-particles. Thus, the peak at -0.3 eV corresponds to the spin-orbit excitations $(f^{5/2} - f^{7/2})$ while the -2 eV structure is associated with the charge excitations $(f^1 - f^0)$. High energy excitations are thus dominated by atomic-like processes in these compounds which cannot be described by band structure calculations.

3.2.4 Selection rules and symmetry

We have seen above that optical absorption by a Bloch state (the first step in the three-step model) is a vertical transition in the first Brillouin zone, at least for UV photons for which the momentum is negligible with respect to the electron momentum. This leads to the conservation of the wave vector in the process. We will show that there are other conditions on the symmetry of the initial state using polarized light for well-chosen geometries. To determine these transition rules on the band states, we must consider their symmetry which is given at each point of the first Brillouin zone by the irreducible representations of the wave vector group. This group is the group of symmetry operations that leave the wave vector unchanged in the first Brillouin zone, i.e., modulo a reciprocal lattice vector. It is a point group that depends on the considered $\vec{k}$ point and is of higher symmetry for particular directions and points of the Brillouin zone. Thus at the Γ point (corresponding to $\vec{k} = 0$) the group of the wave vector is the point group of the crystal.

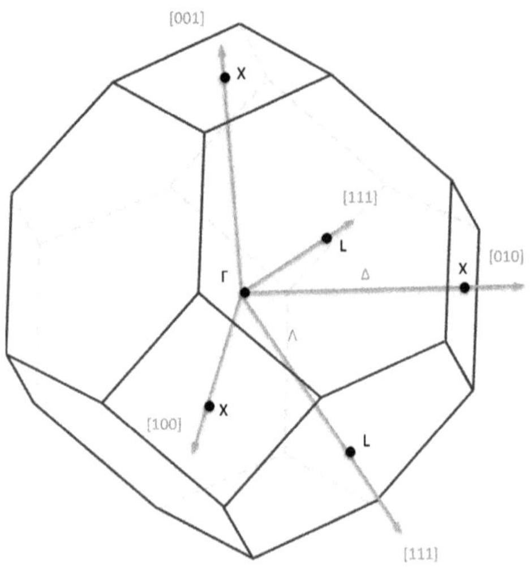

Figure 3.20 *Brillouin zone for a face-centered cubic (fcc) lattice indicating points and directions of high symmetry with the usual group theory nomenclature.*

The selection rules for optical absorption and photoemission are tabulated in the articles [33, 34] for body-centered (bcc) and face centered cubic (fcc) structures for the main directions and points of high symmetry. The principle is the following. Optical absorption or photoemission is governed by the matrix element of the electrical dipole interaction between initial and final states:

$$\langle \phi_i | \vec{p} \cdot \vec{A} | \phi_f \rangle$$

As recalled above, in the UV domain the transition is vertical in the first Brillouin zone and the initial and final states are characterized by the same wave vector. From the point of view of symmetry, $|\phi_i\rangle$ and $|\phi_f\rangle$ are basis functions of irreducible representations of the wave vector group. The matrix element must therefore be invariant under all the symmetry operations of this point group. According to the rules of group theory, this matrix element is non-zero if and only if the product of the three irreducible representations associated with the initial state, the dipolar operator and the final state contains the completely symmetrical representation or if the product of the irreducible representations of the initial state and the dipolar operator contains the irreducible representation of the final state. Let us take the example of the ΓL direction called Λ of a face centered cubic compound (figure 3.20). The corresponding wave vector group is C_{3v} (consisting of the identity E, two rotations of order 3 denoted C_3 and three mirror symmetries denoted σ). The character table of C_{3v}:

C_{3v}	E	$2C_3$	3σ	
Λ_1	1	1	1	A_z ($\parallel$)
Λ_2	1	1	-1	
Λ_3	2	-1	0	(A_x, A_y) ($\perp$)

shows that there are three irreducible representations: Λ_1 (the totally symmetric or trivial representation), Λ_2 (an odd representation with respect to mirrors) and Λ_3 a two-dimensional representation. The components of the vector potential have also been described, describing for A_z linearly polarized photons along Oz, thus parallel to the direction Λ and for (A_x, A_y) linearly polarized photons along Oy and O_z perpendicular to the Λ direction or circularly polarized photons right or left in the Oxy plane. For example, if the final state symmetry is Λ_3, and if we use polarized photons along Ox or Oy (Λ_3), we must consider the product of representations:

$$\Lambda_3 \otimes \Lambda_3 = \Lambda_1 \oplus \Lambda_2 \oplus \Lambda_3$$

which means that the product of the two irreducible representations is a reducible representation decomposing on the three irreducible representations of the group. The transition matrix element is non-zero if the final state has the symmetry of one of the representations appearing in the product $\Lambda_3 \otimes \Lambda_3$. Thus the final states of symmetry Λ_1, Λ_2 and Λ_3 are possible since they appear in the decomposition. For the same initial state of symmetry Λ_3 and polarized photons according to Oz, we must consider the product:

$$\Lambda_3 \otimes \Lambda_1 = \Lambda_3$$

indicating that the final state can only be of Λ_3 symmetry. The table below sets out all the possible cases: in column the irreducible representation of the initial state is indicated, whereas the lines give the final state representations. Inside the table the symbols $\parallel$ and $\perp$ indicate the possible transitions according to the polarizations ($\parallel$ and $\perp$ correspond to the representations Λ_1 et Λ_3 respectively associated with the electric dipolar operator). The empty boxes in this table correspond to the optical transitions forbidden by symmetry.

C_{3v}	Λ_1	Λ_2	Λ_3
Λ_1	$\parallel$		$\perp$
Λ_2		$\parallel$	$\perp$
Λ_3	$\perp$	$\perp$	$\perp,\parallel$

Likewise for the direction (100) called Δ and for the corresponding wave vector group C_{4v} and its 5 irreducible representations (Δ_1,Δ_1',Δ_2,Δ_2',Δ_5), we obtain the following table:

C_{4v}	Δ_1	Δ_1'	Δ_2	Δ_2'	Δ_5
Δ_1	$\parallel$				$\perp$
Δ_1'		$\parallel$			$\perp$
Δ_2			$\parallel$		$\perp$
Δ_2'				$\parallel$	$\perp$
Δ_5	$\perp$	$\perp$	$\perp$	$\perp$	$\parallel$

The detection outside of the crystal introduces an additional constraint. In fact, the final state must be measurable by the analyzer, which means that the wave function must not present a node at the point where it is measured. We will illustrate this concept in the case of the normal emission. The state of the photoelectron is characterized by a wave vector perpendicular to the surface that remains invariant under the operations of the crystal wave vector group. The final state must be symmetric under all symmetry operations. In the opposite case, the wave function has a node and the spectral weight vanishes. The final state must therefore be of symmetry Λ_1 for the normal emission of a surface (111) i.e. along the direction Λ and symmetry Δ_1 for the emission normal of a surface (100) i.e., along the Δ direction.

Let us consider a geometry of emission less particular than the normal emission, that where the electrons are emitted in a mirror plane of the crystal (figure 3.21). The directions of the photon incidence and the photoelectron emission are in a mirror plane of the crystal with a polarization in the plane (p) or perpendicular to the plane (s). The symmetry of the experimental device *crystal and analyzer* is limited to the symmetry plane containing the wave vector and the analyzer. Indeed another element of symmetry of the crystal sends the analyzer in another direction. This point symmetry group has two irreducible representations, one even and the other odd with respect to the mirror. The odd final states have a node in the plane and cannot be detected in the analyzer. Therefore only the even states contribute to the photoemission spectra. Let us consider the geometry of the experiment in the reciprocal space represented by the left part of figure 3.22. The studied surface of a Cu crystal is (110), so that the normal of the surface corresponds to the ΓK direction of the Brillouin zone. The analyzer is in the ΓXL plane, in grey on this figure. Consider initial states along the Λ direction in this mirror symmetry plane, and the final states, even with respect to the emission mirror plane to be detected in the analyzer. Outside the crystal, the direction of propagation is no longer along the Λ direction due to the decrease of the momentum component perpendicular to the surface but remains in the mirror plane. The symmetry of the initial state with respect to the plane will therefore depend on the symmetry of the Hamiltonian and therefore on the polarization. The transition matrix element is non zero, if the parity of the

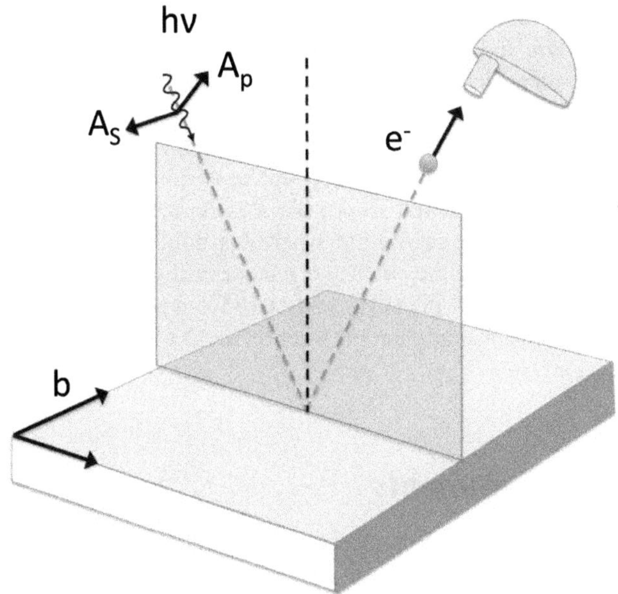

Figure 3.21 *Measurement geometry to probe the parity of the initial state.*

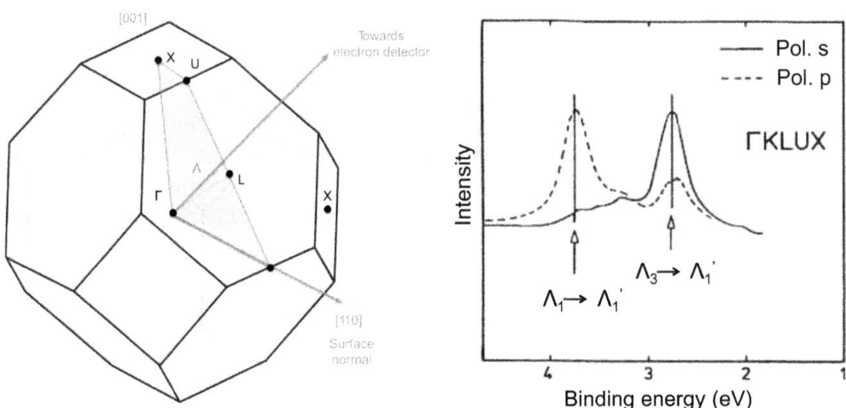

Figure 3.22 *Geometry (left) and Cu (110) photoemission spectra (right) in s and p polarizations illustrating the selection rules according to the symmetry of the two bands. (from [35]). Reprinted from Phys. Rep. Vol. 112, R. Courths, S. Hüfner, "Photoemission experiments on copper", page 53, ©1984, with permission from Elsevier.*

initial state is the same as that of the electric dipole interaction Hamiltonian. So, if the polarization is in the plane (*p* polarization), the initial state must be even with respect to the mirror plane. If the polarization is perpendicular to the plane (*s* polarization), then the initial state must be odd. The rules of group theory allow to specify the symmetry of the initial state. The product of the irreducible representations associated with the initial state and the dipolar interaction Hamiltonian for the polarization under consideration must contain the final state irreducible representation (Λ_1). The s polarization (component A_x) is perpendicular to the direction Λ and corresponds to the odd part of the Λ_3 irreducible representation. The p polarization has to be decomposed according to the A_z (Λ_1 representation) and A_y (even part of the Λ_3 representation) components. Thus a band of Λ_1 symmetry will only be active in photoemission for a p polarization whereas a doubly degenerate band of Λ_3 symmetry will be active for the two s and p polarizations. This is illustrated by the spectra of the figure 3.22 of the (110) surface of Cu which show that the Λ_1 band has no spectral weight in s polarisation whereas the Λ_3 band is visible for the two s and p polarizations.

3.2.5 *Matrix elements*

If the dispersion relations are periodic in reciprocal space, the spectral weight is not. For a given band, the spectral weight is often found in only a few Brillouin zones, sometimes only one. It depends on the matrix element of the electric dipolar operator between the one-electron initial and final states. The overlap of these states is therefore essential to determine the photoemission intensity, and modeling in the one-step description of the photoemission process allows to quantitatively determine the spectral weight distribution of an electronic state in the reciprocal space. As photoemission spectra are often interpreted in the three-step model, the distribution of spectral weight in the different Brillouin zones is then poorly understood. Here we will develop simple approaches to understand this distribution. Several parameters determine this distribution: the nature and the symmetry of the electronic states, the geometrical characteristics of the experiment (angles of incidence and emission, polarization of the photons) and the energy of the photons.

The first example that we will present is the distribution of the spectral weight of the surface state developing on the (111) surfaces of the noble metals (Shockley states). These states have the property of being able to be simply described by a nearly-free-electron approach, with a parabolic dispersion. The plane waves are thus a well-adapted basis for both the initial and final states of the photoemission process of these Shockley states. Being located close to the surface, they are extremely sensitive to any structural change. Thus, a surface reconstruction, which changes the periodicity in the direct space and therefore changes the size of the Brillouin zone in reciprocal space, leads to a modification of the band structure. The figure 3.23(a) shows an STM image of the triangular reconstruction of a monolayer of Ag on Cu(111).

The cell of the reconstruction is $\sim 9\times 9$ with respect to the surface cell of Cu(111). The Brillouin zone of the reconstructed surface is represented in figure 3.23(b) with points of high symmetry Γ, M and K. We show that the M point is equidistant from Γ and the first reciprocal lattice vector $\vec{G}_1$ whereas K is equidistant from 3

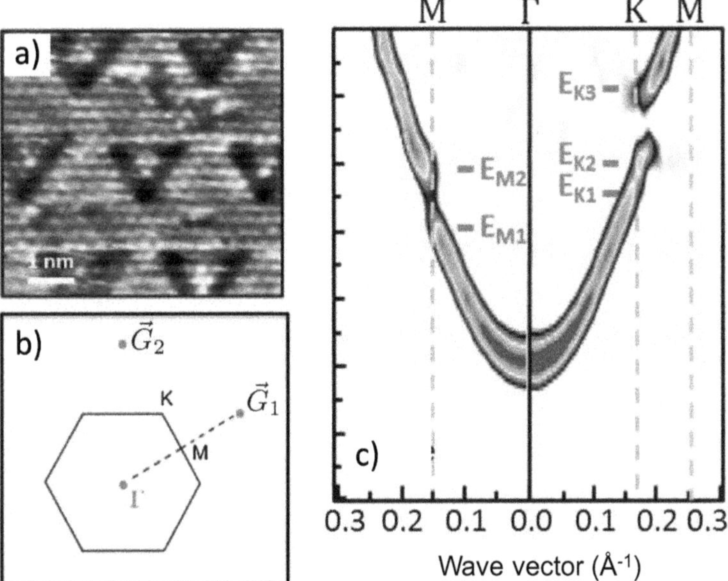

Figure 3.23 *a) STM image of the reconstruction of a monolayer of Ag on Cu (111). b) Brillouin zone of the reconstructed surface. c) Dispersion of surface state in high symmetry directions (from [36]).*

reciprocal lattice vectors (Γ, $\vec{G}_1$ and $\vec{G}_2$). Due to the character of near-free electrons of the surface states, it is expected to open bands prohibited at these points of high symmetry. We can observe in the figure 3.23(c) which presents the ARPES intensity in second derivative in the two directions ΓK et ΓM (the surface state was shifted to high binding energies by deposit of a small amount of potassium adatoms).

It should nevertheless be noted that the spectral weight mainly follows the parabola of free electrons centered in Γ, i.e., it is found in the first Brillouin zone below the first gap and in the second one above the gap etc. The weight of the folded bands remains very low. This behavior can be understood by expressing that an electronic surface state $|\Psi_{\vec{k}}^{(i)}\rangle$ associated with the wave vector $\vec{k}$ in the first Brillouin zone is written in the ΓM direction near the M point:

$$|\Psi_{\vec{k}}^{(i)}\rangle = C_{\vec{k}}^{(i)}|\vec{k}\rangle + C_{\vec{k}-\vec{G}_1}^{(i)}|\vec{k}-\vec{G}_1\rangle \tag{3.68}$$

where $|\vec{k}\rangle$ and $|\vec{k}-\vec{G}_1\rangle$ are states associated with plane waves centered at the points Γ and $\vec{G}_1$ respectively and (i) the band index (in the neighborhood of the K point, we would express the electronic state as a linear combination of three plane waves corresponding of the three equidistant reciprocal space vectors). The normalization of the state is written: $|C_{\vec{k}}^{(i)}|^2 + |C_{\vec{k}-\vec{G}_1}^{(i)}|^2 = 1$. The $C_{\vec{k}}^{(i)}$ and $C_{\vec{k}-\vec{G}_1}^{(i)}$ coefficients

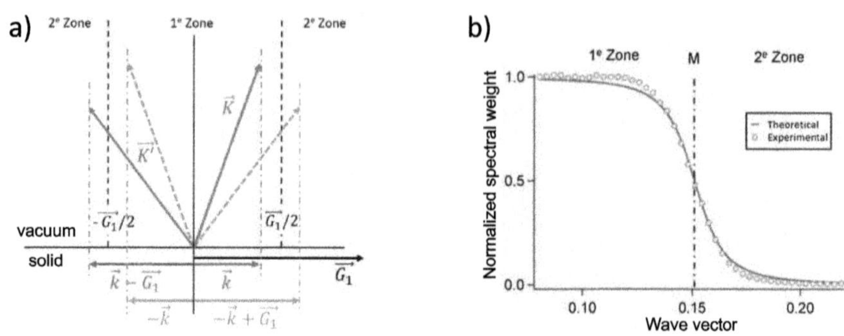

Figure 3.24 *a) Schematic representation of the emission directions from the two symmetric states $|\Psi^{(i)}_{\vec{k}}\rangle$ and $|\Psi^{(i)}_{-\vec{k}}\rangle$. b) Experimental and calculated spectral weight in the ΓM direction with the Fourier component describing the amplitude of the gap.*

and the energy $\varepsilon^{(i)}(\vec{k})$ of the bands can be obtained by a first order perturbation calculation. We obtain for the energy of the two bands:

$$\varepsilon^{(i)}(\vec{k})=\frac{1}{2}(E^0_{\vec{k}}+E^0_{\vec{k}-\vec{G}_1})+\alpha_i\frac{1}{2}\sqrt{(E^0_{\vec{k}}-E^0_{\vec{k}-\vec{G}_1})^2+4|V_{\vec{G}_1}|^2} \tag{3.69}$$

where $E^0_{\vec{k}}$ and $E^0_{\vec{k}-\vec{G}_1}$ are the energies of plane waves ($E^0_{\vec{k}}=\hbar^2k^2/2m$), α_i a coefficient depending on the band index ($\alpha_1=-1$ and $\alpha_2=1$) and $V_{\vec{G}_1}$ the Fourier component of the reconstruction potential associated with the vector $\vec{G}_2$. We therefore have a gap at the point M whose amplitude is:

$$\varepsilon^{(2)}_{\vec{G}_1/2}-\varepsilon^{(1)}_{\vec{G}_1/2}=2|V_{\vec{G}_1}| \tag{3.70}$$

while the ratio of the coefficients is:

$$\frac{C^{(i)}_{\vec{k}}}{C^{(i)}_{\vec{k}-\vec{G}_1}}=\frac{-2V_{\vec{G}_1}}{(E^0_{\vec{k}}-E^0_{\vec{k}-\vec{G}_1})-\alpha_i\sqrt{(E^0_{\vec{k}}-E^0_{\vec{k}-\vec{G}_1})^2+4|V_{\vec{G}_1}|^2}}. \tag{3.71}$$

At the M point, we obviously have $|C^{(i)}_{\vec{k}}|=|C^{(i)}_{\vec{k}-\vec{G}_1}|$ for $i=1,2$, whereas when we move away from this high symmetry point, we have for the first band whose energy is below the gap:

$$|C^{(1)}_{\vec{k}}|>|C^{(1)}_{\vec{k}-\vec{G}_1}|,$$

and, for the second band, above the gap:

$$|C^{(2)}_{\vec{k}}|<|C^{(2)}_{\vec{k}-\vec{G}_1}|.$$

We will assume that the photoemission final state is a plane wave denoted $|\vec{K}\rangle$. The conservation of the parallel component shows that the photoemission from the state $|\Psi^{(i)}_{\vec{k}}\rangle$ leads to the electron emission in the directions given by $\vec{K}$ and $\vec{K}'$ whose components parallel to the surface correspond to the wave vectors of the two plane waves of the initial state:

$$\vec{K}_{\parallel} = \vec{k}$$
$$\vec{K}'_{\parallel} = \vec{k} - \vec{G}_1$$

with a spectral weight respectively proportional to $|C^{(i)}_{\vec{k}}|^2$ and $|C^{(i)}_{\vec{k}-\vec{G}_1}|^2$. It is therefore expected to find spectral weight in the first and second Brillouin zone. Figure 3.24(a) which presents schematically the directions of emission for the two symmetrical states $|\Psi^{(i)}_{\vec{k}}\rangle$ and $|\Psi^{(i)}_{-\vec{k}}\rangle$ shows that the folded band in the second zone at the symmetrical point of $\vec{K}$ with respect to the normal has a weight proportionate to the square of the coefficient $C^{(i)}_{-(\vec{k}-\vec{G}_1)} = C^{(i)}_{\vec{k}-\vec{G}_1}$. According to the equation 3.71, the weight of the folded band is smaller as the Fourier component of the potential $|V_{\vec{G}_1}|$ is small. This parameter determines not only the amplitude of the gap but also the spectral weight of the bands in the different Brillouin zones. Figure 3.24(b) demonstrates that this simple model allows to explain quantitatively the distribution of the spectral weight in the Fourier space for the Shockley states. The spectral weight is thus a measure of the Fourier components of the Bloch function.

A similar approach can also be used to account for the angular distribution of photoelectrons for molecules. Consider a molecule whose electronic states $|\Psi_i\rangle$ are described in the linear combination model of atomic orbitals (Hückel method). Let us assume that the photon energy is sufficiently high that the photoelectrons can be described by plane waves $|\vec{k}_f\rangle$ but not such that we can neglect the photon momentum (vertical transitions). The transition matrix element is written using $\vec{A} \cdot \vec{p} = A\vec{\epsilon} \cdot (-i\hbar\vec{\nabla})$:

$$|M_{if}| \sim |\langle\Psi_i|\vec{\epsilon} \cdot \vec{\nabla}|\vec{k}_f\rangle| = |(\vec{\epsilon} \cdot \vec{k}_f)\ \tilde{\Psi}_i(\vec{k}_f)| \tag{3.72}$$

where $\vec{\epsilon}$ is the vector defining the photon polarization and $\tilde{\Psi}_i$ the Fourier transform of the Ψ_i molecular orbital. The angular distribution of the photoelectrons is therefore given by [37]:

$$I(\vec{k}_f) \sim |\epsilon \cdot \vec{k}_f|^2\ |\tilde{\Psi}_i(\vec{k}_f)|^2. \tag{3.73}$$

Figure 3.25 shows the one-dimensional Fourier transforms (1D, in the direction parallel to the polymer) of occupied monoelectronic states built from the atomic states $C - 2p_z$ of a molecule of benzene, a sexiphenyl and a linear polyphenyl consisting of 20 molecules. For the benzene molecule, the ground level corresponds to the completely binding state (all the π bonds are binding), symmetrical with respect to all the symmetry planes perpendicular to the molecule. Its Fourier transform is

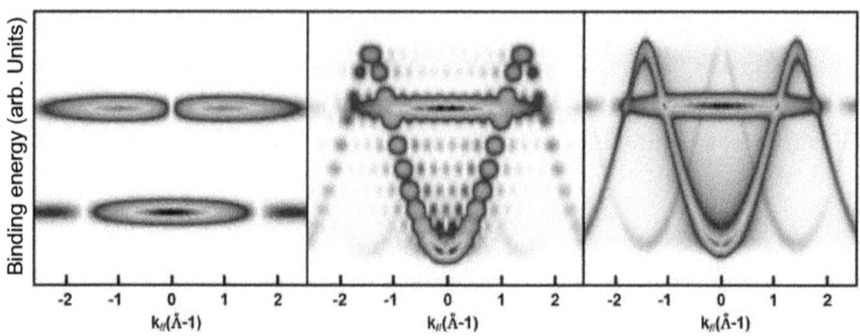

Figure 3.25 *Sum of squares of 1D Fourier transforms ($k_\perp = 0$) of occupied molecular orbitals of the same energy for benzene molecules (left) of sexiphenyl (center) and of a polyphenyl chain of 20 molecules (right). A Gaussian broadening ($\Delta E = 200$ meV) was applied. At the top of the band the two intensity maxima correspond to $\pm 2\pi/a$ (from the G. Vasseur thesis [38]).*

centered on $k = 0$. The first excited level of the molecule is doubly degenerate, with two states that can be chosen symmetric and antisymmetric with respect to the two families of planes perpendicular to the molecule. Since the $k = 0$ direction is at the intersection of these planes, the Fourier transforms of the two states are zero in $k = 0$, as is shown in the left part of the figure representing the sum on the molecular states of same energy of the Fourier transforms squared. For the molecule of sexiphenyl, the number of molecular states is six times greater than the molecule of benzene and the central part of the figure 3.25 shows that, for a given energy, the map of the Fourier transforms presents two extrema, narrow in k, whose k-separation increases with increasing energy to reach $\pm 2\pi/a$ at the band maxima (a is the distance between two phenyl molecules). In addition, an extended horizontal structure in k appears, which corresponds to the non-binding states of the sexiphenyl molecule. These states, which do not contribute to the interphenyl bonds, are antisymmetric with respect to the plane perpendicular to the molecules and containing the axis of the molecule. For the polyphenyl molecule, the Fourier transforms of the molecular states tend towards the Fourier transforms of Bloch states of an infinite chain with a dispersive band and a non-binding (flat) band.

Note that the spectral weight of the folded band is very small, the intensity being highest in the first Brillouin zone ($0 < k < \pi/a$) for the low part of the band and in the second Brillouin zone ($\pi/a < k < 2\pi/a$) for the high part of the band. The two-dimensional Fourier transforms of the HOMO (Highest Occupied Molecular Orbital) and LUMO (Lowest Unoccupied Molecular Orbital) orbitals respectively representing the last occupied and the first unoccupied orbitals of the sexiphenyl are shown in the figure 3.26. The Fourier transform of the HOMO exhibits two structures located at $k_x = \pm 2\pi/a$ and centered on $k_y = 0$. These characteristics can be understood from the symmetry of the molecular orbital shown in part (a) of the figure. Indeed, it is symmetrical with respect to the plane perpendicular to k_y and

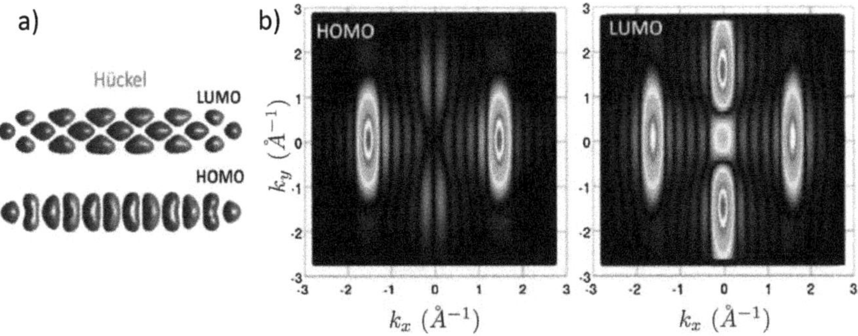

Figure 3.26 *a) a) Representation of the two molecular orbitals (HOMO and HUMO) of the sexiphenyl in the Hückel approach. The red and blue parts represent the positive and negative signs of the atomic orbitals involved. b) Square of the two-dimensional Fourier transform ($k_X = k_{parallel}$, $k_Y = k_{perp}$) representing the spectral weight in reciprocal space (from G. Vasseur thesis [38]).*

antisymmetric with respect to the plane perpendicular to k_x and passing through the center of the molecule. Therefore the Fourier transform must change sign and cancel in the plane $k_x = 0$ whereas it must present a single structure in the direction k_y centered on $k_y = 0$. Moreover, the sign of the molecular orbital oscillates once on a phenyl cycle along k_x leading to the two structures in $k_x = \pm 2\pi/a$. On the other hand, the Fourier transform of LUMO presents additional structures in $k_x = 0$. These structures result from the symmetrical character of the LUMO with respect to the plane perpendicular to k_x and from the presence of two lobes in the k_y direction. To obtain the angular distribution observed experimentally with polarized photons, the polarization of photons must be taken into account by multiplying by the term $|\vec{\epsilon} \cdot \vec{k_f}|^2$.

We have just seen that the angular distribution of photoelectrons, or in other words the spectral weight in reciprocal space, is directly related to the Fourier transforms of molecular orbitals in molecules or Bloch functions in solids. An approach recently developed by S.K. Moser in his thesis, which we will briefly summarize below, allows us to relate the spectral weight to the Fourier transform of the Wannier functions [39]. Remind that the Wannier functions are defined from the Fourier transform of the Bloch functions. These are functions located in a cell of the direct space which have the property of forming a complete base of orthogonal functions. For the band identified by the (n) index, they are written:

$$\Phi^{(n)}_{\vec{R}}(\vec{r}) = \Phi^{(n)}(\vec{r} - \vec{R}) = \frac{1}{\sqrt{N}} \sum_{\vec{k}} e^{-i\vec{k}\cdot\vec{r}} \, \Psi^{(n)}_{\vec{k}}(\vec{r}) \tag{3.74}$$

where $\Psi^{(n)}_{\vec{k}}(\vec{r})$ is the Bloch function of the (n) band for the $\vec{k}$ wave vector belonging to the first Brillouin zone, $\vec{R}$ a vector associated with a cell of the lattice, N the number of cells and $\Phi^{(n)}_{\vec{R}}(\vec{r})$ the corresponding Wannier function. We have written a discrete sum on the wave vectors but we can transform it into an integral on the first Brillouin zone for an infinite crystal. For a one-atom-per-cell crystal described in the tight binding approach, the Wannier function is identified with the atomic orbital from which the given band was constructed. The inverse Fourier transform allows us to write the Bloch states from the Wannier states:

$$\Psi^{(n)}_{\vec{k}}(\vec{r}) = \frac{1}{\sqrt{N}} \sum_{\vec{R}} e^{i\vec{k}\cdot\vec{r}}\, \Phi^{(n)}_{\vec{R}}(\vec{r})\,. \tag{3.75}$$

We now will discuss the first step of the photoemission, i.e. the photon absorption process without the effect of the surface, considering that the electron in the final state can be described by a plane wave ($|\vec{k}_f\rangle$). The transition element is written by decomposing the state of Bloch on the Wannier functions:

$$\begin{aligned}
|M_{\vec{k},\vec{k}_f}| &\sim |\langle \vec{k}_f|\vec{\epsilon}\cdot\vec{\nabla}|\Psi^{(n)}_{\vec{k}}\rangle| \sim |\langle \vec{k}_f|(\vec{\epsilon}\cdot\vec{\nabla}\sum_{\vec{R}} e^{i\vec{k}\cdot\vec{R}}|\Phi^{(n)}_{\vec{R}}\rangle| \\
&\sim \left| -\vec{\epsilon}\cdot\sum_{\vec{R}} e^{i\vec{k}\cdot\vec{R}} \int d^3r\, \Phi^{(n)}(\vec{r}-\vec{R})\left(\vec{\nabla} e^{-i\vec{k}_f\cdot\vec{r}}\right)\right| \\
&\sim \left| i(\vec{\epsilon}\cdot\vec{k}_f)\sum_{\vec{R}} e^{i\vec{k}\cdot\vec{R}} e^{-i\vec{k}_f\cdot\vec{R}} \int d^3r\, \Phi^{(n)}(\vec{r}-\vec{R}) e^{-i\vec{k}_f\cdot\vec{r}} e^{i\vec{k}_f\cdot\vec{R}}\right|
\end{aligned}$$

Since $\vec{k}$ belongs the first Brillouin zone and is defined modulo a reciprocal lattice vector, the sum on $\vec{R}$ leads to $\delta(\vec{k}-\vec{k}+\vec{G})$ and we recognize in the integral the Fourier transform at $\vec{k}_f$ of the Wannier function in the cell containing the origin:

$$|M_{\vec{k},\vec{k}_f}| \sim \left| i(\vec{\epsilon}\cdot\vec{k}_f)\,\delta(\vec{k}-\vec{k}_f+\vec{G}) \times \langle \vec{k}_f|\Phi^{(n)}_0\rangle\right|. \tag{3.76}$$

The photoemission matrix element is therefore proportionate to the Fourier transform of the Wannier function.

The above expression does not take into account the presence of a surface. Indeed, before being detected outside the crystal, the photoelectrons must reach the surface and cross it. Inelastic interactions in matter lead to a finite mean free path (λ) which limits the thickness probed by photoemission since the final state has an evanescent character from the surface. The wave function of the final state is indeed written:

$$\Psi_{\vec{k}_f}(\vec{r}) \sim e^{i\vec{k}_f\cdot\vec{r}}\, e^{-|r_\perp|/\lambda} \tag{3.77}$$

where $|r_\perp|$ is the distance from the surface. It is therefore necessary to modify the matrix element to take into account the presence of the surface and its influence on the electronic states. The modifications are twofold: the sum over $\vec{R}$ must be carried out on a semi-infinite crystal and the evanescent term must be introduced into the wave function. We can consider that the mean free path introduces an imaginary term into the wave vector of the final state, i.e., we must replace $\vec{k_f}$ by $\vec{k_f} - i\frac{1}{\lambda}\vec{e}_\perp$ where $\vec{e}_\perp$ is a unit vector in the normal direction to the surface. The matrix element becomes:

$$|M_{\vec{k},\vec{k_f}}| = \left| i\vec{\epsilon} \cdot (\vec{k_f} + i\frac{1}{\lambda}\vec{e}_\perp)\, \delta(\vec{k}_\parallel - \vec{k}_{f\parallel} + \vec{G}_\parallel) \times \frac{1}{i(k_\perp - k_{f\perp}) + 1/\lambda} \langle \vec{k_f} + i\frac{1}{\lambda}\vec{e}_\perp | \Phi_0^{(n)} \rangle \right| \tag{3.78}$$

we can easily check that at the limit of infinite λ, we find the Fourier transform of the Wannier function (equation 3.76). We have seen that the crossing of the surface leads to a Snell-Descartes refraction (equations 3.34 and 3.39 and figure 3.12):

$$K_{ext\parallel} = \sqrt{\frac{2m}{\hbar^2} E_{kin}}\, \sin\theta, \quad K_{ext\perp} = \sqrt{\frac{2m}{\hbar^2} E_{kin}}\, \cos\theta$$

$$k_{f\perp} + G_\perp = \sqrt{\frac{2m}{\hbar^2}(E_{kin}\cos^2\theta + V_0)}$$

with the conservation of the parallel component: $\vec{K}_{ext\parallel} = \vec{k}_{f\parallel} + \vec{G}_\parallel$. Be careful, the geometry of the experiment and in particular the direction of the photoelectrons is given by $\vec{K}_{ext}$ whereas the matrix element involves the wave vector of the final state in the solid $\vec{k_f}$.

Consider a simple case, a crystal with one atom per cell, and a description in the tight binding model. The Wannier functions can be identified to atomic orbitals. The Fourier transform of an atomic orbital, a radial function times a spherical harmonic, is written as the product of the Fourier transform of the radial function and the Fourier transform of the angular part. The Fourier transform of a given spherical harmonic is the same spherical harmonic, the angles of which determine the direction of the wave vector. But the angular distribution is not only determined by the symmetry of the atomic orbitals but also by the polarization of the photons and the geometry of the experiment. Let us consider linearly polarized photons the incident direction of which in the *Oyz* plane makes an angle α with respect to the normal to the surface that we will call the *Oz* axis. The components of the polarization vector are expressed according to:

$$\vec{\epsilon}_s = \begin{pmatrix} 1 & 0 & 0 \end{pmatrix} \text{ for a } s \text{ polarisation, and}$$

$$\vec{\epsilon}_p = \begin{pmatrix} 0 & \cos\alpha & \sin\alpha \end{pmatrix} \text{ for a } p \text{ polarisation.}$$

leading to the polarization term in the matrix element for the two polarizations:

$$\vec{\epsilon}_s \cdot (\vec{k_f} + i\frac{1}{\lambda}\vec{e}_\perp) = (k_f)_x$$

$$\vec{\epsilon}_p \cdot (\vec{k_f} + i\frac{1}{\lambda}\vec{e}_\perp) = \cos\alpha\ (k_f)_y + \sin\alpha \left[(k_f)_z + i\frac{1}{\lambda}\right].$$

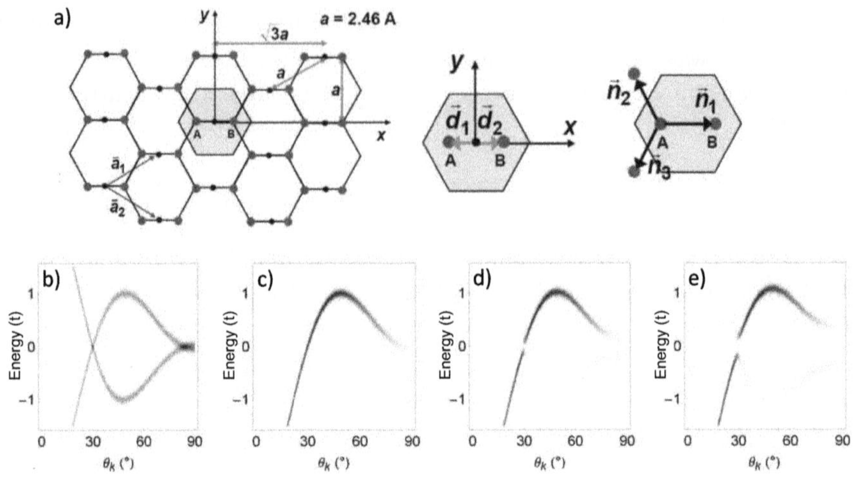

Figure 3.27 a) Graphene structure with A and B sublattices; choice of the origin between the two atoms and definition of the different $\vec{d}_i$ et $\vec{n}_i$ vectors. b) Simulation of the dispersion in the ΓK direction ($\phi_K = 0$) (without matrix elements). b) Simulations of the spectral functions, taking into account matrix elements for different directions; c): $\phi_K = 0$, d) : $\phi_K = 1.5°$, e) : $\phi_K = 3°$ (from SK Moser [39]).

Let's illustrate this approach of matrix elements on the example of graphene. The atomic structure of this carbon plane is organized in a honeycomb lattice with two carbon atoms per cell (figure 3.27(a)). With the choice of the origin between the two C-atoms, the position of the atoms is given by $\pm\frac{a}{2\sqrt{3}}(1,0)$. Bloch states of $2p_z$ symmetry described by the tight binding model are written as a linear combination of atomic orbitals with different phase terms for both types of carbon atoms [40]:

$$\Psi_{\vec{k}}(\vec{r}) = \frac{1}{\sqrt{N}}\sum_j \left[e^{i\vec{k}\cdot\vec{R}_j^A} c^A(\vec{k})\Phi(\vec{r}-\vec{R}_j^A) + e^{i\vec{k}\cdot\vec{R}_j^B} c^B(\vec{k})\Phi(\vec{r}-\vec{R}_j^B)\right] \quad (3.79)$$

where $\vec{R}_j^{A/B}$ represents the positions of the A/B atoms in the cell marked by j. The tight binding Hamiltonian is:

$$H = -t\sum_{i,j}(|\Phi_j^A\rangle\rangle\langle\Phi_i^B| + h.c.) \quad (3.80)$$

with the notation $\langle\vec{r}|\Phi_j^B\rangle\rangle = \Phi(\vec{r}-\vec{R}_j^B)$. The $c^A(\vec{k})$ et $c^B(\vec{k})$ coefficients are obtained by solving the Schrödinger equation. We obtain the following 2 equation system:

$$\epsilon(\vec{k})c^A(\vec{k}) = -t(e^{-i\vec{k}\cdot\vec{n}_1} + e^{-i\vec{k}\cdot\vec{n}_2} + e^{-i\vec{k}\cdot\vec{n}_3})\ c^B(\vec{k}) = -tf(\vec{k})\ c^B(\vec{k})$$
$$\epsilon(\vec{k})c^B(\vec{k}) = -t(e^{+i\vec{k}\cdot\vec{n}_1} + e^{+i\vec{k}\cdot\vec{n}_2} + e^{+i\vec{k}\cdot\vec{n}_3})\ c^A(\vec{k}) = -tf^*(\vec{k})\ c^A(\vec{k})$$

where the energies of the two bands are:

$$\varepsilon^{\pm}(\vec{k}) = \pm t|f(\vec{k})| = \pm t\sqrt{3 + 2\cos(\sqrt{3}k_x a) + 4\cos(\sqrt{3}k_x a/2)\ \cos(3k_y a)} \tag{3.81}$$

Define by $-\vartheta_{\vec{k}}$ the phase of $f(\vec{k})$: $f(\vec{k}) = |f(\vec{k})|\ e^{-i\vartheta_{\vec{k}}}$. We can express the coefficients:

$$\begin{pmatrix} c^{A+}(\vec{k}) \\ c^{B+}(\vec{k}) \end{pmatrix} = \frac{1}{\sqrt{2}}\begin{pmatrix} 1 \\ -e^{i\vartheta_{\vec{k}}} \end{pmatrix} \text{ and } \begin{pmatrix} c^{A-}(\vec{k}) \\ c^{B-}(\vec{k}) \end{pmatrix} = \frac{1}{\sqrt{2}}\begin{pmatrix} 1 \\ e^{i\vartheta_{\vec{k}}} \end{pmatrix} \tag{3.82}$$

The Bloch states of energy $\varepsilon^{\pm}(\vec{k})$ are written:

$$\Psi^{\pm}_{\vec{k}}(\vec{r}) = \frac{1}{\sqrt{2N}}\sum_j \left[e^{i\vec{k}\cdot\vec{R}^A_j}\ \Phi(\vec{r} - \vec{R}^A_j) \mp e^{i\vec{k}\cdot\vec{R}^B_j}\, e^{i\vartheta_{\vec{k}}}\ \Phi(\vec{r} - \vec{R}^B_j)\right] \tag{3.83}$$

Since $\vec{R}^B_j - \vec{R}^A_j = \vec{d}_2 - \vec{d}_1 = \vec{\delta}_{AB}$, we can write:

$$\Psi^{\pm}_{\vec{k}}(\vec{r}) = \frac{1}{\sqrt{2N}}\sum_j e^{i\vec{k}\cdot\vec{R}^A_j}\left[\Phi(\vec{r} - \vec{R}^A_j) \mp e^{i\vec{k}\cdot\vec{\delta}_{AB}}\, e^{i\vartheta_{\vec{k}}}\ \Phi(\vec{r} - \vec{R}^B_j)\right] \tag{3.84}$$

We have seen above that the matrix element is proportionate to the Fourier transform of the Wannier function (equation 3.76). We thus have the matrix elements for the two bands:

$$|M^{\pm}_{\vec{k},\vec{k}_f}| \sim |\langle\vec{k}_f|\Phi^{\pm}_{\vec{0}}\rangle| = |\langle\vec{k}_f|\big(|\Phi^A_{\vec{0}}\rangle \mp e^{i\vec{k}\cdot\vec{\delta}_{AB}}\, e^{i\vartheta_{\vec{k}}}\ |\Phi^B_{\vec{0}}\rangle\big)|$$

where $|\Phi^A_{\vec{0}}\rangle = |\Phi^B_{\vec{0}}\rangle$ is the $2p_z$ orbital of the carbon atom. Figure 3.27 (b)–(e) presents simulations of the photoelectron distribution for a photon energy of about 50 eV (an arbitrary Gaussian broadening has been applied). The distribution (b) is computed for $\phi_{\vec{k}} = 0$ without matrix elements and thus represents the band structure in the ΓK direction, i.e. the two $\varepsilon^{\pm}(\vec{k})$ functions in that direction. The observed intersection point close to the $\theta_{\vec{k}} \sim 30°$ corresponds to the K point of the Brillouin zone (Dirac point). The distribution presented in c) corresponds to the same direction of the reciprocal space ($\phi_{\vec{k}} = 0$) but taking into account the matrix elements. We show that only the $\varepsilon^-(\vec{k})$ band is observed in the first zone and the $\varepsilon^+(\vec{k})$ band in the second zone. When moving away from the ΓK direction, $\phi_K = 1.5°$ for d) and $\phi_K = 3°$ for e) the spectral weight of the folded bands increases. This example shows again that the spectral weight is very strongly affected by the matrix elements.

3.2.6 *Temperature dependence*

3.2.6.1 Temperature dependence of core level spectra

The temperature dependence of the core photoemission spectra has been studied by Larson and Pendry [41]. It is due to the atom vibration activated by the temperature, the electron-phonon coupling and results in a modification of intensity and width of the spectral structures. The selection rules can also be relaxed because of the change of symmetry. Indeed, vibration symmetry modes can have a lower symmetry than the ground state. We will not deal in detail with the effect of temperature in all types of materials and for all electron-phonon interactions but only give the most general characteristics. The effect of atomic vibrations on photoemission spectra is similar to the temperature dependence of X-ray diffraction. In coherent diffraction experiments, the Bragg peaks show a decrease in intensity as the temperature increases by a Debye-Waller factor $e^{-2W(T)}$. A similar dependence is observed in photoemission (even in the valence band) because of the vibration of each ion around its equilibrium position: $\vec{R}(t) = \vec{R}^0 + \Delta\vec{R}_0(t)$. When taking into account this oscillation in the photoemission intensity $I_T \propto \langle f|H|i\rangle$. There are two terms that appear [11]:

$$I_T(E, h\nu) = \left[e^{-|\Delta k|^2 \Delta R_0^2} I_D + (1 - e^{-|\Delta k|^2 \Delta R_0^2}) I_{ND}\right] \tag{3.85}$$

The second term is associated with non-direct transitions, i.e., the phonon assisted transitions, and gives an incoherent background independent of T. The temperature dependence is in the first term of the direct transitions, with an exponential variation which is the Debye factor where $W = \frac{1}{2}|\Delta k|^2 \Delta R_0^2$. According to the calculation of the average displacement (ΔR_0), we obtain:

$$W = \frac{3\hbar^2 |\Delta k|^2}{2Mk_B\Theta^2} T \tag{3.86}$$

where k_B is the Boltzmann constant, Θ is the Debye temperature, M is the atomic mass, and Δk is a vector of the reciprocal lattice.

3.2.6.2 Temperature dependence of the valence band spectra

We have just seen that the electron-phonon interaction leads to a modification of the core level photoemission spectra. Effects are also expected on the valence states and can be described by the many body approach developed above. Indeed, the quasi-particles are dressed by virtual phonons resulting in a broadening of the spectral function or even the appearance of satellite structures. Similarly, self-energy leads to a renormalization of the bands in a narrow energy domain in the vicinity of the Fermi level of the order of the characteristic frequencies of the phonons. We shall describe these effects qualitatively in a simple approach.

The electron-phonon coupling yields the scattering of the valence electrons by the lattice with creation or destruction of a phonon (proper mode of vibration of a crystalline lattice). It can be described by the Fröhlich Hamiltonian:

$$H_{e-p} = \sum_{\vec{q},\vec{k}} V(\vec{q},\vec{k})\, a^{\dagger}_{\vec{k}+\vec{q}} a_{\vec{k}} (b^{\dagger}_{-\vec{q}} + b_{\vec{q}}) \tag{3.87}$$

describing the destruction of an electron with $\vec{k}$ wave vector and the creation of an electron with $\vec{k}+\vec{q}$ wave vector. The $\vec{q}$ transferred wave vector comes from the destruction of a $\vec{q}$ wave vector phonon or the creation of a $-\vec{q}$ wave vector phonon. According to Pauli's principle, this mechanism can only involve states in the vicinity of the Fermi level in an energy domain of the order of the phonon energy since the energy change of the electron must correspond to the phonon energy and that the electron final state must be unoccupied.

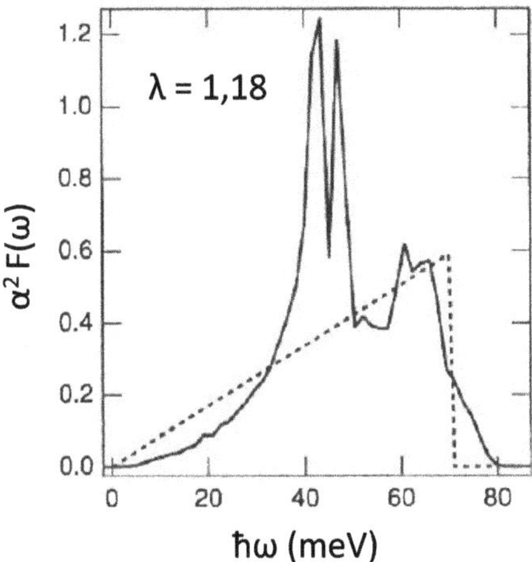

Figure 3.28 *The Eliashberg function calculated for a coupling parameter value* $\lambda = 1.18$ *(solid line) and a simple approximated linear function (dashed line) (from [42]).*

We have seen that the spectral function is completely determined by the self-energy. It is therefore necessary to determine the self-energy corresponding to the electron-phonon interaction. We can make the approximation where it does not depend explicitly on the wave vector so that it can be written [43]:

$$\Sigma_{e-p}(\omega) = \int d\epsilon \int_0^{\omega_m} d\tilde{\omega}\, \alpha^2 F(\tilde{\omega}) \times \left[\frac{1 - f(\epsilon, T) + N(\tilde{\omega}, T)}{\omega - \epsilon - \tilde{\omega} + i\delta} + \frac{f(\epsilon, T) + N(\tilde{\omega}, T)}{\omega - \epsilon + \tilde{\omega} + i\delta} \right] \tag{3.88}$$

where $f(\epsilon, T)$ and $N(\tilde{\omega}, T)$ are the distributions of Fermi-Dirac and Bose-Einstein, respectively. The $\alpha^2 F(\tilde{\omega})$ quantity called the Eliashberg function represents the

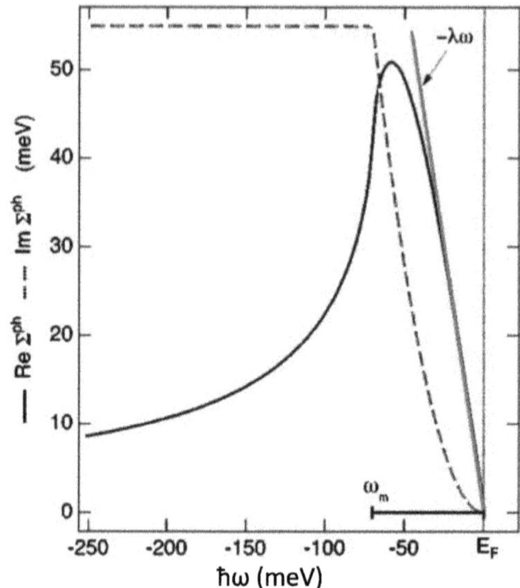

Figure 3.29 *Real and imaginary parts of self-energy calculated at $T = 0$ with an approximated Eliashberg function (from [44]). Reprinted figure with pemission from M. Hengsberger, R. Frésard, D. Purdie, P. Segovia, et Y. Baer, Phys. Rev. B 60, 10796 (1999). ©1999 by the American Physical Society.*

product of the phonon state density $F(\tilde{\omega})$ by the electron-phonon coupling constant α^2. This function for the surface of Be (0001) is represented in figure 3.28. The self-energy does not depend to any significant extent on the fine details of the Eliashberg function, which can be approximated by a linear function represented in dashed lines in figure 3.28:

$$\alpha^2 F(\tilde{\omega}) \propto \frac{\lambda \tilde{\omega}}{2\omega_m} \quad \text{with} \quad \lambda = 2 \int_0^{\omega_m} \frac{\alpha^2 F(\tilde{\omega})}{\tilde{\omega}} d\tilde{\omega} \qquad (3.89)$$

where λ is a dimensionless number called coupling parameter and ω_m a cutoff frequency, in the order of the Debye frequency.

With this simple approach, the real and imaginary parts of the self-energy calculated with this approximated form of the Eliashberg function are presented for $T = 0$ (valid in the $k_B T \ll \omega_m$ limit) in figure 3.29. They present a significant variation in an energy domain of the order of ω_m in the vicinity of E_F. The real part grows linearly with a slope equal to $-\lambda$ and the imaginary part increases monotonically. Beyond this range, the imaginary part is constant while the real part decreases slowly towards 0. The variation of the real part of the self-energy leads to a renormalization of the band dispersion around the Fermi surface as illustrated schematically in figure 3.30. It can be seen that in a range of the order of phonon energy around the Fermi level, the

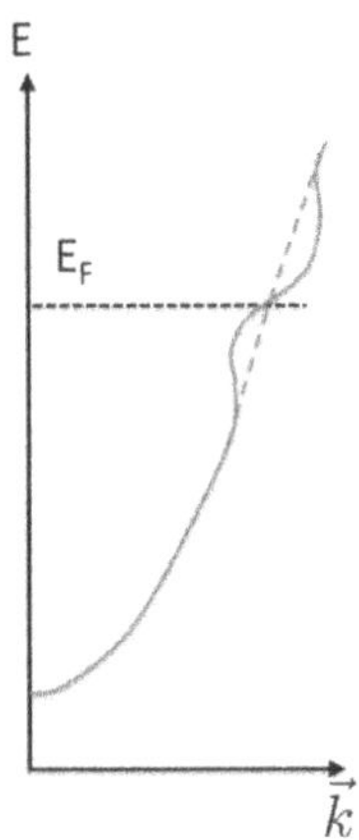

Figure 3.30 *Schematic renormalization of the band dispersion. In an energy range of the order of the phonon energy ($\hbar\omega_m$), the band is renormalized, while outside this range, the expected dispersion for a non-interacting system is expected.*

dispersion is reduced (greater effective mass and flatter band) while at higher energy the bare dispersion (without electron-phonon coupling) is found. With increasing temperature, the presence of phonons in the initial state leads to an increase in self-energy. In the high temperature limit ($k_B T \gg \hbar\omega_m$), the temperature change of the imaginary part of the self-energy at E_F is linear:

$$\Sigma''(\omega = 0, T) = \pi \lambda k_B T.$$

This formalism has been successfully applied to describe the photoemission spectrum of systems where electron-phonon interaction is important. Figure 3.31 shows the photoemission spectra measured on the (0001) Be surface in a high symmetry direction of the surface Brillouin zone. We distinguish very well the two domains where the dispersion is different. At high energy, one observes wide structures (imaginary part of the important self-energy) which disperse rapidly whereas, in the vicinity of the Fermi surface, narrow structures are observed due to the small self-energy imaginary part and with a very small dispersion (strongly renormalized band).

The mechanism discussed above for band renormalization corresponds to direct transitions. But where the photoemission on the core levels is concerned, there is a non-direct contribution which derives from the presence of phonons in the initial state and which corresponds to a photoemission process with phonon absorption. Phonons have a small energy compared to electrons but a wave vector that can be large. Therefore, a phonon-assisted photoemission process does not conserve the electron wave vector. A photoelectron emitted with a given wave vector can come from an initial state with any wave vector of the Brillouin zone. With increasing temperature, the probability of direct transitions decreases and that of indirect

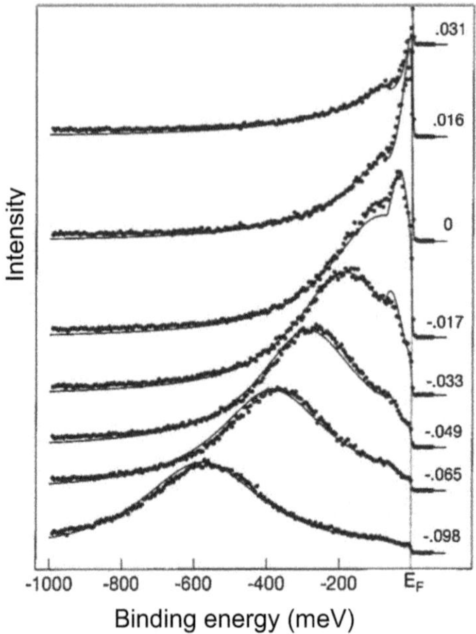

Figure 3.31 *Experimental photoemission spectra on the (0001) Be surface and fits (solid lines) with the model described above (from Hengsberger* et al. *[44]. Figure reprinted with permission from M. Hengsberger, R. Frésard, D. Purdie, P. Segovia, et Y. Baer, Phys. Rev. B 60, 10796 (1999). ©1999 by the American Physical Society).*

transitions increases so that at very high temperatures, the spectrum is an average over the whole Brillouin zone and then becomes proportionate to the density of states. The Debye-Waller factor also allows you to understand the photon energy dependence. Indeed, in the XPS regime, the photons wave vector is no longer negligible, for example, its order of magnitude for the Al K_α line ($h\nu$ =1486 eV) is about 20Å^{-1} which leads to a Δk increase in the Debye-Waller factor $W = \frac{1}{2}|\Delta k|^2 \Delta R_0^2$. The weight of the direct transitions in this energy range is then only a few percent of the total intensity at room temperature for standard materials like Cu.

3.3 Detailed analysis of core levels

As we have seen above, the core levels, because of their atomic nature, allow us to conduct a chemical analysis of the surfaces (composition) but also to gain information as to the nature of the chemical bonds and the degree of oxidation. We will see in this part that a finer analysis can also provide information on the electronic structure. Indeed, the shape of the spectral line and in particular its asymmetry is a function of the density of states at the Fermi level. Moreover, additional structures sometimes appear and the position and intensity of these satellites make it possible to obtain information about the electronic structure of the ground state.

3.3.1 *Core level line shape in metals*

The experimental line shape is the convolution product of the intrinsic line with a function (usually a Gaussian) describing the experimental resolution. When the intrinsic line is only due to the lifetime of the created hole, it is a lorentzian. In this case, the experimental spectrum is then a Voigt function resulting from the convolution product of this intrinsic lorentzian with the Gaussian due to the resolution (cf. previous chapter). However, in metals there is an asymmetrical line that reflects a new mechanism associated with the presence of a metallic band with a finite density of states. This mechanism corresponds to the screening of the core hole by electron-hole pairs near the Fermi level. The core hole yields a localized potential that induces the formation of band excitations to screen the positive charge associated with the hole. This is an extremely complex many body problem similar to the singularity at the X-absorption thresholds in metals theoretically studied by Nozières and Dominicis in 1969. They have shown, by a partial wave phase shift approach, that in the vicinity of the absorption threshold, the response of the electron gas leads to to a spectral absorption function X of the form [45]:

$$A(\omega) \propto \left(\frac{\xi}{\omega - \omega_0}\right)^{\alpha_\ell} \tag{3.90}$$

where ξ represents an energy cutoff of the order of the Fermi energy, ω_0 the threshold energy and α_ℓ the exponent of the singularity which is expressed from the phase shifts δ_ℓ:

$$\alpha_\ell = \frac{2\delta_\ell}{\pi} - \alpha \quad \text{with} \quad \alpha = \sum_{l=0} 2(2l+1)\left(\frac{\delta_\ell}{\pi}\right)^2 \tag{3.91}$$

It is recalled that phase shifts must satisfy Friedel's rule:

$$\sum_{l=0} 2(2l+1)\frac{\delta_\ell}{\pi} = +1 \tag{3.92}$$

corresponding to the associated charge of the hole to be screened.

The two contributions in α_ℓ had already been proposed separately. The first one $(2\delta_\ell/\pi)$, generally positive, was introduced by Mahan [46] into an exciton (hole-electron interaction) approach of the X-absorption leading to the divergence of the absorption at the threshold. The second one (α) was introduced by Anderson [47] who had noticed the orthogonality of the many body wave functions of the initial and final states (orthogonality catastrophe). This contribution tends to cancel the absorption coefficient at the threshold.

This formalism, developed to describe the X absorption thresholds, was adapted by Doniach and Šunjić to describe the photoemission line shape. Unlike X-absorption, the XPS photoelectron leaves the solid so that the excitonic process does not contribute. The singularity of the spectral function is written [48]:

$$F(E) \propto \frac{\sin(\alpha/2)}{\Gamma(\alpha)(E - E_0)^{1-\alpha}} \tag{3.93}$$

where Γ is the gamma function, E_0 the energy of the core level and α the same exponent as for X absorption spectroscopy. We must also take into account the life time of the core hole and convolute $F(E)$ with a lorentzian of width 2γ, which gives for the intensity of the spectrum:

$$I_{DS}(E) = \frac{\Gamma(1-\alpha)}{[(E - E_0)^2 + \gamma^2]^{(1-\alpha)/2}} \cos\left(\frac{\pi\alpha}{2} + \theta(E)\right) \tag{3.94}$$

with

$$\theta(E) = (1-\alpha)\tan^{-1}[(E - E_0)/\gamma]\,. \tag{3.95}$$

This form of line, called Doniach-Šunjić's functions shown in figure 3.32. If $\alpha = 0$, we find a symmetrical line with lorentzian shape corresponding to the life time is obtained.

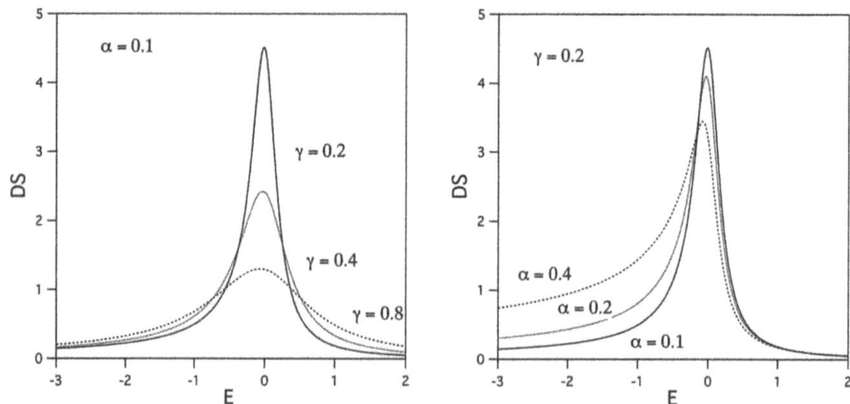

Figure 3.32 *Evolution of the Doniach-Šunjić's function with the asymmetry parameter (α) and the core-hole lifetime (γ).*

The Doniach-Šunjić's function however, causes problems because of its divergent integral. This led Mahan to propose a function of similar form but with a finite integral [49]. The Mahan function is:

$$F_0(E) = \frac{1}{\Gamma(\alpha)} \frac{e^{(E-E_0)/\xi}}{|(E - E_0)/\xi|^{1-\alpha}} \Theta(E_0 - E) \tag{3.96}$$

where ξ is a cutoff parameter of the order of the occupied bandwidth.

The α parameter of the Doniach-Šunjić function measures the line asymmetry and describes the core hole screening by electron-hole pairs at the Fermi surface. It therefore depends on the density of states at the Fermi level. The figure 3.33 illustrates this dependence of the asymmetry parameter on the density of states at the Fermi level in the $Pd_{1-x}Ce_x$ alloys. It exhibits the evolution of the valence band UPS spectra as a function of the cerium concentration. The Pd-*d* band is incompletely filled and corresponds to a large density of states at the Fermi level.

The corresponding asymmetry parameter of the Pd-$3d^{5/2}$ line presented in figure 3.33(b) is large. When the Ce- concentration (x) increases, there is a progressive filling of the Pd-d band leading to a decrease of the density of states at E_F. This dependence is accompanied by a drastic decrease in the asymmetry parameter. Finally, beyond x =0.125, the d band is full and is shifted towards high binding energies. The density of states at E_F, due to the sp states, then remains nearly constant and the asymmetry parameter presents a plateau for a very low α value. This shows that the α asymmetry parameter gives qualitative information about the density of states at the Fermi level; a very strong asymmetry of a core level line shape reflects a large density of states at the Fermi level.

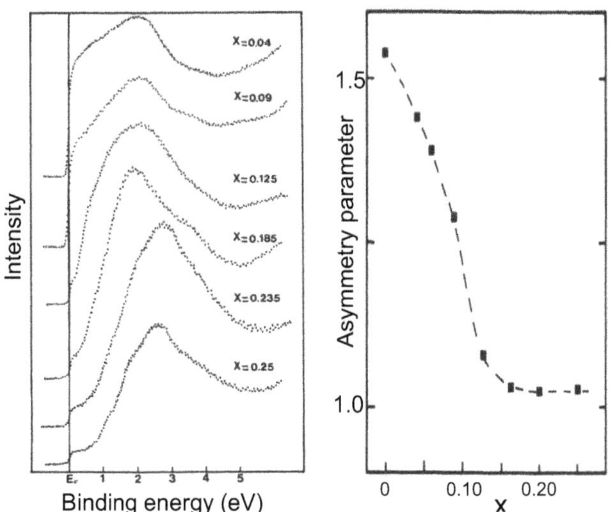

Figure 3.33 a) UPS spectra of the valence band in $Pd_{1-x}Ce_x$ for several x values. b) Asymmetry parameter of the Pd $3d^{5/2}$ line, from [50]. Figure reprinted with permission from D. Malterre, G. Krill, J. Durand, et G. Marchal, Phys. Rev. B 38, 3766 (1988). ©1988 by the American Physical Society.

3.3.2 *Multiplet effects*

A complex spectral shape may appear when the photoabsorber atom has an incompletely filled localized outer layer. For example, this is the case for ionic transition metal compounds or rare earth based alloys. For these materials, the electrons of the outer sub-layer of d or f symmetry are very localized and have a significant atomic character. In the photoemission final state, these localized electron will interact strongly with the core hole. This coupling leads to additional structures in the photoemission spectra. This splitting of a spectral line into several structures is called a multiplet effect; it results from the coupling of the angular momenta of the two incomplete sub-layers and each of the structures is associated with a value of the total angular momentum of the photo-excited atom. Describing the detailed calculation of these multiplet states is outside the ambitions of this book. This is a complex problem, but we would like to explain its principle, at least in the atomic limit. But first of all, we must describe the initial state that contains an outer incomplete sublayer, for example $n\ell^p$,corresponding to p electrons in the atomic sublayer associated with the main (n) and azimuth (ℓ) quantum numbers. The degeneracy of the electron configuration associated with this sublayer is lifted by electron-electron interactions, spin-orbit interactions and possibly crystal-field interactions. The orders of magnitude of these different interactions depend on the type of solids, but Coulomb and exchange interactions are generally dominant. When we neglect the crystalline field effects, it is an atomic physics problem and the energy levels are multiplets characterized by the different angular momenta: the total orbital momentum L of the sublayer, its total spin momentum S and total angular momentum J ($\vec{J} = \vec{L} + \vec{S}$). The ground level is usually given by Hund's rules of atomic physics and is the initial state of the photoemission process. We can recall the nomenclature and spectroscopic notation used for multiplets in the Russel-Saunders model (or LS coupling). A multiplet associated with L, S, J is designated by the notation ${}^{2S+1}A_J$ where A is a capital letter giving the value of the orbital angular momentum according to the following convention:

value of L	notation of A
$L = 0$	S
$L = 1$	P
$L = 2$	D
$L = 3$	F
$L = 4$	G
$L = 5$	H
$L = 6$	I
etc.	

where the exponent determines the spin degeneracy equal to $2S + 1$, whereas the subscript gives the J value. For example, the term spectroscopic:

$$^4G_{11/2}$$

corresponds to the multiplet $L = 4$, $S = 3/2$ and $J = 11/2$.

In the photoemission process, an electronic hole is created in an internal sub-layer, for example the $n'\ell'$ sub-layer. This hole is associated with an orbital angular $L' = \ell'$ and spin $S' = 1/2$ momenta. The energy levels in the final state are multiplets corresponding to the values of the different angular momenta (orbital, spin and total) of the two incomplete (core and outer) sub-layers. The photoemission spectrum corresponds to the transitions between the initial state given by the ground state multiplet derived from Hund's rules and all the multiplets corresponding to the final states of the photoemission process. The intensity of a contribution is given by the matrix element of the electric dipole interaction operator between the initial and final multiplets. Let's take the example of a cerium compound and therefore a $4f^1$ sub-layer. The ground multiplet is the $^2F_{5/2}$ state ($L = 3$, $S = 1/2$ and $J = 5/2$). Consider the photoemission on a core level of d symmetry. The coupling of the two incomplete d and f sub-layers leads to a spin singlet ($S=0$) or triplet ($S=1$) and a total orbital momentum between 3+2 and 3−2. We therefore have the spectroscopic terms:

$$^3P, ^1P, ^3D, ^1D, ^3F, ^1F, ^3G, ^1G, ^3H, ^1H$$

The spin-orbit interaction yields a splitting of each spectroscopic term as a function of the J values (between$|L - S|$ and $L + S$).
The energy separation of the different terms results from the electron-electron interactions and can be calculated in the LS coupling model. The calculation method was developed by Slater in the framework of the LS coupling model and can be found in Condon and Shortley's book [51]. Let us recapitulate on the the main steps. The interaction between two monoelectronic states (ϕ_p and ϕ_q) involves a direct Coulomb term:

$$J(\phi_p, \phi_q) = \langle \phi_p \phi_q | \frac{e^2}{r_{12}} | \phi_p \phi_q \rangle = \int \phi_p^*(\vec{r}_1 s_1) \phi_q^*(\vec{r}_2 s_2) \frac{e^2}{r_{12}} \phi_p(\vec{r}_1 s_1) \phi_q(\vec{r}_2 s_2) d\tau_1 d\tau_2$$

and an exchange term:

$$K(\phi_p, \phi_q) = \langle \phi_p \phi_q | \frac{e^2}{r_{12}} | \phi_q \phi_p \rangle = \int \phi_p^*(\vec{r}_1 s_1) \phi_q^*(\vec{r}_2 s_2) \frac{e^2}{r_{12}} \phi_q(\vec{r}_1 s_1) \phi_p(\vec{r}_2 s_2) d\tau_1 d\tau_2$$

where the index of the atomic wave functions (p and q) denotes the quantum n, ℓ, m numbers. To determine these different integrals, the electron-electron interaction term has to be written in a multipolar expansion:

$$\frac{e^2}{r_{12}} = \frac{e^2}{r_>} \sum_{k=0}^{\infty} \left(\frac{r_<}{r_>} \right)^k P_k(\cos\alpha)$$

where α is the angle between $\vec{r}_1$ and $\vec{r}_2$, $r_<$ and $r_>$ the smallest and largest values of r_1 and r_2. The $P_k(\cos\alpha)$ are the Legendre polynomials that obey the identity:

$$P_k(\cos\alpha) = \frac{4\pi}{2k+1} \sum_{m=-k}^{k} Y_k^{-m}(\theta_1, \varphi_1) Y_k^m(\theta_2, \varphi_2).$$

That leads to the expression of the Coulomb term:

$$J(\phi_p, \phi_q) = \sum_{k=0}^{\infty} F^k(n^p \ell^p, n^q \ell^q) \times a^k(\ell^p m^p, \ell^q m^q) \tag{3.97}$$

and exchange term:

$$K(\phi_p, \phi_q) = \delta(s_p, s_q) \sum_{k=0}^{\infty} G^k(n^p \ell^p, n^q \ell^q) \times b^k(\ell^p m^q, \ell^p m^q) \tag{3.98}$$

where $\delta(s_p, s_q)$ indicates that the spins of the two states must be the same to obtain a non-zero exchange term. The a^k and b^k values depend only on symmetry and are tabulated for the most common electronic configurations in the reference books, for example, [51]. The b^k are zero unless:

$$|\ell^p - \ell^q| \leqslant k \leqslant \ell^p + \ell^q$$

where k must obey:

$$k + \ell^p + \ell^q = 2g \quad \text{for} \quad g \text{ positive integer.}$$

For the a^k, only the terms, with parity equal to $\ell^p + \ell^q$, less than or equal to $2min(\ell^p, \ell^q)$ are non-zero. The F^k and G^k are integrals on the radial parts of the wave functions. The conditions on the a^k and b^k values show that the multipolar expansion is truncated. For 2 electrons belonging to the same electronic sublayer, we have $F^k = G^k$. Thus for 2 p electrons, only F^0 and F^2 are non-zero, for 2 d electrons, we must consider F^0, F^2 and F^4, and for 2 f electrons, F^0, F^2, F^4 and F^6. To simplify the expressions, we define the quantities $F_k = F^k/D_k$ and $G_k = G^k/D_k$ where D_k is an integer dependent on k. We can therefore express the terms of Coulomb and exchange as a function of F_k and G_k and thus calculate the energies of the multiplets. For example, for the $n'd\ nf$ configuration discussed above, we obtain:

$$\begin{aligned}
{}^1P,\ {}^3P:&\ E = F_0 + 24F_2 + 66F_4 \pm (\ G_1 + 24G_3 + 330G_5)\\
{}^1D,\ {}^3D:&\ E = F_0 + 6F_2 - 99F_4 \mp (3G_1 + 42G_3 - 165G_5)\\
{}^1F,\ {}^3F:&\ E = F_0 - 11F_2 + 66F_4 \pm (6G_1 + 19G_3 + 55G_5)\\
{}^1G,\ {}^3G:&\ E = F_0 - 15F_2 - 22F_4 \mp (10G_1 - 35G_3 - 11G_5)\\
{}^1H,\ {}^3H:&\ E = F_0 + 10F_2 + 3F_4 \pm (15G_1 + 10G_3 + \ G_5)
\end{aligned}$$

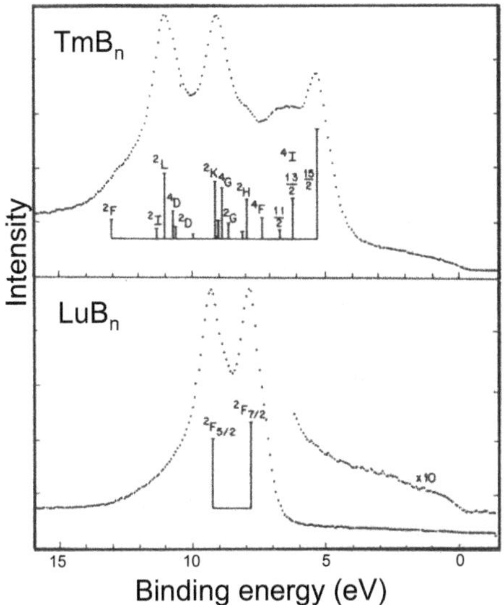

Figure 3.34 *Valence band spectra of LuB_n and TmB_n, from [52]. B_n indicates a phase mixture. Reproduction from "Photoemission in Solids II", edited by L. Ley and M. Cardona, 1979, chapter entitled "Unfilled Inner Shells: Rare Earths and Their Compounds" by M. Campagna, G. K. Wertheim, and Y. Baer, Fig. 4.27, © by Springer-Verlag Berlin Heidelberg 1978, with kind permission from Springer Science and Business Media.*

Each multiplet will contribute to the photoemission spectrum of the core levels ($3d$ or $4d$), the weight of each contribution being given by the dipolar Hamiltonian matrix element between the initial and the final states corresponding to the different multiplets. This multiplet approach can also be used to describe the $4f$ valence spectra of rare earth compounds because of quasi-atomic character of these $4f$ states. Only the $4f$ sub-layer has to be considered and the energies calculation is simplified because we have $F_k = G_k$. The figure 3.34 shows the spectra of the valence band for several rare earth compounds. For the lutetium compound, the $4f$ sublayer is full in the initial state and contains a single hole in the final state. Then the final multiplets corresponds to a single hole in the $4f$ shell with only a spin-orbit splitting: $^2F_{5/2}$ and $^2F_{7/2}$.

3.3.3 *Satellite structures*

In many systems characterized by highly localized valence electrons (transition metal oxides, rare earth compounds, ...) the core level photoemission spectra exhibit a complex structure with satellites. To explain the physical origin of these structures, we will now introduce the Anderson impurity model.

3.3.3.1 The single impurity Anderson model

The Anderson impurity model [53] was used to describe a magnetic impurity in a 'normal' metal. It allows us to explain the magnetic or non-magnetic character of the impurity (the so-called Kondo effect). We will use it here to describe the core level spectroscopy of cerium compounds in particular (the same approach can be used for similar systems such as transition metal oxides).

Cerium is the first rare earth element and corresponds to an atomic configuration with one 4f electron. In the rare earth compounds, the 4f electrons, although belonging to the valence states, are very localized and may be, in a first approximation, to be described in an atomic approach. Indeed, they do not contribute to the chemical bond and their properties are dominated by intra-atomic Coulomb interactions. We note that the repulsion of two f electrons is of the order of 5–7 eV. Consequently, the number of f electrons per atom is generally integer and their properties, for example the magnetic properties, are mainly atomic-like. However, in some rare earths, f states can hybridize with conduction states, i.e., an electron can jump from a localized f state to a delocalized band state. This mechanism has important consequences for the lower atomic number rare-earth elements (cerium), near the middle atomic numbers (europium, samarium) or higher atomic numbers (ytterbium) of the lanthanide series. The physical properties then result from the competition between Coulomb repulsion and hybridization. We can first approximate these systems by the Hamiltonian of the single impurity of Anderson which is written:

$$H = \sum_{\vec{k}} \varepsilon_{\vec{k}} a^{\dagger}_{\vec{k}} a_{\vec{k}} + \sum_{m} \varepsilon_f a^{\dagger}_{fm} a_{fm} + U_{ff} \sum_{mm'} n_m n_{m'} + \sum_{\vec{k},m} V_{\vec{k},m} \left(a^{\dagger}_{\vec{k}} a_{fm} + a^{\dagger}_{fm} a_{\vec{k}}\right) \tag{3.99}$$

where the first term describes the energy of the conduction band states, the second the energy of the f states (they are 14 times degenerate ($m = 1, \ldots 14$), the third is the repulsion of Coulomb between f states ($n_m = a^{\dagger}_{fm} a_{fm}$ is the occupation operator of the mth f state and the last term describes the hybridization between f states and conduction states. If the last term is zero, the f and conduction states are not coupled and the number of f electrons is integer and only depends of the values of the ε_f and U_{ff} parameters. The eigenstates correspond to the different $|f^n\rangle$ configurations. For example, taking the f^0 configuration as the energy origin, the energy of the different configurations can be easily obtained:

$$\begin{aligned}
E(f^1) &= \varepsilon_f \\
E(f^2) &= 2\varepsilon_f + U_{ff} \\
E(f^3) &= 3\varepsilon_f + 3U_{ff} \\
E(f^n) &= n\varepsilon_f + \frac{n(n-1)}{2} U_{ff}
\end{aligned}$$

since for n electrons, we have $n(n-1)/2$ electron pairs that interact. The typical values of these parameters for the Ce compounds are $\varepsilon_f \simeq -1.5$ eV and $U_{ff} \simeq 6$ eV so that the most stable state is the f^1 configuration and the first excited state the f^0 configuration (figure 3.35).

In the final states of the photoemission on a Ce core level, there is a core hole which will interact with the f electrons that are very localized. With the hole being positively charged, the Coulomb interaction term between the hole and f states is attractive ($U_{fc} < 0$). Experiment shows that $U_{fc} \simeq -10$ eV. The Hamiltonian in the final state can be written to take into account the interaction of the core hole with the f electrons:

$$H_{fin} = H + \varepsilon_c + U_{fc} \sum_m n_c n_m \tag{3.100}$$

where ε_c is the energy of the relevant core level. There is an additional term U_{fc} corresponding to the interaction term which modifies the energy of the electronic $|c, f^n\rangle = a_c|f^n\rangle$ configurations (a_c is the annihilation operator of a core electron and c index indicates the presence of a core hole). Thus the energy of the (c, f^1) configuration is lowered by U_{fc}, that of the configuration (c, f^n) is lowered by nU_{fc}.

$$\begin{aligned} E(c, f^0) &= 0 \\ E(c, f^1) &= \varepsilon_f + U_{fc} \\ E(c, f^2) &= 2\varepsilon_f + U_{ff} + 2U_{fc} \end{aligned}$$

It can be seen that due to the interaction with the core hole, the configuration energies are more separated than in the initial state and that the final configuration state of lower energy is $|c, f^2\rangle$. The physical origin of this mechanism is simple to understand, a core hole is much better screened by a localized f electron than by a delocalized conduction electron. Figure 3.35 presents the energy of the different configurations (in the $V_{\vec{k},m} = 0$ limit) for the initial and the final states. Only the f^1, f^0 and f^2 configurations contribute to the ground state.

A finite hybridization will couple the configurations together so that the ground state can be written symbolically as a quantum admixture of configurations[5]:

$$|\Psi_{G.S.}\rangle = a_1|f^1\rangle + a_0|f^0\rangle + a_2|f^2\rangle \tag{3.101}$$

The largest coefficient is of course a_1 and the f^1 configuration is dominant for the ground state of Ce. The order of magnitude of the other coefficients can be obtained from the second-order perturbation theory which yields that the mixture of states is proportional to the matrix element of the hybridization between f^n and f^{n+1} states squared divided by the difference of the unperturbed state energies. We can write that $a_0 \propto |V|^2/\varepsilon_f$ and $a_2 \propto |V|^2/(\varepsilon_f + U_{ff})$ from which we deduce $a_1 > a_0 > a_2$. Likewise excited states can also be written as a configuration admixture but with

[5] This notation indicates only the $4f$ configurations; it is clear that the total number of electrons, localized and delocalized, is the same for the different configurations.

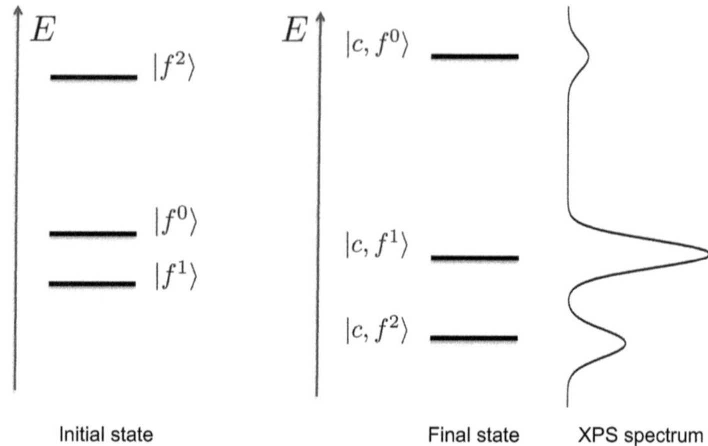

Figure 3.35 *Energies of the different configurations of Ce for initial and final (with the core hole) systems in the limit of zero hybridization ($V_{\vec{k},m} = 0$). The following values were used: $\varepsilon_f = -1.5$ eV, $U_{ff} = 6$ and $U_{fc} = -9$ eV. Schematic spectrum of a core level for finite hybridization. Three structures corresponding to the three possible final states are observed in the presence of a core hole.*

a dominant configuration. However, this description with discrete states associated with each configuration is very simplified. Indeed, the continuum of band states should be explicitly involved. The resolution of Anderson's Hamiltonian is one of the most complex problems of condensed matter physics and requires the use of very elaborate many body methods. Nevertheless, as far as we are concerned here, a description by a configuration mixture is sufficient to qualitatively understand the spectroscopic properties of these systems.

In the final state, the hybridization mixes the different (c, f^n) electronic configurations but as the energy separation between the different configurations are much larger than in the initial state, the mixture of the configurations is smaller. Let us designate $|\Psi_{fin}^{f2}\rangle$ the lowest energy state. It can be written:

$$|\Psi_{fin}^{f2}\rangle = b_2^{(2)}|c,f^2\rangle + b_1^{(2)}|c,f^1\rangle + b_0^{(2)}|c,f^0\rangle \quad \text{with} \quad b_2^{(2)} > b_1^{(2)} > b_0^{(2)} . \quad (3.102)$$

The other two final states noted $|\Psi_{fin}^{f1}\rangle$ and $|\Psi_{fin}^{f0}\rangle$ are expressed in the same way with the $b_n^{(1)}$ and $b_n^{(0)}$ coefficients. The photoemission intensity is proportionate to the transition probability between the ground state ($|\Psi_{G.S.}\rangle$) and the three $|\Psi_{fin}^{fp}\rangle$ final states ($p = 0,1,2$). We therefore expect three contributions in the core level spectrum. The intensity of the structure associated with $|\Psi_{fin}^{fp}\rangle$ dominated by the (c, f^p) configuration is simply obtained from the matrix transition element (a_c operator)

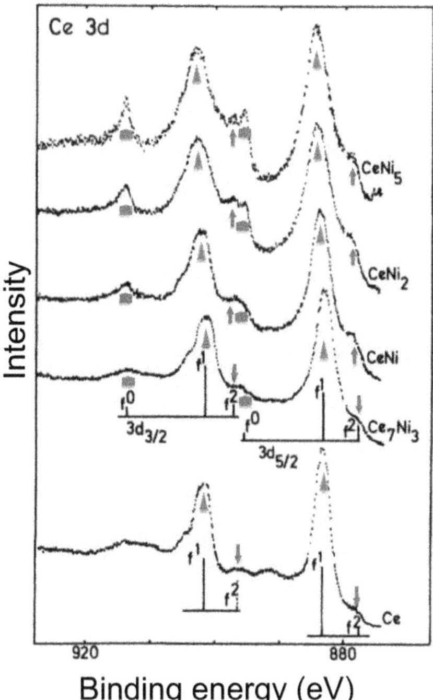

Figure 3.36 *Photoemission spectra of the 3d Ce core level in several intermetallic compounds. We note the presence of two satellites, one at low energy corresponding to the final well-screened configuration dominated by (c, f^2) and one at high energy corresponding to the core hole shielding by the conduction electrons (dominated by the (c, f^0) configuration) (from [54]). Figure reprinted with permission from J. C. Fuggle, F. U. Hillebrecht, Z. Zonierek, R. Lässer, Ch. Freiburg, O. Gunnarsson, and K. Schönhammer, Phys. Rev. B 27, 7330 (1983). ©1983 by the American Physical Society.*

squared:

$$I(c, f^p) = |\langle \Psi_{fin}^{fp} | a_c | \Psi_{G.S.} \rangle|^2 = |a_2 b_2^{(p)} + a_1 b_1^{(p)} + a_0 b_0^{(p)}|^2 . \qquad (3.103)$$

We present in figure 3.36, the Ce-$3d$ photoemission spectra for several intermetallic compounds. The above approach makes it possible to understand the complex structure of these spectra. In fact, because of the spin orbit coupling of the Ce-$3d$ level, which separates the $3d^{5/2}$ and $3d^{3/2}$ contributions by about 18 eV, these spectra are composed of six structures: three for each spin-orbit component. The analysis of these spectra allows to determine the various parameters of the Hamiltonian. From the position of the structures, the energies parameters of the Anderson hamiltonian (ε_f, U_{ff}, U_{fc}) can be obtained while their intensities are functions of the a_n and $b_n^{(p)}$ coefficients which depend on the hybridization and the energies of the different configurations.

3.3.4 Selection rules for core level photoemission

Another consequence of matrix element properties is the selection rules that allow only certain transitions between initial and final states. In a one-electron approach, the wave functions of the core level are characterized by quantum numbers associated with angular momentum. Similarly, the photoelectron wave function has a spherical symmetry with respect to the photoabsorber atom so that one can write:

$$\begin{aligned}\phi_{n_i,\ell_i,m_i}(\vec{r}) &= R_{n_i,\ell_i}(\vec{r})\,Y_{\ell_i}^{m_i}(\theta,\phi)\\ \phi_{n_f,\ell_f,m_f}(\vec{r}) &= R_{n_f,\ell_f}(\vec{r})\,Y_{\ell_f}^{m_f}(\theta,\phi)\end{aligned} \tag{3.104}$$

with $R_{n,\ell}$ representing the radial part and $Y_\ell^m(\theta,\phi)$ the angular part (spherical harmonics). The transition probability is proportional to

$$\int d\Omega\, Y_{\ell_i}^{m_i *}(\theta,\phi)(\vec{\epsilon}\cdot\vec{r})\,Y_{\ell_f}^{m_f}(\theta,\phi) \tag{3.105}$$

If the polarization $\vec{\epsilon}$ is parallel to Ox, Oy or Oz, $\vec{\epsilon}\cdot\vec{r}$ is proportional to $(Y_1^{-1}-Y_1^1)r$, $(Y_1^{-1}+Y_1^1)r$ and Y_1^r. From the properties of the integral of the product of three spherical harmonics (cf. for example, [55]), the transition probability is non-zero if:

$$\begin{aligned}\Delta\ell &= \ell_f-\ell_i=\pm1\\ \Delta m_\ell &= m_f-m_i=-1,0,+1\end{aligned} \tag{3.106}$$

which constitutes the dipole selection rules. The transition with $\Delta m = 0$ corresponds to the linearly polarized light according to Oz. The other two transitions are allowed with circularly polarized or linearly polarized light along Ox and Oy. The transition $\Delta\ell = 0$ is forbidden by parity. Indeed, the dipolar transition operator is odd and can only couple states of different parity[6].

In the presence of the spin-orbit coupling, the stationary states of the system are eigenstates of the total angular momentum $\vec{j}=\vec{\ell}+\vec{s}$, they are denoted $|\ell,s,j,m_j\rangle$ from the ℓ,s,j,m_j quantum numbers. We obtain the selection rules [6]:

$$\begin{aligned}\Delta j &= 0,\pm1\\ \Delta m_j &= 0,\pm1\\ \Delta\ell &= \pm1\end{aligned} \tag{3.107}$$

knowing that the transition with $\Delta j = 0$ is forbidden if $j_i = j_f = 0$.

When it is necessary to describe the system by the many body approach (for example by multiplets), we have to consider the angular momenta of the emitting atom L, S et J. In the most general case, only the total kinetic moment J is involved in the

[6] We recall that the parity of a spherical harmonic is $(-1)^\ell$.

selection rules:

$$\begin{aligned} \Delta J &= 0, \pm 1 \\ \Delta M_J &= 0, \pm 1 \end{aligned} \tag{3.108}$$

with $\Delta J = 0$ forbidden for $J_i = 0$.

In the framework of the LS coupling, we have the additional rules:

$$\begin{aligned} \Delta L &= 0, \pm 1 \\ \Delta S &= 0 \\ \Delta M_L &= 0, \pm 1 \end{aligned} \tag{3.109}$$

with $\Delta L = 0$ forbidden for $L_i = 0$[7] .

3.3.5 Cross section

The photoionization cross section σ measures the probability of ionizing a N-electron system for a given photon energy $(h\nu)$. In a description of independent electrons, the cross section can be simply expressed as a matrix element between one-electron functions. Indeed in the electric dipole approximation, the cross section can be written:

$$\sigma(h\nu) \propto \left| \int \psi_i^*(\vec{r}_1, \vec{r}_2, ..., \vec{r}_N) \times \sum_i^N \vec{r}_i \, \psi_f(\vec{r}_1, \vec{r}_2, ..., \vec{r}_N) \, d\vec{r}_1 \, d\vec{r}_2 ... d\vec{r}_N \right|^2 \tag{3.110}$$

where $\vec{r}_i$ represent the coordinates of each electron. It is therefore necessary to know the wave functions of the initial (ψ_i) and final (ψ_f) states, which are antisymmetrized products of one-electron wave functions (Slater determinants). On the other hand, if the relaxation processes of the one-electron wave functions under the effect of the hole are neglected, the final state is given by the antisymmetric product of $N-1$ electron wave functions identical to those of the initial state. The cross section is therefore:

$$\sigma \propto \int \phi_i^*(\vec{r}_1) \; \vec{r}_1 \; \phi_f(\vec{r}_1) d\vec{r}_1 \tag{3.111}$$

where $\phi_i(\vec{r}_1)$ and $\phi_f(\vec{r}_1)$ are the wave functions of the electron before and after the photon absorption. When we consider the photoemission on an atom, the initial and final one-electron states are solutions of the same equation with spherical symmetry. If the central potential on the electrons of the $n\ell$ sub-layer of energy $\epsilon_{n\ell}$ is $V_{n\ell}(r)$, the radial part $R_{n\ell}(r)$ of the initial wave function is compliant with the equation:

[7] The initial and final states must have a different parity but the parity is not associated with L. It is given by $(-1)^{\sum_n \ell_n}$.

$$\left(-\frac{\hbar^2}{2m}\frac{d^2}{dr^2}+V_{n\ell}(r)-\epsilon_{n\ell}+\frac{\ell(\ell+1)\hbar^2}{2mr^2}\right)R_{n\ell}(r)=0 \tag{3.112}$$

This equation is very similar to the radial equation for the hydrogen atom where the last term is associated with the rotational kinetic energy depending on the orbital angular momentum of the electron. Since this term depends on r, it is called centrifugal potential energy and it is a repulsive term unlike the central potential ($V_{n\ell}(r)$) which is attractive. The radial functions in the final state (solutions for energy scattering states ϵ) will be denoted $R_{\epsilon\ell}(r)$. The cross section can be calculated by averaging all initial states of orbital momentum m and summing over all final states. We obtain [56]:

$$\sigma_{n\ell}=\frac{4\pi^2\alpha a_0^2}{3}\frac{N_{nl}}{2\ell+1}h\nu\left[\ell B_{\ell-1}^2(\epsilon)+(\ell+1)B_{\ell+1}^2(\epsilon)\right] \tag{3.113}$$

where $\alpha=1/137$ is the fine structure constant, a_0 the Bohr radius ($a_0=5,29\times 10^{-9}$ cm to express the cross section in cm^2), $N_{n\ell}$ is the number of electrons in the $n\ell$ sublayer, and the $B_{\ell\pm1}$ factors are given by:

$$B_{\ell\pm1}(\epsilon)=\int_0^\infty R_{n\ell}(\vec{r})\,\vec{r}\,R_{\epsilon,\ell\pm1}(\vec{r})dr \tag{3.114}$$

where ϵ, the kinetic energy of the photoionized electron, verifies $\epsilon=h\nu-\epsilon_{n\ell}$. In the electric dipole approximation (odd vector operator), only states with $\ell\pm1$ angular momenta can be reached from an initial state of ℓ angular momentum. Moreover for a linear polarization the differential photoionization section of the $n\ell$ sub-layer is [56–61]:

$$\frac{d\sigma_{n\ell}(h\nu)}{d\Omega}=\frac{\sigma_{n\ell}(h\nu)}{4\pi}[1+\beta_{n\ell}(h\nu)P_2(\cos\gamma)] \tag{3.115}$$

where $P_2(\cos\gamma)=\frac{1}{2}(3\cos^2\gamma-1)$ is the $\ell=2$ Legendre polynomial and γ is the angle between the polarization and the direction of the photoelectron. $\beta_{n\ell}(h\nu)$ called the asymmetry parameter can take values between -1 and 2:

$$\beta=\frac{\ell(\ell-1)B_{\ell-1}^2+(\ell+1)(\ell+2)B_{\ell+1}^2-6\ell(\ell+1)B_{\ell+1}R_{\ell-1}\cos(\delta_{\ell+1}-\delta_{\ell-1})}{(2\ell+1)[lB_{\ell-1}^2+(\ell+1)B_{\ell+1}^2]} \tag{3.116}$$

where the $\delta_{\ell\pm1}$ are the phase shifts of the spherical waves for the two possible $\ell\pm1$ final states.

The photoionization cross section depends on the atomic layer of the initial state (Fig. 3.37), and it can be shown that the states of zero angular momentum have a large cross section at low energy. On the other hand, for high angular momenta, the centrifugal potential increases and play the role of a potential barrier for the final state which decreases the matrix element ($\langle\phi_f|H_{int}|\phi_i\rangle$) and therefore the cross section. Cross sections can be obtained from experimental data [63, 64] or from calculations (Fig. 3.38). Yeh and Lindau [62] tabulated the atomic cross sections calculated for

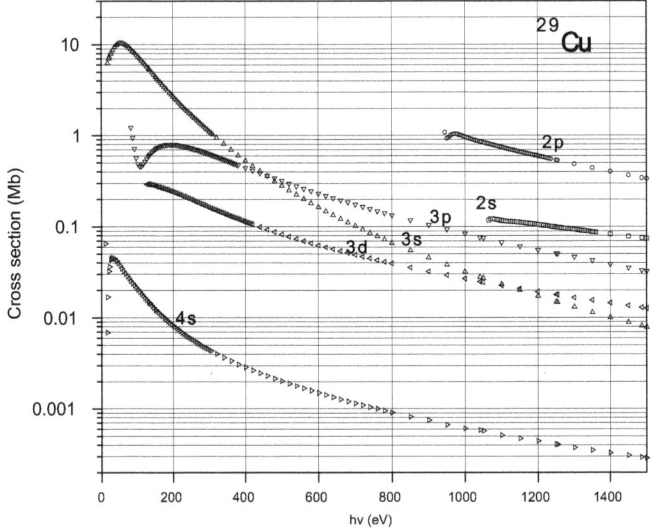

Figure 3.37 *Energy dependence of the ross section of Cu (from [62]). The cross-section varies over several orders of magnitude as a function of hν. Reprinted from Atomic Data and Nuclear Data Tables Vol. 32, J. J. Yeh, I. Lindau, "Atomic subshell photoionization cross sections and asymmetry parameters: 1 ≤ Z ≤ 103", page 49, ©1984, with permission from Elsevier.*

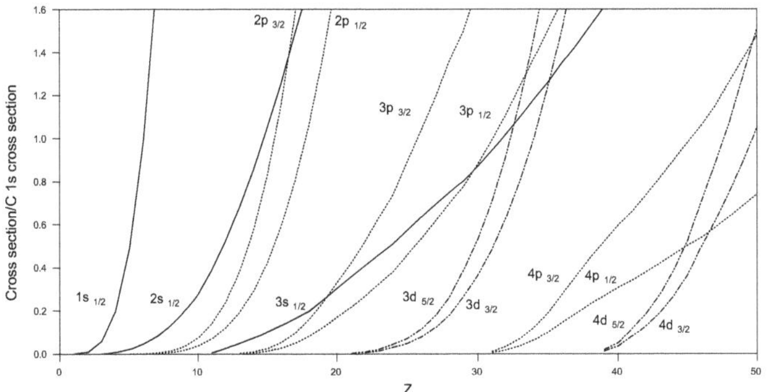

Figure 3.38 *Cross section dependence as a function of Z. From [64].*

each sub-layer of atoms. There are also relativistic calculations for full sub-layers [65]. Calculations show that in each sub-layer there is a minimum of the cross-section when the initial state has a node in its radial part (Cooper minimum [56]).

The causes of the deviations between the calculations and the experimental determinations of the cross section can vary. The cross-sections are often calculated for isolated atoms and may differ from those of atoms in solids [66, 67]. Correlation effects (electron interactions) can also introduce changes to atomic cross sections [68, 69].

3.4 Related processes

In previous sections we saw how photoemission spectroscopy can measure the electronic properties of a system. In the following pages, we will briefly describe other spectroscopies pertaining to related electron excitation processes that can also provide information on electronic properties.

3.4.1 *Auger spectroscopy*

The Auger effect is an electronic emission process involved in the mechanisms of relaxation of a core hole. The latter is created during a primary process where a core electron (A sublayer) is excited by a high energy photon or electron. The core hole can relax according to two different processes leading to the emission either of an electron or a photon. Indeed, an electron of a sub-layer (B) of energy higher than that involved in the primary process fills the core hole. The energy gained during this transition can allow the emission of a third electron (C) belonging to another sub-layer or to the valence band (Auger process) or the emission of a photon (fluorescence). The cross section of the Auger transitions is important at low energy while the fluorescence dominates at high energy.

An Auger process therefore involves three energy levels A, B and C and the final state is characterized by two holes on the B and C sub-layer. The kinetic energy of the Auger electron does not depend on the energy of the particles (photons or electrons) that have been involved in the primary process to create the core hole. The energy conservation allows us to write:

$$E_{kin}^{Auger} = E_A^Z - E_B^Z + E_C^{Z+1} - U - \Phi \tag{3.117}$$

where E_A^Z, E_B^Z and E_C^{Z+1} are the binding energies of the three sub-layers involved, Φ the work function and U the hole-hole interaction and relaxation energy. The exponent Z characterizes the charge of the ion of the emitting atom and we must consider that for the C level we have an additional hole yielding the $Z + 1$ charge. The hole-hole interaction energy has three terms:

$$U = R_x^{int} + R_x^{ext} - E_{BC} \tag{3.118}$$

E_{BC} is the interaction energy between the holes in B and C, R_x^{int} is the intra-atomic relaxation energy corresponding to an isolated atom and R_x^{ext} is the relaxation due to the electrons of the other atoms. These different terms are important (E_{BC} can be larger than 20 eV and R_x larger than 10 eV [70]) but are of different sign so a first approximation is to neglect U and write:

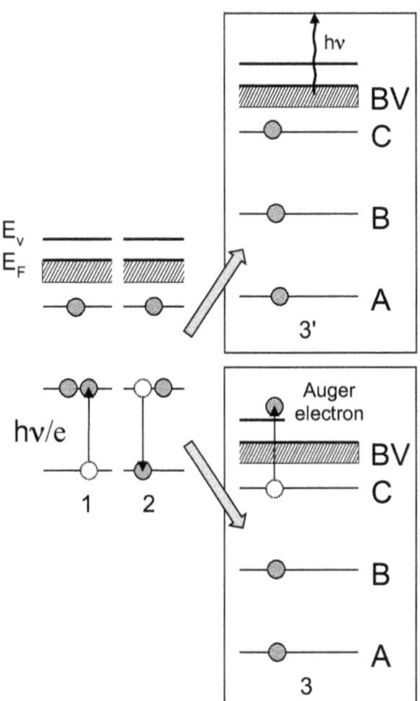

Figure 3.39 *Schematic diagram of the Auger transition and fluorescence (1). An electron in an inner layer is excited by photons or electrons of high energy (primary process). An electron of a higher level takes the place of the created hole (2). Excess energy is released either by emission of an electron, i.e., in an Auger transition (3), or by emission of a photon, i.e. by fluorescence (3').*

$$E_{kin}^{Auger} = E_A^Z - E_B^Z + E_C^{Z+1} - \Phi \tag{3.119}$$

Since we have two holes that affect the energies of the B and C sub-layers, an empirical expression has been proposed taking into account average energies of the B and C sub-layers with one and two holes [71]:

$$E_{kin}^{Auger} = E_A^Z - \frac{1}{2}(E_B^Z + E_B^{Z+1}) - \frac{1}{2}(E_C^Z + E_C^{Z+1}) - \Phi \,. \tag{3.120}$$

Auger transition energies are characteristic of the probed atom (cf. Fig. 3.39). This chemical specificity is the basis of the Auger electron spectroscopy (AES - *Auger Electron Spectroscopy*), which is very useful in surface physics for the chemical characterization of surfaces.

The nomenclature of Auger transitions involves the letters corresponding to the layers of the three intervening levels: (ABC). We use, as for X absorption thresholds, the capital letters K, L, M, N, ... to name the atomic layers of principal quantum number $n = 1, 2, 3, 4,\ldots$ As a hole on an inner layer is considered, multiplet effects are

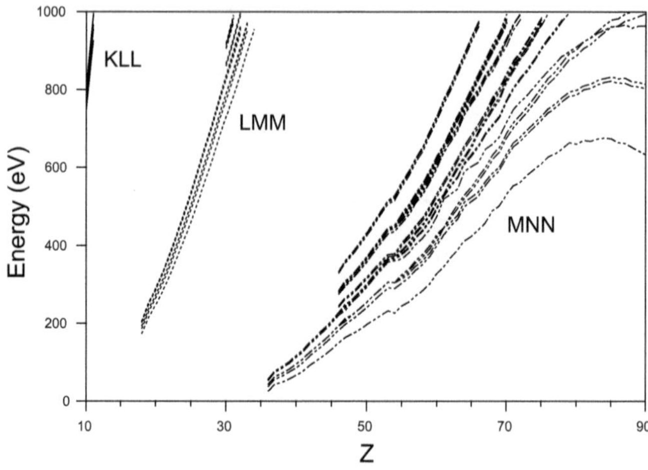

Figure 3.40 *Energy of the KLL, LMM and MNN Auger transitions according to Z. From [72].*

generally small and a one-particle description is valide. We must therefore consider a hole of angular momentum ℓ and spin s so that due to the spin-orbit coupling, we have two types of final states associated with the total angular momentum of the hole: $j = \ell \pm s$. Each energy level is called by the corresponding letter, with an index (1, 2, 3, 4, ...) designating the total angular momenta $j = \frac{1}{2}, \frac{3}{2}, \frac{5}{2}, \frac{7}{2}, ...$ (array 3.4). If the splitting associated with spin-orbit is not resolved, a double index is used, such as $KL_{2,3}L_{2,3}$.

For Auger transitions involving the valence band, the letter "V" is used. Transitions that involve levels with the same principal quantum number are called Coster-Krönig transitions (the A et B sub-layers belong to the same layer) or Super Coster-Krönig (the A, B and C sub-layers belong to the same layer) [73]. These types of transition have a high probability.

A spectrum associated with a given Auger transition has several contributions associated with different states of neighboring energies. The line shape of the Auger transitions is therefore relatively complex and satellites are often observed whose origin is the interaction between the holes in the final state, the possible plasmons etc. The width of the Auger spectra depends on the different levels involved in the transition. Thus, if the process involves the emission of an electron from the valence band (e.g. KLV), the spectrum is broadened since the electron can come from the bottom or the top of the band. In processes involving two electrons of the valence band (e.g., KVV), the broadening is twice [74].

The energy of Auger transitions is tabulated in many reference works [75–79]. In these tables, the energies of the Auger transitions are often related to the differential mode $EdN(E)/dE$ where the tabulated Auger energy corresponds by convention to

Table 3.4 *Notation for the Auger transitions*

n	l	j	Level
1	0	1/2	K
2	0	1/2	L_1
2	1	1/2	L_2
2	1	3/2	L_3
3	0	1/2	M_1
3	1	1/2	M_2
3	1	3/2	M_3
3	2	3/2	M_4
3	2	5/2	M_5
4	0	1/2	N_1
4	1	1/2	N_2
4	1	3/2	N_3
4	2	3/2	N_4
4	2	5/2	N_5
4	3	5/2	N_6
4	3	7/2	N_7

that of the high energy minimum [80]. This energy is not the same to the direct mode energy, without differentiation. This mode corresponds to the experimental method of acquisition called 'lock-in' where, in order to eliminate most of the background noise, the signal is modulated at a given frequency and the derivative is measured directly at the modulation frequency.

3.4.2 Photoelectron diffraction

Photoelectrons propagate inside a crystal before escaping from the solid. Elastic or inelastic scattering may occur that can be described in a particle approach. Moreover, an additional scattering by the periodic lattice is allowed because of the wave nature of the photoelectron. In this section we will explain how we can describe this photoelectron diffraction mechanism. It is a final state effect that is naturally included in the one-step model, but it must be added in the three-step phenomenological model. The reader can find a detailed description of the photoelectron diffraction (PED) in many publications [81–93].

The propagation of a photoelectron in a crystal is similar to the band structure that describes stationary states of an electron in a periodic potential. The opening of a band gap at the boundaries of the Brillouin zone corresponds to a Bragg diffraction. In addition, the description of the photoemission in a one-step model is based on the inverse-LEED theory (Low Energy Electron Diffraction) in [7, 12, 16, 18, 20, 94–99]. Indeed, in a LEED experiment, an incoming electron beam is diffracted by the periodic lattice. The photoemission process corresponds to the same states (initial and final) but by reversing the direction of time with out-coming electrons emitted by the crystal. Technically, a multiple scattering formalism is used in multi-layers or aggregates [100–103]. An infinite system is not necessary since the inelastic average free path of the photoelectron makes it possible to limit the size of the aggregates to be considered.

As in LEED, photoelectron diffraction provides structural information. Multiple scattering formalism shows that the photoelectron intensity and therefore the transition matrix elements depend on the emission angle and the photon energy. The dependence of the intensity as a function of the wave vector $\vec{k}$ can be written:

$$I(\vec{k}) \propto |\langle\psi(\vec{k})|\vec{\epsilon}\cdot\vec{r}|\psi_L\rangle|^2 \qquad (3.121)$$

where $\psi_L(\vec{r})$ is the wave function of the localized initial state and $\psi(\vec{r},\vec{k})$ is the wave function of the final state. This final state wave function is composed of the wave function ϕ_0 emitted directly by the photoabsorbing atom and waves diffused by all the other atoms j of the crystal:

$$\psi(\vec{r},\vec{k}) = \phi_0(\vec{r},\vec{k}) + \sum_j \phi_j \qquad (3.122)$$

Therefore, when we introduce the expression of the final state in the photoemission intensity (cf. equation 3.121), the sum of wave functions will give rise to interferences at the origin of photoelectron diffraction mechanism. This mechanism is always present in the usual photoemission process, but it is exploitable only if one analyzes quantitatively the intensities of the lines. Moreover, it is not necessary to have a long-range order as in the X-ray diffraction. Indeed, since the detector is sufficiently far, the wave functions of the various components of the final state are spherical waves. Their amplitude then decreases in $1/r$ and the finite mean free path of the photoelectrons leads to an exponential decay in the solid.

The photoelectron diffraction will give structural information around the emitting atom. On the other hand, since the initial state $\psi_L(\vec{r})$ is localized, this structural information can be obtained for elements of different chemical nature. These states can be accessed by detecting a core level or an Auger transition. This is called XPD (*X-ray Photoelectron Diffraction*) or AED (*Auger Electron Diffraction*). The photoelectron coming from a core level is usually described by a state with spherical symmetry (average of states in a given sublayer). In the case of an Auger emission, such a description is less direct: there is a mixture of final states with different angular momenta. Nevertheless, if the electron is described by a high number of spherical harmonics, the spherical wave approximation remains valid [104–106] and the description is similar to that of the XPD.

Now let us describe the wave function of the final state. The scattering of ϕ_0 by an atom j is described by a complex factor $f(\theta_j) = |f_j(\theta_j)| \exp i\varphi_j(\theta_j)$ called scattering factor which depends on the angle θ_j between the emitter and the atom j. $|f_j(\theta_j)|$ and $\varphi_j(\theta_j)$ determine, respectively, the amplitude and the phase of the scattered wave. The wave function of the final state can be then written:

$$\psi(\vec{r}, \vec{k}) \propto \frac{e^{i\vec{k}\cdot\vec{r}}}{r} + \sum_j f_j(\theta_j)\frac{e^{ik|\vec{r}-\vec{r}_j|}}{|\vec{r}-\vec{r}_j|} \tag{3.123}$$

with the scattering factor:

$$f_j(\theta) = \frac{1}{2ik}\sum_{\ell=0}^{\infty}(2\ell+1)[\exp(2i\delta_\ell) - 1]P_\ell(\cos\theta) \tag{3.124}$$

where the δ_ℓ parameters are partial phase shifts and the P_ℓ the Legendre polynomials. For large r (i.e., at the detector), the phase change φ_j is given by:

$$kr_j(1 - \cos\theta_j) + \psi_j(\theta_j) \tag{3.125}$$

because of the difference in path between the main wave and the scattered wave. This term contains structural information in photoelectron diffraction. The anisotropy of the signal, i.e., the variation of the intensity detected with the detection angle, is given by the quantity:

$$\chi(\vec{k}) = \frac{I(\vec{k}) - I_0(\vec{k})}{I_0(\vec{k})} \tag{3.126}$$

where the wave vector of the electron $\vec{k}$ contains the dependence as a function of the photon energy and the emission angle ($I(\vec{k})$ is normalized to the flux of photons) and $I_0(\vec{k})$ gives the slow variation of $I(\vec{k})$ due, for example, to the variation of the photoionization cross section as a function of energy. The photoelectron diffraction models make it possible to determine the function$\chi(\vec{k})$.

Energy ranges

The scattering of electrons in solids strongly depends on their kinetic energy. There are two types of extreme behavior depending on whether the energy is greater or less than a few hundred eV. Low energies define the backscattering regime. This regime is characterized by a high surface sensitivity and multiple scattering processes. Data interpretation requires sophisticated simulations [88]. On the other hand, for high kinetic energies, the behavior is simpler because forward scattering processes dominate [87, 90]. The intensity maxima appear in the directions defined by the emitter and the scattering atoms. This property simplifies the interpretation of the data and makes it possible to find very quickly the directions in which the first neighbors of the emitting atom are located. These two schemes are detailed below.

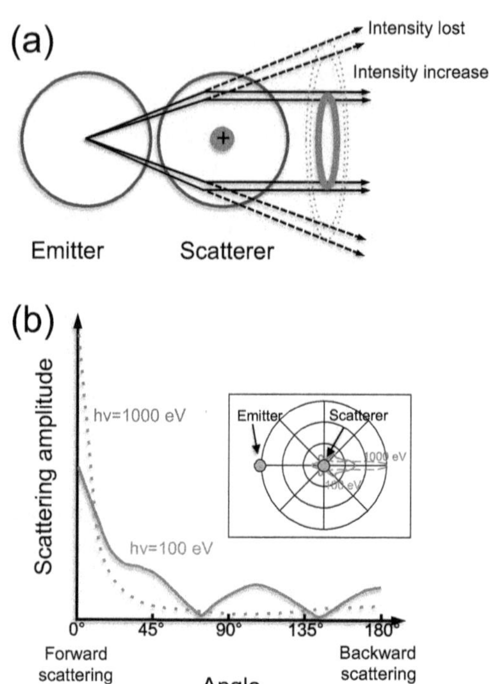

Figure 3.41 *Photoelectron diffraction. a) Diagram showing the forward focusing of the emitting atom. b) Energy regimes for photoelectron diffraction. The scattering amplitude varies significantly with energy. At low kinetic energy of the photoelectron, the angular distribution is more homogeneous with a maximum for backscattering. At high kinetic energy a pronounced maximum is observed for forward scattering. The insert shows a polar diagram with the qualitative behavior for 100 and 1000 eV.*

Forward scattering can be understood with semiclassical theory [107]. Figure 3.41(a) illustrates the deviation of the electronic wave by the atomic potential of a diffusing atom in the emitter-scatterer direction. Due to the cylindrical symmetry all the points of a ring centered on this direction are in phase, leading to a constructive interference. The forward preferential scattering results from the angular dependence of the scattering factor $f(\theta_S, E)$ which describes elastic scattering as a function of energy. It gives the amplitude of the scattered wave ϕ_S for the scattering angle θ_S (Fig. 3.41(b)). For large photoelectron kinetic energies (greater than a few hundred eV), the scattering factor has a forward maximum in a cone of 10 to 20 degrees [108]. This cone decreases with increasing energy [109]. The intensity anisotropy will therefore essentially reflect the atomic directions around the absorber atom. On the other hand, the intensity of the signal is affected by the nature of the scatterer atoms; the scattering amplitude of atoms generally increases with the nuclear charge [110]. The intensity of the XPD peak also depends on the distance between the emitter and the scatterer [111] and if the distance between the two atoms is large, the kinetic energy must increase to observe a maximum [110].

The relatively direct correlation between diffraction maxima and scatterer directions is the strength of this technique. Nevertheless, multiple scattering or screening effects can sometimes complicate the analysis. The intensity due to the higher orders of scattering appears in concentric circles around the internuclear axis. The angular position of the rings depends on the path difference between the forward wave and the scattered wave, which makes it possible to identify multiple scattering effects by varying the kinetic energy. In general, their intensity is lower than that of the principal maximum. Nevertheless, they can generate large maxima if several rings are superimposed, which leads to a strong intensity outside the internuclear directions [112]. Another effect is the defocusing of initially propagating electrons along a chain of atoms. The intensity and the angular width of the scattering maximum are affected, especially for weak kinetic energies. Another curious effect is screening, which removes maxima from the main directions. This phenomenon occurs when there are two relatively close atoms, one of which is closer to the emitter. For some geometries this first atom focuses the outgoing wave that does not reach the second [112].

We have seen that backscattering occurs when the kinetic energy of the photoelectron is low. Multiple scattering phenomena therefore occur more than at higher energy and the form factor $f(\theta_S, E)$ has several maxima in addition to the forward scattering maximum. In particular, a new maximum appears in the opposite direction of the initial direction of the outgoing photoelectron. This backscattering is greater at low kinetic energy and for the high atomic numbers of the scatterer [113].

Comparison with other structural techniques

The most important structural techniques that use photons or electrons as probes are low energy electron diffraction (LEED), photoelectron diffraction (PED), the study of fine structures in X-ray absorption spectra (EXAFS) and X-ray diffraction (XRD)[114]. Each of these techniques has specificities that we will briefly recall now.

Due to a small mean free path, the surface structure can be studied by PED and LEED. The main differences are the chemical sensitivity for PED and the need for a long-range order for LEED. On the other hand, PED is only sensitive to short-range order, since the average free path only probes the first two-three spheres of close neighbors [115]. The distant spheres of 15-20 Å contribute marginally to the diffraction intensity.

The PED and the EXAFS techniques have a lot in common. EXAFS measures the absorption coefficient as a function of the photon energy around the energy of a threshold. Like PED, it is thus a technique which gives access to the atomic structure around an absorbing atom (chemical sensitivity). The essential difference between the two techniques is the angular dependence. The absorption coefficient averages all the scattering phenomena and therefore the emission directions of the photoelectron. EXAFS measures the total cross-section (angle-integrated) while the photoelectron diffraction measures the angle-resolved cross section. The signal in PED is much weaker, but with a modulation that can be an order of magnitude larger [116]. On the other hand, EXAFS determines the nearest neighbor distances with higher accuracy.

The difference between XRD and PED is the nature of the probe particles. The photon-matter interaction is much weaker than the electron-matter interaction so that the XRD multiple scattering processes are generally negligible. Models to obtain the structure from a X-ray diffraction pattern are therefore much simpler than with electrons so that the accuracies achieved by XRD are much better (0.01 Å) [117]. In addition XRD has no chemical selectivity unless using photons whose energy corresponds to an absorption threshold of one of the elements (anomalous X-ray diffraction).

3.4.3 Resonant photoemission

The resonant photoemission of valence states was first observed on Ni (001) [118]. It consists of varying the energy of photons $h\nu$ close to an X-ray absorption threshold associated with a core level of a chemical species in the sample. At the threshold energy, the process of photoemission of a valence state comes into competition with the absorption of the photon by an electron of the core level. This last process leads to an intermediate state with a core hole that leads to a relaxation mechanism reminiscent of the Auger state, i.e., a valence band electron fills the core hole with the emission of another valence band electron. Take the example of Ni to illustrate this effect. Consider that in the initial state[8], Ni has 9 d electrons. If the photons correspond to the energy of the $3p$-threshold, we can have the absorption process which will lead to an intermediate state with 10 d electrons and a $3p$ hole:

$$3p^6 3d^9 4s + h\nu \rightarrow 3p^5 3d^{10} 4s\,. \tag{3.127}$$

The relaxation process corresponds to the emission of one electron:

$$3p^5 3d^{10} 4s \rightarrow 3p^6 3d^8 4s + e^- \tag{3.128}$$

where e^- indicates the electron leaving the solid. It is found that the resulting final state is the same as that resulting from the direct photoemission of a valence state:

$$3p^6 3d^9 4s + h\nu \rightarrow 3p^6 3d^8 4s\,. \tag{3.129}$$

It can be seen that the two processes: direct photoemission and absorption followed by relaxation have the same initial and final states (Fig. 3.42). So there is an interference between the different channels and we must sum the two amplitudes and then take the square to obtain the intensity.

[8] This is a simplified approach. The Ni d states form an energy band with an average number of electrons d per atom of about 9.6.

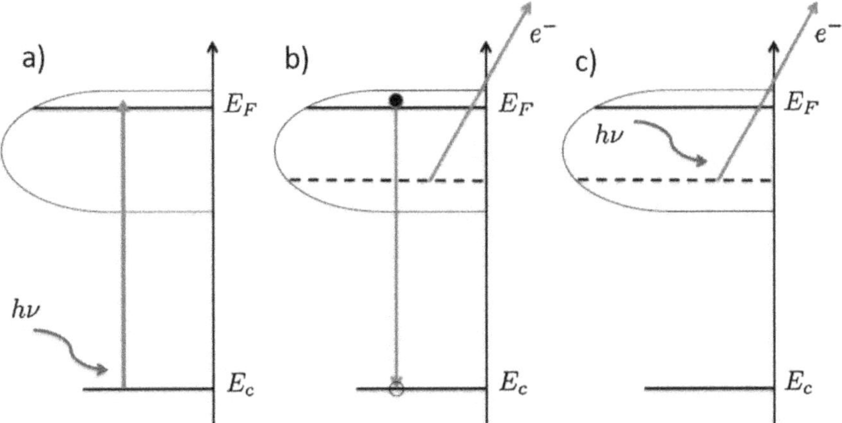

Figure 3.42 *Principle of resonant photoemission. a) Absorption of a photon by a core electron leading to an intermediate state. This state corresponding to an electron in unoccupied states has a local character due to the interaction with the core hole. b) Deexcitation of the core hole with the emission of an electron from the valence band. c) Direct photoemission of the same electron from the valence band. Both processes interfere to give rise to a resonance of the transition probability.*

The probability of emission of the electron depends on the energy and has a significant modulation around the energy of the threshold. Indeed, far from the threshold only the process of direct photoemission occurs whereas in the vicinity of the threshold, the 2 processes interfere and lead to a maximum of the effective photoemission cross section of the states for constructive interference or to a minimum for destructive interference. The dependence of the photoemission cross-section as a function of photon energy can be modeled by the Fano model [119]. In this approach, we consider an initial state $|i\rangle$, a discrete intermediate state $|\phi\rangle$ with energy E_ϕ, and a continuum $|\psi_E\rangle$. In the Fano model, the discrete intermediate state is coupled to the continuum by a hybridization term V_E, so that the eigenstates $|\Psi_E\rangle$ are hybrid states:

$$|\Psi_E\rangle = a|\phi\rangle + \int dE' b_{E'}|\psi_{E'}\rangle \tag{3.130}$$

The a (energy dependent) and $b_{E'}$ coefficients depending on the hybridization term defined by:

$$\langle\psi_E|H|\phi\rangle = V_E \,. \tag{3.131}$$

This term is fundamental to determining the transition probability. Indeed, if it is zero, there is no relaxation from the intermediate state. A non-zero value of the hybridization term transforms the discrete state into a resonance. Two effects are expected, on the one hand an energy shift of the resonance compared to that of the intermediate state of the quantity $F(E_R)$:

$$E_R = E_\phi + F(E_R) \tag{3.132}$$

where

$$F(E) = \mathcal{P} \int dE' \frac{|V_{E'}|^2}{E - E'} \tag{3.133}$$

where $\mathcal{P}$ is the Cauchy principal part. On the other hand, the resonance has a finite lifetime and therefore a spectral width (half-width equals $\Gamma/2 = \pi |V_E|^2$)[9]. The resonance is characterized by the energy dependence of the a parameter:

$$|a(E)|^2 = \frac{|V_E|^2}{(E - E_\phi - F(E))^2 + \pi^2 |V_E|^4} \tag{3.134}$$

which shows that the discrete state is diluted in the continuum by forming a resonance[10]. The state has a lifetime ($\hbar/2\pi |V_E|^2$).

As for any resonance phenomenon (for example an RLC electric circuit), a phase shift appears on both sides of the resonance. Here it corresponds to the phase of the continuum state induced by hybridization. The phase shift is:

$$\Delta = -\arctan \frac{\pi |V_E|^2}{E - E_\phi - F(E)} \, . \tag{3.135}$$

It exhibits a change of $\sim\pi$ in the energy range corresponding to the width of the resonance. We can express the a and b_E parameters as a function of Δ:

$$a = \sin \Delta / \pi V_E \tag{3.136}$$

and

$$b_{E'} = \frac{V_{E'} \sin \Delta}{\pi V_E^* (E - E')} - \cos \Delta \delta(E - E') \, . \tag{3.137}$$

The objective is to calculate the transition probability as a function of the photon energy and thus of the final state energy. We have to calculate the matrix element of the dipolar transition operator T between the initial state and the final state ($|\langle \Psi_E | T | i \rangle|^2$):

$$\langle \Psi_E | T | i \rangle = \frac{1}{\pi V_E^*} \langle \Phi | T | \, i \rangle \sin \Delta - \langle \psi_E | T | \, i \rangle \cos \Delta \tag{3.138}$$

by introducting the "modified" discrete state:

$$|\Phi\rangle = |\phi\rangle + \mathcal{P} \int dE' \frac{V_{E'}}{E - E'} |\psi_{E'}\rangle \, . \tag{3.139}$$

[9] The energy shift and broadening correspond respectively to the real and imaginary parts of the self-energy of the intermediate state.

[10] The resonance has a strictly Lorentzian form if $V_E = \text{const}$.

The transition amplitude is thus a sum of 2 terms. The rapid variation of the Δ phase in the vicinity of the resonance induces a rapid variation of $\langle \Psi_E | T | i \rangle$. Indeed, $\langle \Phi | T | i \rangle$ and $\langle \psi_E | T | i \rangle$ interfere with opposite phases on each side of the resonance energy. Moreover, there is a value of the phase for which the transition amplitude is canceled (anti-resonance). This phase Δ_0 at energy E_0 is obtained from:

$$\tan \Delta_0 = -\frac{E_0 - E_\phi - F(E_0)}{\pi |V_{E_0}|^2} = \frac{\pi V_{E_0^*} \langle \psi_{E_0} | T | i \rangle}{\langle \Phi | T | i \rangle} . \tag{3.140}$$

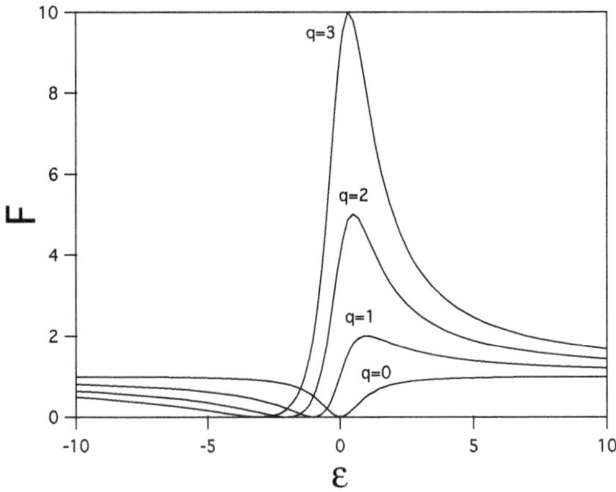

Figure 3.43 *Dependence of the Fano lineshape as a function of the asymmetry parameter q.*

The behavior of the transition probability can be characterized by plotting the normalized transition probability by the transition probability to unperturbed states, i.e., by the ratio:

$$\frac{|\langle \Psi_E | T | i \rangle|^2}{|\langle \psi_E | T | i \rangle|^2} . \tag{3.141}$$

We will express the dependence as a function of an energy parameter, this parameter (ϵ) corresponds to the energy relative to the resonance and normalized by the half-width, ie:

$$\epsilon = -\cot \Delta = \frac{E - E_\phi - F(E)}{\pi |V_E|^2} = \frac{E - E_\psi - F(E)}{\Gamma/2} . \tag{3.142}$$

The equation 3.138 shows that there are two transition channels, one associated with the unperturbed states $|\psi_E\rangle$, the other with the "modified" discrete state $|\Phi\rangle$. We introduce the dimensionless parameter called asymmetry parameter:

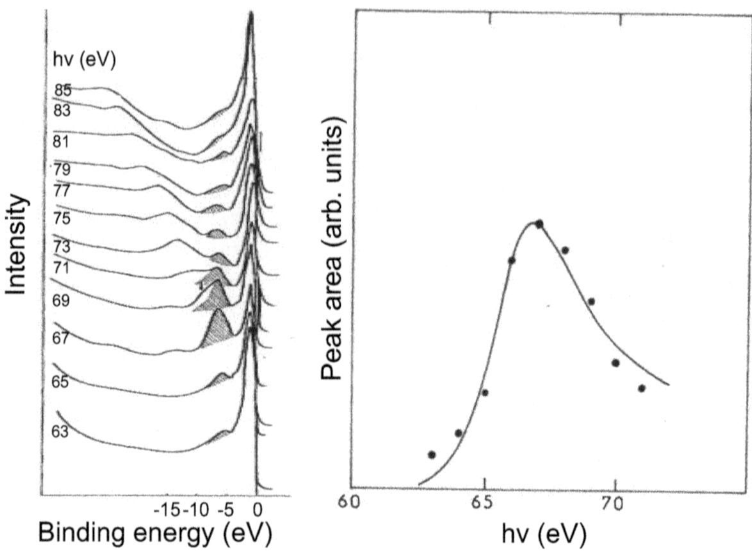

Figure 3.44 *Resonant photoemission on the Ni. a) Spectra of the valence band for several energies showing the resonance of the structure at 6 eV. b) Intensity of the 6eV-structure as a function of the photon energy. The intermediate state is an impurity state localized by the potential of the core hole. The solid line represents an adjustment with the Fano formula ($\Gamma = 2$ eV; $E_R = 66$ eV and $q = 2.5$ from [118]. Reprinted figure with permission from C. Guillot, Y. Ballu, J. Paigné, J. Lecante, K. P. Jain, P. Thiry, R. Pinchaux, Y. Pétroff, and L. M. Falicov, Phys. Rev. Lett. 39, 1632 (1977). ©1977 by the American Physical Society.*

$$q = \frac{\langle \Phi | T | \, i \rangle}{\langle \psi_E | T | \, i \rangle \Gamma / 2} \,. \tag{3.143}$$

For small q, it is the transitions to the continuum that dominate while for large q are those to the discrete state.

It is easy then to show that the transition probability has the following behavior:

$$S_q(\epsilon) = \frac{|\langle \Psi_E | T | \, i \rangle|^2}{|\langle \psi_E | T | \, i \rangle|^2} = \frac{(q + \epsilon)^2}{\epsilon^2 + 1} \,. \tag{3.144}$$

The function $S_q(\epsilon)$ is called the Fano profile (Fig. 3.43). For $q = 0$, the profile is symmetrical and vanishes at the resonance energy ($\epsilon = 0$). This particular point corresponds to $\langle \Phi | T | \, i \rangle = 0$ and $\cos \Delta = 0$ hence a zero transition probability (equation 3.138). When q increases, the profile exhibits an anti-resonance followed by a resonance all the more marked as q is large. When we reverse the sign of q, we obtain the same type of profile but with an inverted ϵ value ($S_{-q}(\epsilon) = S_q(-\epsilon)$).

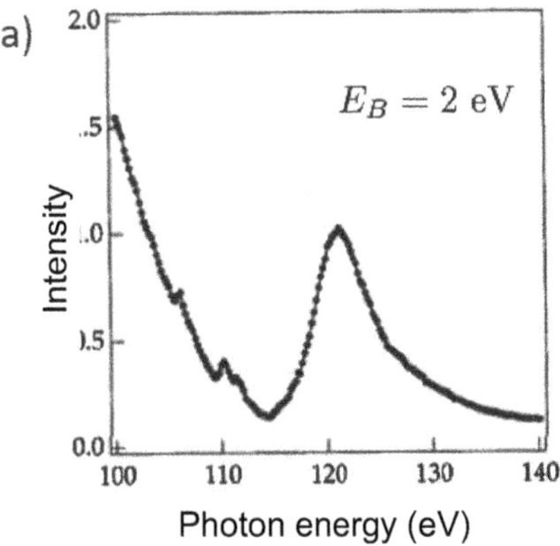

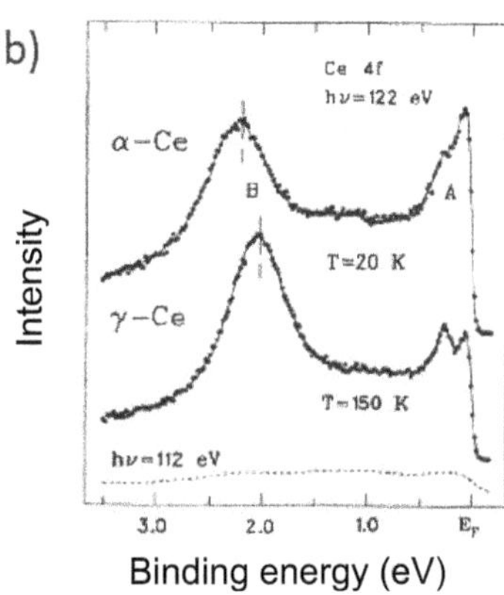

Figure 3.45 *Resonant photoemission on cérium. (a) Resonant behavior of the 4f intensity (E_B = 2eV) at the 4d threshold (from [120]). Figure reprinted with permission from N. Witkowski, F. Bertran, T. Gourieux, B. Kierren, D. Malterre, and G. Panaccione, Phys. Rev. B 56, 12054 (1997). ©1997 by the American Physical Society. (b) Spectra at anti-resonance (112 eV) and resonance (122 eV) for Ce. Between 20K and150 K, Ce metal exhibits a phase transition ($\gamma - \alpha$) modifying the f states (from [121]). Figure reprinted with permission from E. Weschke, C. Laubschat, T. Simmons, M. Domke, O. Strebel, and G. Kaindl, Phys. Rev. B 44, 8304 (1991). ©1991 by the American Physical Society.*

The Fano profile is a generic behavior that reproduces the resonant character of the transition probability when we have a discrete state coupled to a continuum and a transition characterized by the interference between two channels[11]. Fano developed this model to describe the resonant inelastic scattering of electrons by the $2s2p$ level of He. The Fano profile is also observed on tunnel spectroscopy curves near an adatom on a surface measured by a scanning tunneling microscope. Two channels interfere: the discrete intermediate state is an atomic state of the adatom while the continuum corresponds to the electronic states of the surface (bands). The evolution of the photoemission cross section of the valence band when the photon energy reaches an absorption threshold also follows a Fano profile. Figure 3.44 shows the resonance of a structure (at the binding energy $E_B = 6$ eV) in the Ni valence spectrum when the photon reaches the energy of the $3p$ core level. The intensity of this structure as a function of energy has a resonant character.

A similar behavior is observed at the $4d$ threshold of cerium, a rare earth corresponding to the $4f^1$ configuration. As we have already seen, the $4f$ states are very localized in rare earths and do not form an energy band. They are quasi-atomic states located spatially and energy of which is close to the Fermi level. According to the dipolar

[11] This approach remains qualitatively correct even when the intermediate state is a set of states close in energy and sufficiently localized. This is generally the case for resonant photoemission.

selection rules, the absorption of a photon by a $4\,d$ electron allows us to populate the f states (intermediate state). In the following relaxation process, a $4f$ electron fills the $4d$ core hole while the second f electron is emitted:

$$4d^{10}4f^{1} + h\nu \rightarrow 4d^{9}4f^{2} \rightarrow 4d^{10}4f^{0} + e^{-}\,. \tag{3.145}$$

Interference with the direct photoemission process is expected. Figure 3.45(a) shows the evolution of the spectral weight of the $4f$ states as a function of the photon energy. A strong resonance is observed around 120 eV preceded by a minimum as predicted in Fano's theory. Figure 3.45(b) shows the Ce spectra recorded at the photon energies corresponding, respectively, to the minimum and maximum of the $4f$ signal. The spectrum at 112 eV (antiresonance) has a very low intensity and essentially probes the non f states. On the other hand, for a photon energy corresponding to the resonance (122 eV), the intensity is large and reveals the signature of the f states. The difference between the two spectra makes it possible to estimate the $4f$ spectral function. This resonant photoemission technique is particularly useful when the f states are superimposed on a large density of states of other symmetries that mask the $4f$ signal in a standard photoemission experiment.

3.4.4 *Two-photon processes*

Photoemission is a single-photon process, the probability that an electron absorbs two photons with common sources is extremely low. However, under certain conditions, using very intense sources such as lasers, it is possible to observe two-photon processes. This technique is called two-photon photoemission with the acronym 2PPE (for *two photon photoemission*)[122–124].

Two-photon transitions can only be observed in energy domains where single-photon processes are forbidden, otherwise they would be completely masked (very low probability). This two-photon spectroscopy allows us to study the unoccupied states between the Fermi level E_F and the vacuum level E_V. It has been used extensively to study image states of metal surfaces and electron dynamics of semiconductor surfaces. The image states are, like surface states, located in energy in the forbidden band of volume states, and spatially located in the outer vicinity of the surface. The dynamics of excitation and relaxation can be studied by 2PPE by the use of femtosecond laser pulses [125, 126]. The principle is based on a pump-probe device: the first energy photon $h\nu_1$ from a laser pulse excites an electron to an intermediate state between E_F and E_V (pump) while the second photon with energy $h\nu_2$ belonging to a pulse of another laser (probe) extracts it from the solid. Figure 3.46 illustrates the principle of two-photon photoemission with the absorption of the pump photon (transition from from the initial state E_i to the intermediate state E_{int} located in the gap). We must have:

$$h\nu_1 < \phi\,. \tag{3.146}$$

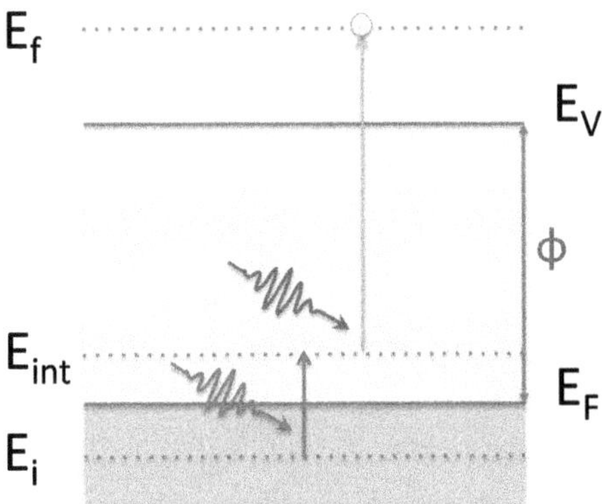

Figure 3.46 *Principle of the two-photon photoemission: An electron of energy E_i below the Fermi level absorbs a photon $h\nu_1$ and populates an intermediate state of energy E_{int}. The second photon with energy $h\nu_2$ allows the electron in the intermediate state to leave the solid and be detected.*

The absorption of the probe photon allows the electron to escape from the solid with the energy E_f higher than the energy of the vacuum E_V. We must therefore have:

$$h\nu_1 + h\nu_2 > \phi \,. \tag{3.147}$$

These simple considerations result from energy conservation. However, the situation is more subtle because the intermediate state can be a real electronic state, such as, for example, a image state, or a virtual state intervening as the intermediate state of a single two-photon process. These differences lead to two types of 2PPE spectrum dependence according to the photon energy of the pump. Indeed, when there is a true intermediate electronic state, the 2-photon photoemission can be considered as the result of 2 independent processes. The absorption of the pump photon makes it possible to reach the energy state E_{int} of spectral width Γ_{int}, the excitation probability depending on the energy of the photon. The absorption of the photon probe allows to reach the final state of energy $E_{int} + \nu_2$ independently of a modulation $\Delta h\nu_1$ of the energy of the pump photon. This modulation will affect the transition probability. On the other hand, if the initial state is a virtual state, the final state will be at energy $E_i + h\nu_1 + h\nu_2$ so that a modulation of the pump energy of $\Delta h\nu_1$ will lead to a shift of the same magnitude of the peak position in the spectrum. These considerations show that a structure in a 2PPE spectrum can be associated with the initial state, this is the case when the intermediate state is virtual, or in the intermediate state when the pump energy is tuned.

Figure 3.47 illustrates this behavior on the two-photon photoemission of the (111) surface of Cu with a single source of photons ($h\nu_1 = h\nu_2 = h\nu$). On this type of surface, there is an intermediate state in the gap (image state). By changing the energy by $\Delta h\nu$, we expect a shift of $2\Delta h\nu$ for a single process probing the initial state and only $\Delta h\nu$ for a process at two independent stages passing through the intermediate state. The figure shows two structures presenting these 2 types of dependence and crossing for $\Delta h\nu = 0$ defined for the resonance excitation $E_{int} - E_i$.

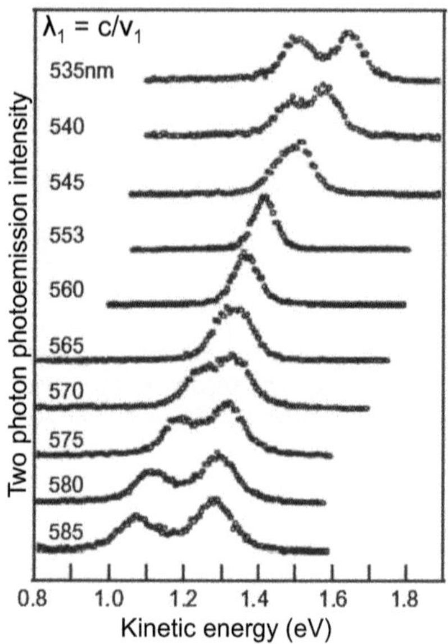

Figure 3.47 *Two photon photoemission spectra as a function of the photon energy $h\nu_1 = h\nu_2$ (from [127]). Reprinted from Surf.Sci. Vol. 374, W. Wallauer, Th. Fauster, "Two-photon excitation processes and linewidths of surface and image states on Cu(111)", page 44, ©1997, with permission of Elsevier.*

As mentioned above, it is possible to study the dynamics of electronic properties by studying the response as a function of the delay between pump and probe excitations. With current femtosecond lasers, the temporal resolution is of the order of a few femtoseconds. Two-photon time resolved photoemission was used to study the lifetime of image states of noble metal (111) surfaces. Because of their location outside the solid, their lifetime is longer than the bulk electronic excitations. We will not describe this sophisticated technique that requires complex modeling of temporal correlations. The measurements consist of studying the fixed kinetic energy intensity of the photoelectron as a function of the delay between the pump and probe excitations. The analysis shows that it is thus possible to estimate the lifetime using pulses longer than the lifetime.

Finally, it should be noted that processes with more than two photons are also possible if the intensity of the radiation is sufficiently high. For identical photons, the absorption of n photons can lead to the emission of an electron if:

$$n \times h\nu > \Phi. \tag{3.148}$$

3.4.5 *Inverse photoemission*

Inverse photoemission is the complementary technique of photoemission, because it probes unoccupied states and provides the spectral function $A(\vec{k},\omega)$ for positive energies [128–130]. The inverse photoemission process can be considered as the time-reversal image of the photoemission process. It therefore consists of sending electrons in unoccupied states (excited states), and in detecting the radiation resulting from the relaxation of these states. The pioneering group in the development of inverse photoemission was Ulmer [131] and the technique was later refined by Lang and Baer [132].

Several variants of the technique have been developed. Inverse photomission spectroscopy (IPES : *Inverse photoelectron Spectroscopy*) is used when incoming electrons are monoenergetic and the photons are detected as a function of their energy. On the other hand, one can work with constant photon energy and vary the energy of the incident electrons. The technique is then called BIS (*Bremsstrahlung Isochromat Spectroscopy*).

As we have seen above, inverse photoemission and photoemission give access to the one-electron excitations of an N electron system. An electron is removed or added so that in the final state the system has $N+1$ and $N-1$ electrons, respectively. For an atom, this involves ionization energy and electronic affinity.

One can wonder about the value of the inverse photoemission matrix element since in the initial state there is no photon and in a classical treatment, the electromagnetic field is null and $\vec{A}=0$. To solve this paradox, it is necessary to consider the quantum character of the electromagnetic field and even in the vacuum of the field (0 photons), the fluctuations of the vacuum can induce a transition [130, 133]. The differential cross section for inverse photoemission is:

$$\frac{d\sigma}{d\Omega_{IPES}} = \frac{\omega}{mc^2}\cdot\frac{\alpha}{hk}|\langle \mathrm{oc}|\vec{e}\cdot\vec{p}|\mathrm{inoc}\rangle|^2 \tag{3.149}$$

and for photoemission:

$$\frac{d\sigma}{d\Omega_{PES}} = \frac{k}{m}\cdot\frac{\alpha}{h\omega}|\langle \mathrm{inoc}|\vec{e}\cdot\vec{p}|\mathrm{oc}\rangle|^2 \tag{3.150}$$

The relation between the two cross sections is:

$$R = \frac{(d\upsilon/d\Omega)_{IPES}}{(d\sigma/d\Omega)_{PES}} = \frac{\omega^3}{c^2k^2} \tag{3.151}$$

which in terms of the wavelengths of the electron λ_e and the photon λ_{ph} is rewritten as:

$$R = \frac{\lambda_e}{\lambda_{ph}}^2 \tag{3.152}$$

For the UPS regime the photoemission is 10^5 times more favorable than inverse photoemission and for X-rays the factor becomes 10^3, but the cross section for the valence band decreases: the inverse photoemission therefore has a yield rather weak.

Bibliography

[1] B. Feuerbacher, B. Fitton, and R. Willis, *Photoemission and the electronic properties of surfaces* (Wiley, Chichester, 1978).

[2] *Angle resolved photoemission - Theory and current application*, edited by S. D. Kevan (Elsevier, Amsterdam, 1992).

[3] L. Hedin and S. Lundqvist, *Effects of electron-electron and electron-phonon interactions on the one electron states of solids*, Vol. 23 of *Solid State Physics* (Academic, New York, 1970).

[4] C.-O. Almbladh and L. Hedin, *Beyond the one electron model/many body effects in atoms, molecules and solids. Handbook on synchrotron radiation* (North Holland, Amsterdam, 1983), Chap. 8, pp. 607–904.

[5] J. Braun, Rep. Prog. Phys. **59**, 1267 (1996).

[6] *Mecanique Quantique*, edited by C. C. Tannoudji, B. Diu, and F. Laloë (Hermann, Paris, 1973).

[7] G. Mahan, Phys. Rev. B **2**, 4334 (1970).

[8] J. A. Bearden and A. F. Burr, Rev. Mod. Phys. **39**, 125 (1967).

[9] *Photoemission in solids*, edited by M. Cardona and L. Ley (Springer, Berlin, 1978), Vol. 1.

[10] J. C. Fuggle and N. Mårtensson, J. Electron Spectrosc. Relat. Phenom. **21**, 275 (1980).

[11] S. Hüfner, *Photoelectron spectroscopy* (Springer, Heidelberg, 1995).

[12] J. Pendry, Surf. Sci. **57**, 679 (1976).

[13] J. Pendry, J. Phys. C **8**, 2413 (1975).

[14] J. Pendry and D. Titterington, Commun. Phys. **2**, 31 (1977).

[15] C. Berglund and W. Spicer, Phys. Rev. **136**, A1030 (1964).

[16] B. Feuerbacher and R. Willis, J. Phys. C **9**, 169 (1976).

[17] H. Puff, Phys. Status Solidi **1**, 636 (1961).

[18] P. Feibelman and D. Eastman, Phys. Rev. B **10**, 4932 (1974).

[19] A. Liebsch, Phys. Rev. Lett. **32**, 1203 (1974).

[20] D. Spanjaard, D. Jepsen, and P. Marcus, Phys. Rev. B **15**, 1728 (1977).

[21] T. Miller, W. McMahon, and T.-C. Chiang, Phys. Rev. Lett. **77**, 1167 (1996).

[22] E. Hansen, T. Miller, and T.-C. Chiang, Phys. Rev. B **55**, 1871 (1997).

[23] G. Malmström and J. Rundgren, Comput. Phys. Commun. **19**, 263 (1980).

[24] M. Seah and W. Dench, Surf. Int. Anal. **1**, 2 (1979).

[25] J. Stöhr *et al.*, Phys. Rev. B **17**, 587 (1978).

[26] P. Thiry *et al.*, Phys. Rev. Lett. **43**, 82 (1979).

[27] Y. Petroff and P. Thiry, Appl. Optics **19**, 3957 (1980).

[28] R. Courths, H. Schulz, and S. Hüfner, Solid State Commun. **29**, 667 (1979).

[29] R. Courths, S. Hüfner, and H. Schulz, Z. Phys. B **35**, 107 (1979).

[30] I.-W. Lyo and E. W. Plummer, Phys. Rev. Lett. **60**, 1558 (1988).

[31] J. Northrup, M. Hybertsen, and S. Louie, Phys. Rev. Lett. **59**, 819 (1987).

[32] F. Patthey, W. D. Schneider, Y. Baer, and B. Delley, Phys. Rev. Lett. **58**, 2810 (1987).

[33] J. Hermanson, Solid State Commun. **22**, 9 (1977).

[34] W. Eberhardt and F. Himpsel, Phys. Rev. B **21**, 5572 (1980).
[35] R. Courths and S. Hüfner, Phys. Rep. **112**, 53 (1984).
[36] G. Vasseur *et al.*, Phys. Rev. B **89**, 121409 (2014).
[37] P. Puschnig *et al.*, Science **326**, 702 (2009).
[38] G. Vasseur, PhD. (Université de Lorraine, Nancy, 2014).
[39] K. Moser, PhD. (EPFL, Lausanne, 2014).
[40] C. Bena and G. Montambaux, New J. of Phys. **11**, 095003 (2009).
[41] C. Larsson and J. B. Pendry, J. Phys. C **14**, 3089 (1981).
[42] Hegensberger, PhD. (Univ. Neuchâtel, ADDRESS, 2000).
[43] G. Grimvall, Phys. Kond. Mat. **9**, 283 (1969).
[44] M. Hengsberger *et al.*, Phys. Rev. B **60**, 10796 (1999).
[45] P. Nozières and C. de Dominicis, Phys. Rev. **178**, 1097 (1969).
[46] G. Mahan, Phys. Rev. **163**, 612 (1967).
[47] P. Anderson, Phys. Rev. Lett. **18**, 1049 (1967).
[48] S. Doniach and M. Šunjić, J. Phys. C: Solid State Phys. **3**, 285 (1970).
[49] G. Mahan, Phys. Rev. B **11**, 4814 (1975).
[50] D. Malterre, G. Krill, J. Durand, and G. Marchal, Phys. Rev. B **38**, 3766 (1988).
[51] *The theory of atomic spectra*, edited by E. Condon and G. Shortley (Cambridge University Press, Cambridge, 1935).
[52] *Photoemission in solids*, edited by L. Ley and M. Cardona (Springer, Berlin, 1979), Vol. 2.
[53] P. Anderson, Phys. Rev. **124**, 41 (1961).
[54] J. C. Fuggle *et al.*, Phys. Rev. B **27**, 7330 (1983).
[55] G. Arfken, *Mathematical methods for physicists* (Academic Press, San Diego, 1985).
[56] J. Cooper, Phys. Rev. **128**, 681 (1962).
[57] M. Seaton and G. Peach, Proc. Phys. Soc. London **79**, 129 (1962).
[58] J. Cooper, Phys. Rev. **165**, 126 (1968).
[59] H. A. Bethe and E. Salpeter, in *Handbuch der Physik* (Springer, Berlin, 1957), Vol. 35, Chap. Quantum mechanics of one- and two-electron atoms, sects. 59, 69.
[60] U. Fano and J. Cooper, Rev. Mod. Phys. **40**, 441 (1968).
[61] J. Cooper and R. Zare, J. Chem. Phys. **48**, 942 (1968).
[62] J. Yeh and I. Lindau, Atomic Data and Nuclear Data Tables **32**, 1 (1985).
[63] S. Goldberg, C. Fadley, and S. Kono, J. Electron Spectrosc. Relat. Phenom. **21**, 285 (1981).
[64] J. Scofield, J. Electron Spec. Rel. Phenom. **8**, 129 (1976).
[65] I. Band, Y. I. Kharitonov, and M. Trzhaskovskaya, Atomic Data and Nuclear Data Tables **23**, 443 (1979).
[66] I. Abbati *et al.*, Phys. Rev. Lett. **50**, 1799 (1983).
[67] J. Rossbach, Nucl. Instrum. Methods A **28**, 3031 (1983).
[68] *Breakdown of the One-Electron Pictures in Photoelectron Spectra, Structure and Bonding*, edited by G. Wendin (Springer, New York, 1981), Vol. 45.
[69] M. Hecht and I. Lindau, Phys. Rev. Lett. **47**, 821 (1981).
[70] K. Kim, S. Gaarenstroom, and N. Winograd, Chem. Phys. Lett. **41**, 503 (1976).
[71] M. Chung and L. Jenkins, Surf. Sci. **21**, 253 (1970).
[72] F. Larkins, Atomic Data Nuclear Data Tables **20**, 311 (1977).
[73] D. Coster and R. Krönig, Physica **2**, 13 (1935).
[74] R. Weissmann and K. Müller, Surf. Sci. Rep. **105**, 251 (1981).
[75] D. Briggs and M. Seah, *Practical Surface Analysis* (John Wiley & Sons, Chichester, 1983).

[76] L. Davis *et al.*, *Handbook of Auger Electron Spectroscopy* (Physical Electronics Division, Perkin-Elmer Corporation, USA, 1978).
[77] G. McGuire, *Auger Electron Spectroscopy Reference Manual* (Plenum Press, New York, 1979).
[78] Y. Shiokawa, T. Isida, and Y. Hayashi, *Auger Electron Spectra Catalogue: a Data Collection of Elements* (Anelva, Tokyo, 1979).
[79] T. Sekine *et al.*, *Handbook of Auger Electron Spectroscopy* (JEOL, Tokyo, 1982).
[80] H. Bishop and J. Rivière, Surf. Sci. **17**, 462 (1969).
[81] G. Margaritondo and J. E. Rowe, J. Vac. Sci. Technol. **17**, 561 (1980).
[82] D. A. Shirley, Crit. Rev. Solid State Mat. Sci. **10**, 373 (1982).
[83] D. P. Woodruff, Le vide **38**, 189 (1983).
[84] S. A. Chambers, Surf. Sci. Rep. **16**, 261 (1992).
[85] H. P. Bonzel, Prog. Surf. Sci. **42**, 219 (1993).
[86] D. P. Woodruff and A. M. Bradshaw, Rep. Prog. Phys. **57**, 1029 (1994).
[87] J. Osterwalder *et al.*, Surf. Sci. **331-333**, 1002 (1995).
[88] C. S. Fadley *et al.*, Surf. Rev. Lett. **4**, 421 (1997).
[89] P. Aebi *et al.*, Surf. Sci. **402-404**, 614 (1998).
[90] J. Osterwalder *et al.*, Prog. Surf. Sci. **64**, 65 (2000).
[91] G. Grenet *et al.*, Surf. Interface Anal. **14**, 367 (1989).
[92] J. J. Barton, S. W. Robey, and D. A. Shirley, Phys. Rev. B **34**, 778 (1986).
[93] V. Fritzsche, Surf. Sci. **265**, 187 (1992).
[94] I. Adawi, Phys. Rev. A **134**, 788 (1964).
[95] W. Schaich and N. Ashcroft, Solid State Commun. **8**, 1959 (1970).
[96] W. Schaich and N. Ashcroft, Phys. Rev. B **3**, 2452 (1971).
[97] C. Caroli, D. Lederer-Rozenblatt, B. Roulet, and D. Saint-James, Phys. Rev. B **8**, 4552 (1973).
[98] A. Liebsch, Phys. Rev. B **13**, 544 (1976).
[99] W. Bardyszewski and L. Hedin, Physica Scripta **32**, 439 (1985).
[100] C. S. Fadley, Prog. Surf. Sci. **16**, 275 (1984).
[101] A. Kaduwela, D. Friedman, and C. Fadley, J. El. Spec. Rel. Phenom. **57**, 223 (1991).
[102] V. Fritzsche, J. Phys.: Cond. Matter **2**, 1413 (1990).
[103] F. J. G. de Abajo, M. A. V. Hove, and C. S. Fadley, Phys. Rev. B **63**, 75404 (2001).
[104] D. Woodruff, Surf. Sci. **53**, 538 (1975).
[105] L. McDonnell, D. Woodruff, and B. Holland, Surf. Sci. **51**, 249 (1975).
[106] H. Helfering, E. Lang, and K. Heinz, Surf. Sci. **93**, 398 (1980).
[107] H. C. Poon, D. Snider, and S. Y. Tong, Phys. Rev. B **33**, 2198 (1986).
[108] D. A. Wesner, F. P. Coenen, and H. P. Bonzel, Phys. Rev. B **39**, 10770 (1989).
[109] H. C. Poon and S. Y. Tong, Phys. Rev. B **30**, 6211 (1984).
[110] W. F. Egelhoff, Crit. Rev. Solid State Mat. Sci. **16**, 213 (1990).
[111] D. Naumovic *et al.*, Phys. Rev. B **47**, 7462 (1993).
[112] M. Seelmann-Eggebert and H. J. Richter, Phys. Rev. B **43**, 9578 (1991).
[113] T. Greber, J. Wider, E. Wetli, and J. Osterwalder, Phys. Rev. Lett. **81**, 1654 (1998).
[114] J. B. Pendry, Surf. Sci. Rep. **19**, 87 (1993).
[115] C. S. Fadley, Surf. Sci. Rep. **19**, 231 (1993).
[116] D. P. Woodruff *et al.*, Surf. Sci. **201**, 228 (1988).
[117] M. A. V. Hove *et al.*, Surf. Sci. Rep. **19**, 191 (1993).
[118] C. Guillot *et al.*, Phys. Rev. Lett. **39**, 1632 (1977).
[119] U. Fano, Phys. Rev. **124**, 1866 (1961).
[120] N. Witkowski *et al.*, Phys. Rev. B **56**, 12054 (1997).

[121] E. Weschke *et al.*, Phys. Rev. B **44**, 8304 (1991).
[122] S. Ogawa and P. Petek, Surf. Sci. **363**, 313 (1996).
[123] M. Aeschlimann *et al.*, J. Chem. Phys. **102**, 8606 (1995).
[124] R. Haight, Surf. Sci. Rep. **21**, 275 (1995).
[125] H. Petek and S. Ogawa, Prog. Surf. Sci. **56**, 239 (1997).
[126] M. Wolf, Surf. Sci. **377-379**, 343 (1997).
[127] W. Wallauer and T. Fauster, Surf. Sci. **374**, 44 (1997).
[128] V. Dose, Prog. Surf. Sci. **5**, 337 (1985).
[129] F. Himpsel, Commun. Condens. Mater. Phys. **12**, 199 (1986).
[130] N. Smith, Rep. Prog. Phys. **51**, 1227 (1988).
[131] H. Claus and K. Ulmer, Z. Phys. **173**, 462 (1963).
[132] J. Lang and Y. Baer, Rev. Sci. Instrum. **50**, 221 (1979).
[133] P. Johnson and J. Davenport, Phys. Rev. B **31**, 7521 (1985).

Experimental techniques

In this chapter we introduce the experimental aspects of the photoemission technique by describing how to perform the measurements. In particular, we shall briefly discuss the need for ultra-high vacuum, the photon sources and the electron detectors.

4.1 Ultra-high vacuum

Photoemission spectroscopy and the ultra-high vacuum are connected for several reasons that we will now clarify. First of all, UV light and soft XR propagation requires a good vacuum because air, as other gases, absorb these radiations. Furthermore, for pressures above 10^{-5}–10^{-6} mbar, photoelectrons are also strongly absorbed. But the main reason is the high surface sensitivity of photoemission in the UV and soft XR domains. The surface has to remain clean during the experiment. In order to maintain a clean surface, the number of particles that adsorb at the surface during the measurement time has to remain much smaller than the atoms in a monolayer. The number g of particles per unit of surface and time to reach the sample can be determined from the kinetic theory of gases:

$$g = \frac{1}{4} n\overline{v} = \frac{p}{\sqrt{2\pi\, mkT}} \tag{4.1}$$

where n is the molecular density, $\overline{v}$ the average velocity, p the pressure, M the molecular mass and kT the Boltzmann term. However, contamination depends on

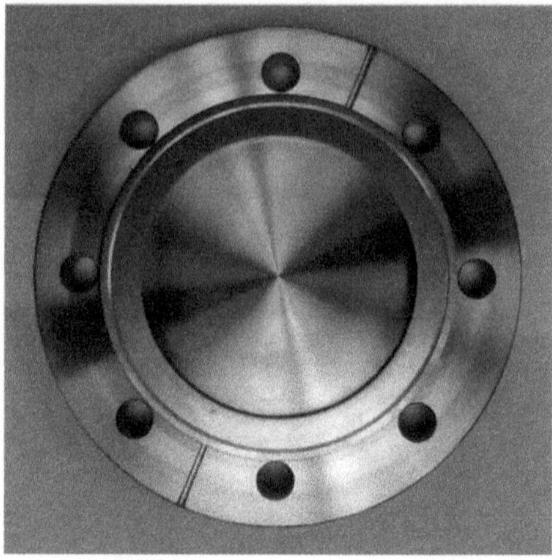

Figure 4.1 *Ultra-high vacuum flange in stainless steel with its copper gasket.*

the sticking coefficient (between 0 and 1) of the residual vacuum molecules onto the surface. This coefficient varies with the surface temperature and the surface reactivity. Let us now recall the orders of magnitude involved. At room temperature with a sticking coefficient of 1, a monolayer is formed in one second at a pressure of 10^{-6} mbar. If the surface is to remain clean during the experiment time (typically one hour), a vacuum of the order of 10^{-10} mbar is necessary. Such a quality of vacuum requires the so-called *ultra-high vacuum* technology.

This technology imposes severe experimental constraints. For instance, it requires special metal seals used to close out the experimental chamber where the electron source and the electron detector are located. Specific pumping systems are also needed and the choice of the materials inside the vacuum chamber is restricted. Materials compatible with ultra-high vacuum are those with low degassing rates, such as most metals: aluminum, copper, molybdenum or stainless steel. Plastics or elastomers are usually avoided and even some metal alloys such as bronze, because of the highly volatile zinc. Borosilicates are used for the standard windows and Be for those to be transparent to X-ray radiation. The connections between different parts of the ultra-high vacuum chambers are made with special flanges using gasket usually made of Cu (Fig. 4.1). When the flange is tightened, the seal is ensured by a part of the flanges called the knife which sinks on both sides of the Cu gasket.

It then becomes necessary to pump considerably on this hermetic chamber to reach ultra-high vacuum. Different pumps are used to reach the ultimate pressure (Fig. 4.2): rotary, turbo-molecular, ionic, cryogenic, sublimation pumps... However, pumping alone is not enough. Just by pumping, the final pressure is only of the order of 10^{-8} mbar, far from the 10^{-10} mbar necessary to carry out photoemission experiments. This limitation arises from the degassing of the chamber walls. There, polar molecules, essentially the water molecules that are present in the air, adhere to the walls and are

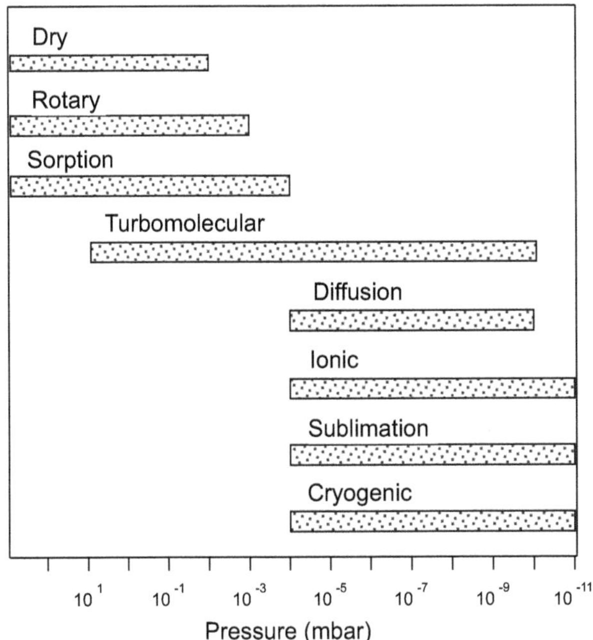

Figure 4.2 *Operation range of the different pumps.*

very poorly pumped at room temperature. To remove these molecules, the chamber must be annealed under vacuum (baked out) at temperatures between 150-180° C for about 48 hours. When the chamber is cold, a vacuum of 10^{-10} mbar can be reached. The need of baking out the chamber adds additional constraints to the building materials and the various equipment parts of an ultra-high vacuum chamber: they have to withstand the annealing at 150° C.

4.2 Micromechanics

In order to carry out photoemission experiments, particularly in angle-resolved photoemission, it is necessary to correctly orient the sample with respect to the analyzer. Nowadays, analyzers are generally fixed and the sample is mounted on a precision manipulator. These manipulators possess at best six degrees of freedom, three translations and three rotations (Fig. 4.3). One rotation is generally around the normal axis of the sample and the two other rotations have their axes in the surface of the sample (often the vertical and the horizontal axis).

The main difficulty in the manipulator design is not only the mechanical precision under ultra-high vacuum but also the cryogenic cooling. It is often necessary to measure a very low temperature (typically 10 K), implying the optimization of the

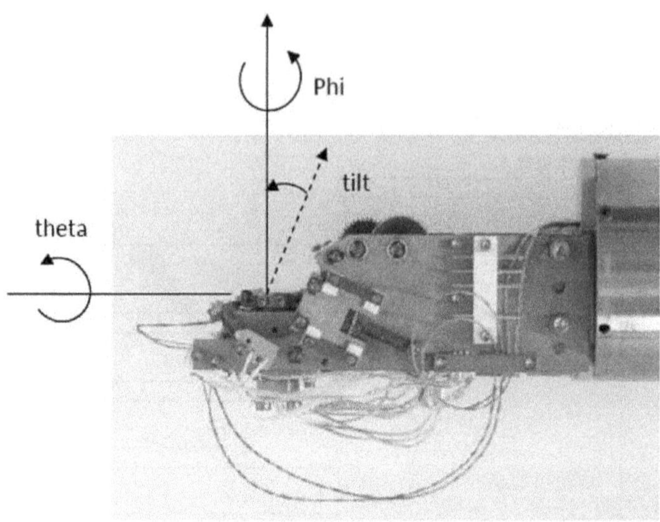

Figure 4.3 *Head of the photoemission manipulator of the Cryoscan company (Nancy, France). This manipulator has the azimuthal rotation ϕ around the sample normal and both the polar rotation θ and the* tilt *along two horizontal axes that pass through the sample surface.*

thermal contact between the sample and the cold finger of the manipulator, while ensuring a precise motion of the sample.

4.3 Photon sources

Photoemission requires a photon source to excite electrons from their initial state. The precise determination of the initial state energy requires monochromatic light. Photon sources for this spectroscopic technique are discharge lamps for UPS, X-ray tubes for XPS, and synchrotrons for both. Let us briefly describe the principle of each of these sources.

4.3.1 *Discharge lamps*

The design of modern discharge lamps dates back to the 1960s [1]. The lamps operate on the principle of the discharge in a low pressure gas, of the order of 10^{-5} mbar (Fig. 4.4). Helium is the most commonly used gas in these lamps for several reasons. It is a monoatomic gas and it has the simplest electronic structure. The number of transitions is thus reduced, giving only a few spectral lines. The absence of vibration and rotation in the monoatomic molecules also leads to narrow lines. The most intense spectral line, called HeI, corresponds to a photon energy of 21.22 eV, well-suited for the study of the valence band of materials. Another advantage of He is

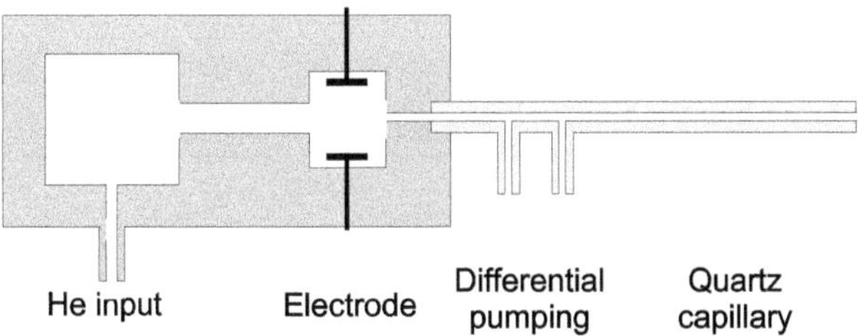

Figure 4.4 *Schematic diagram of a discharge lamp. The discharge in a rarefied gas leads to a plasma. Photons of well-defined energies are produced by the desexcitation of the gas atoms. A capillary and a differential pumping allow you to maintain an excellent pressure in the measuring chamber.*

its small weight. Because of it, He preserves the discharge chamber better when He ions bombard their walls. The mechanism of photon generation in a discharge is as follows. An electron beam between the cathode and the anode excites the He atoms in between and leads to a plasma generation. When the excited atoms desexcite, they emit UV photons the energy of which corresponds to the energy difference between the excited and the ground state. The energies that can be obtained by discharges in rare gas lamps are listed in table 4.1. Since there is no material transparent to the He emission energy, it is not possible to place a window between the discharge lamp and the photoemission chamber. Differential pumping devices using capillaries can nevertheless maintain a very good vacuum in the measurement chamber.

A higher flux discharge is obtained when the plasma is created by microwave-induced Electron Cyclotron Resonance (ECR [2]) (Fig. 4.5). A generator (klystron) generates microwaves confined in a cavity, with a maximum resonance in the axis. The electric field of the microwaves has a frequency ω of linear polarization, which can be decomposed into two circularly polarized waves of opposite helicities. Magnets placed in the cavity add a continuous magnetic field in the cavity axis, which induces the rotation of the electrons with the cyclotron frequency $\omega_c = eB/m$. If the electric and magnetic fields are in resonance ($\omega = \omega_c$), the electric field helicity that is rotating in the same direction as the electrons induces a force $-eE$ in the transverse direction of non-zero temporal average, which leads to a continuous increase of the electron energy. The opposite helicity gives rise to a periodic force of zero average. The main advantage of these sources is the flux, which can be almost three orders of magnitude higher than that of a conventional UV discharge lamp.

For all types of discharge lamps, the spectral width of the emitted light depends mainly on the lifetime of the excited state (intrinsic width), the degree of self-absorption (absorption of photons by the plasma) and the Doppler widening (frequency variation due to the emitter displacement, *i.e.*, the gas atoms) [3, 4]. However, the intrinsic width is generally very small (<10 μeV pour He). The Doppler effect contributes to the broadening according to:

$$\Delta E = 7.2 \times 10^{-7} \sqrt{T(\mathrm{K})/M}\, E \tag{4.2}$$

Table 4.1 *Discharge lamp lines.*

Line	Energy (eV)	Intensity with respect to the most intense line (%)
He I_α	21.22	100
He I_β	23.09	1-2
He II_α	40.81	100
He II_β	48.37	<10
Ne I_α	16.67	15
Ne I_β	16.85	100
Ar I	11.83/11.62	
Kr I	10.64/10.03	
Xe I	9.57/8.44	

where T is the temperature and M is the atomic mass of the gas [3], so the Doppler effect is smaller for heavy atoms. Self-absorption depends on the plasma pressure and it reduces the intensity mainly at the center of the line, where the absorption process is the most resonant, leading to the broadening of the line [4]. To reduce self-absorption, the pressure in the discharge chamber must be reduced. However, the lamp becomes then less stable and the emission can stop. A compromise between the emission intensity and the stability of the lamp must therefore be found. Another concern is that plasmas produce several spectral lines so monochromators are often used to select a given line and therefore a single photon energy.

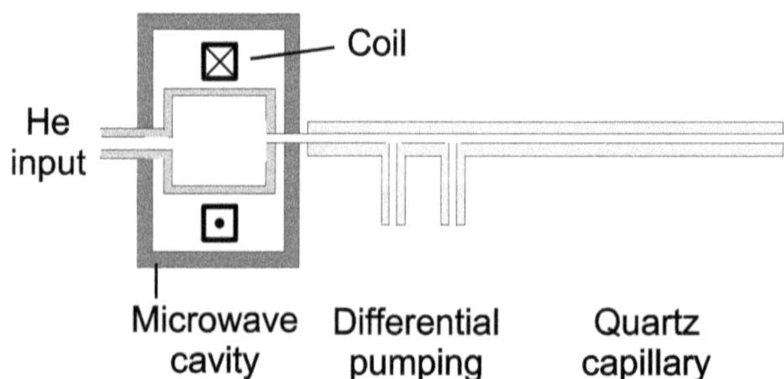

Figure 4.5 *Schematic diagram of a plasma UV lamp. The plasma is generated by ECR excited with microwaves.*

4.3.2 *X-ray tubes*

X-rays are necessary to study deep core levels. In the laboratory, X-rays are produced with cathode ray tubes. Their operation principle is simple. Electrons are emitted by a heated filament (cathode) by Joule effect. An electric potential between cathode

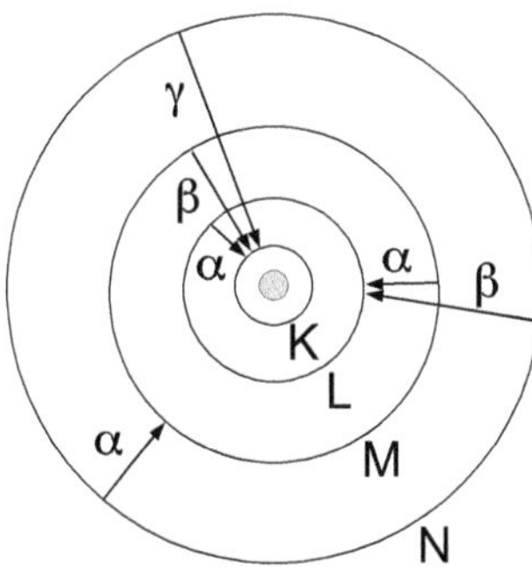

Figure 4.6 *Nomenclature of spectral lines produced during electronic transitions.*

and anode is stabilized. This potential accelerates electrons to an energy high enough to excite a core level of the metal atoms constituting the anode (Fig. 4.7). The atom relaxes when an electron of an upper electronic layer fills the hole created by the electron beam, therefore emitting a X-ray photon. The nomenclature of the X-ray emission lines is given by a Latin capital letter (K, L, M, ...) accompanied by a Greek letter. The capital letter characterizes the deep hole layer (K for $n = 1$, L for $n = 2$, etc.) while the Greek letter indicates the difference between the principal quantum numbers of the two layers involved in the desexcitation ($\alpha \leftrightarrow \Delta n = 1$, $\beta \leftrightarrow \Delta n = 2$,...) as shown in Fig. 4.6. The energy of the X-ray photons depends on the chemical nature of the anode. The material of the anode is chosen so that the energy of the lines is sufficient to explore the core levels of most of the elements and the spectral line-width is narrow, since the latter determines the energy resolution if the X-rays are not monochromatic.

The anode material must be easy to manufacture, stable during electronic bombardment and have a convenient heat dissipation. Mg and Al meet these conditions and therefore most commercial anodes are proposed with a double Mg and Al anode (usually thin films deposited on a copper support cooled by a water circulation. The precise line shape of the Kα transitions can be found in [5] and some typical energies

Table 4.2 *X-ray lines of cathode ray tubes.*

Line	Energy (eV)	Intensity with respect to the most intense line (%)
Al Kα	1486.6	100
Al Kβ	1557.5	1.7
Mg Kα	1253.6	100
Mo Lα	2293.2	100
Mo Lβ	2394.8	44
Cu Lα	929.7	100
Cu Lβ	949.8	25.3

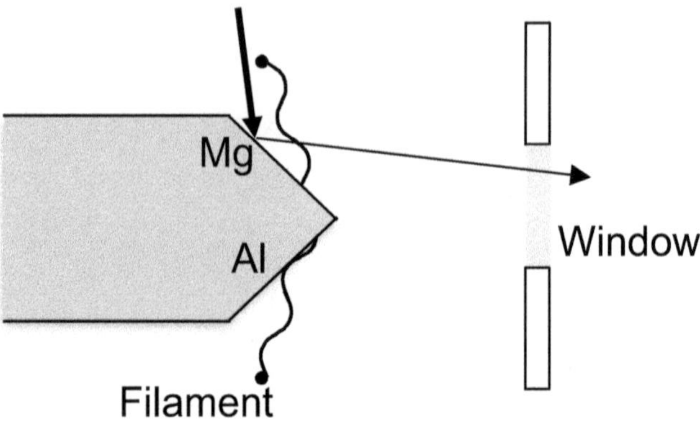

Figure 4.7 *Schematic of an X-ray tube. Electrons emitted by a filament are accelerated by the potential between the anode and the cathode. The X-rays generated at the cathode go through a Be window before arriving to the sample.*

are listed in Table 4.2. The table also shows the Cu line, which appears if the Mg and Al thin films (about ten microns thick) are degraded and their Cu support is exposed to the electron flow. A Kα emission line is large and asymmetric because it consists of several transitions of slightly different energies (K $\alpha_{1,2}$), because the energy depends not only on the principal quantum number n. This asymmetry is greater for Al than for Mg [6]. In addition to the emission lines associated with the quantum transitions between the atomic levels of the anode, an X-ray tube produces a continuous background associated with the deceleration of the incident electrons (Bremsstrahlung), characteristic of accelerated charges.

The X-ray flux depends on several parameters, in particular on the electron acceleration energy and on the filament current, which determines the number of electrons leaving the filament. An adequate flux is obtained with an acceleration energy of an order of magnitude greater than the energy of the line to be excited (typically 15 keV to excite the Kα line of Mg or Al). The flux can also be enhanced by increasing the current. However, this leads to an increase in the temperature of the anode which can be detrimental for the operation, although to a certain extent the heating can be controlled by using a rotating anode, that changes continuously the anode region exposed to the electron beam.

4.3.3 *Laser photoemission: ultra-high resolution and dynamical studies*

Recently, laser photon sources have been applied to photoemission, in order to perform time-resolved experiments and two-photon photoemission. Laser photon sources are more intense than synchrotrons in the visible and also in peak power. The advantages of lasers are threefold. First, lasers have a very well-defined energy

Table 4.3 *Mg and Al anode satellites [7].*

Mg	$\alpha_{1,2}$	α_3	α_4	α_5	α_6	β
ΔE (eV)	0	8.4	10.1	17.6	20.6	48.7
Height (%)	100	8.0	4.1	0.6	0.5	0.5
Al	$\alpha_{1,2}$	α_3	α_4	α_5	α_6	β
ΔE (eV)	0	9.8	11.8	20.1	23.4	69.7
Height (%)	100	6.4	3.2	0.4	0.3	0.6

and allow us to carry out ultra-high resolution photoemission experiments (of the order of $\sim$ 300 μeV). Second, the pulsed nature of certain lasers allows to carry out experiments resolved in time, with a resolution of the order of up to ten femtoseconds, beyond the possibilities of the synchrotrons. Finally, the high intensities of the lasers make it possible to carry out two-photon photoemission experiments: a first photon populates an excited state which is then probed by a second photon. However, in order to extract photoelectrons, a higher photon energy than that of the original laser (typically 1.5 eV) is required to overcome the work function (about 3-5 eV). It is therefore necessary to obtain higher harmonics of the fundamental frequency of optical lasers. This is made possible thanks to non-linear crystals such as the barium β-borate which allows you to double the frequency of the incoming photon. The most commonly used laser for this type of experiment is the Ti:sapphire which allows to reach 6 eV by two successive doublings or 6.3 eV by mixing frequencies. Figure 4.8 shows this type of mounting for time-resolved photoemission measurements, where the system is first excited with the infrared photon (with the fundamental

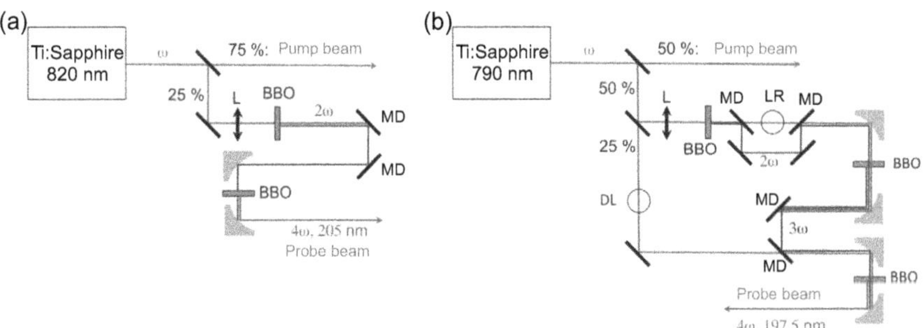

Figure 4.8 *Examples of a laser setup for photoemission. 1.5 eV photons of a Ti: sapphire laser are not energetic enough to extract photoelectrons. It is therefore necessary to use non-linear crystals to reach at least 6 eV either by frequency doubling (a) or by frequency mixing (b) (according to [8]). Reprinted from J. Faure et al., Rev. Sci. Instrum. 83, 043109 (2012), with the permission of AIP Publishing.*

frequency ω of the laser) and subsequently studied with a delayed laser pulse. Figure 4.9 (a) shows precisely the variation of the bismuth spectrum as a function of the delay between a first pulse at ω and a second pulse at 4ω. The system is excited at $t = 0$ and then unoccupied electronic states are populated.

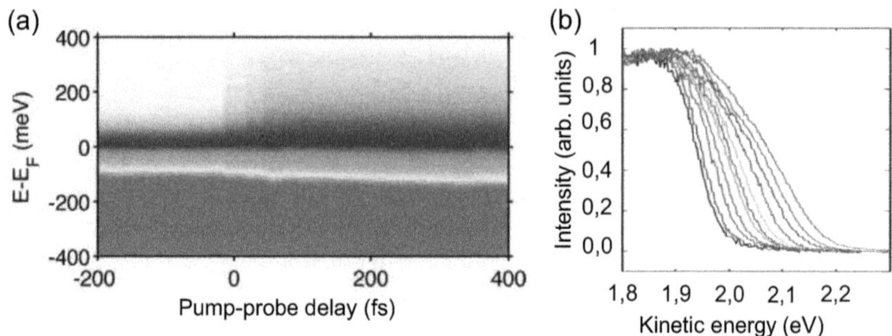

Figure 4.9 *(a) Normal emission photoemission spectrum of Bi as a function of the delay between excitation and probe beams. (b) Spatial charge effect on polycrystalline copper. The position of the Fermi level shifts and expands as a function of the number of photons per pulse, which varies between* 2×10^5 *and* 6×10^6 *photons/pulse (after [8]). Reprinted from J. Faure et al., Rev. Sci. Instrum. 83, 043109 (2012), with the permission of AIP Publishing.*

This type of analysis is powerful although it is necessary to take into account the highly out-of-equilibrium nature of the excited state. Indeed, pulsed lasers can extract a considerable number of electrons in a narrow spatial domain. These electrons leave the surface in a very dense cloud which favors their Coulomb interaction, by shifting and widening the peaks (Fig. 4.9b). This effect is known as the spatial charge effect and it can be minimized by reducing the number of photons per pulse or by increasing the spatial size of the beam. Since the space charge effect limits the number of photons per pulse, in order to keep reasonable statistics during the measurement, it is necessary to use high repetition rates (beyond 10 kHz). As an example, the space charge effect when measuring copper can be avoided for fluxes below 5×10^5 photons/pulse.

Lasers therefore offer new possibilities for photoemission although with some limitations due to the low photon energy of lasers. Indeed, a photon of 6 eV allows you to explore only one energy domain of the order of 2 eV below the Fermi level (the work function being of the order of 4 eV in most of solids). Moreover, at this energy, the accessible space is limited to about 0.5 Å^{-1} which generally does not allow you to explore the whole Brillouin zone. Is it possible to increase energy by further doubling the frequency? This is not possible in solid crystals because beyond a certain photon energy no crystal is transparent. The solution consists in producing high harmonics of the fundamental frequency with gases, as in the work of Haight and Peale in 1994 [9].

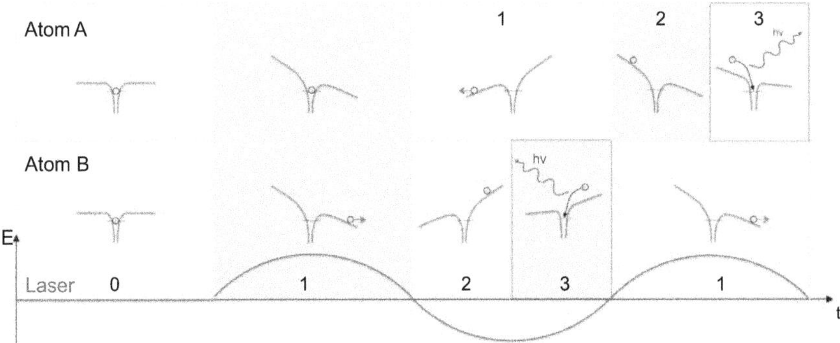

Figure 4.10 *Three-step model of the high harmonics emission. Very intense lasers can change the atomic potential. In the first stage, the potential distortion allows the electron to escape from the atom by tunnel effect. In the second stage, the electron is accelerated by the laser. In the third step, the electron emits a photon by recombination with the ion.*

High Harmonic Generation (HHG) is described by the Corkum semiclassical model [10] which explains the response of an atom to a very intense electric field. This response can be decomposed into three steps (Fig. 4.10). In a first step, the electric field modifies the atomic potential and a weakly bound electron escapes from the atom by tunnel effect. The electron is free, with zero velocity. When the electric field E of the laser changes its sign, the electron gains kinetic energy. With a subsequent change of the field sign, the electron recombines with its ion and emits a photon. Recombination can occur whenever the phase of the field E is between $\pi/2$ and π or between $3\pi/2$ and 2π. This temporal distribution of the emission has a Fourier transform composed of odd harmonics of the frequency ω. It is thus possible to generate high harmonics of the laser frequency ω.

Since the efficiency of this process is very low (between 10^{-4} and 10^{-7}), this technique requires very powerful lasers, reaching intensities of the order of 10^{14} W/cm^2- 10^{15} W/cm^2 depending on the rare gas. The photon number is of the order of 10^{10} photons/s. This is lower than with conventional lasers, but on the other hand it is possible to have very high photon energies (e.g., 91.5 eV corresponding to the 59ème harmonic with Ne) and extremely short pulses (a few hundred attoseconds).

Finally, it is necessary to select the desired harmonic. This selection is possible by means of diffraction gratings or interferometric mirrors (Fig. 4.11). Diffraction gratings that were used in the early studies [9] are very versatile because they allow an easy tuning of the photon energy. However, they had the disadvantage of reducing the photon flux (their efficiency is of the order of 10%). Another option for selecting a particular photon energy is the use of interferometric mirrors, for example with Mo/Si multilayers [13], which allow to have higher photon fluxes without widening the photons pulse.

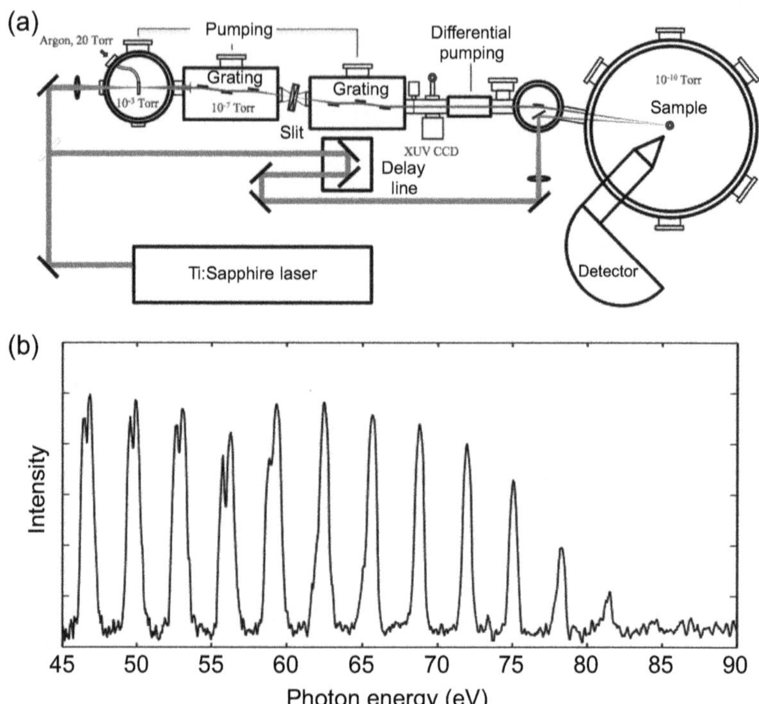

Figure 4.11 *(a) Example of a higher harmonic generation device with a gas. A laser is focused on a rare gas, allowing to obtain high harmonics of the initial laser frequency. Diffraction gratings or interferometric mirrors select afterwards the desired harmonic. For time-resolved photoemission, the fundamental frequency is sent to the sample to excite the system (from [11]). Reprinted from G. L. Dakovski et al., Rev. Sci. Instrum. 81, 073108 (2010), with the permission of AIP Publishing. (b) Harmonics intensity generated by a Ne gas (from [12]). Reprinted from Nucl. Instrum. Meth. A Vol. 615, B. H. Christensen, M. K. Raarup, P. Balling, "Photoemission with high-order harmonics: A tool for time-resolved core-level spectroscopy", page 114, ©2010, with permission from Elsevier.*

4.3.4 *Synchrotron radiation and beam lines*

A non-relativistic oscillating particle emits a monochromatic and quasi- isotropic dipole radiation in the equatorial plane of the dipole, like a radio antenna. As for relativistic particles subjected to a transverse acceleration, their radiation is focused in the direction tangent to the trajectory and the emitted spectrum is polychromatic. We will briefly recall the physical origin of these properties.

Let us consider an electric dipole oscillating along Oz with an amplitude z_0 and an angular frequency ω. The dipole moment is defined by $\vec{d_0} = qz_0\vec{u}_z$ where $\vec{u}_z$ is the following unit vector Oz. At large distances, the field is dominated by the radiative part which varies in 1/r. It is written as follows:

$$\vec{E}(\vec{r},t) \approx \frac{q}{4\pi\varepsilon_0}\frac{a}{c^2}\frac{\sin\theta}{r}\,\vec{u}_\theta$$

where a is the acceleration of the electric charge of the dipole at the delayed time $(t - r/c)$, θ is the angle with respect to the dipole direction and $\vec{u}_r$, $\vec{u}_\theta$ and $\vec{u}_\phi$ are the unit vectors associated with the spherical coordinates. The harmonic character of the movement allows us to write:

$$\vec{a}(t - r/c) = -z_0\omega^2\, e^{-i\omega(t-r/c)}\vec{u}_z$$

and the magnetic field can therefore be obtained:

$$\vec{B}(\vec{r},t) = \frac{\vec{u}_r \times \vec{E}}{c} \approx \frac{q}{4\pi\varepsilon_0}\frac{a}{c^3}\frac{\sin\theta}{r}\,\vec{u}_\phi$$

The electric field is therefore located in the plane defined by the dipole and the observation point, while the magnetic field is located in the direction perpendicular to this plane. The Poynting vector at a given point $\vec{r}$ is expressed from the electric and magnetic fields created by the dipole:

$$\vec{S}(\vec{r},t) = \frac{\vec{E}(\vec{r},t) \times \vec{B}(\vec{r},t)}{\mu_0} = \frac{2q^2}{32\pi^2\varepsilon_0}\frac{a^2}{c^3}\frac{\sin^2\theta}{r^2}\,\vec{u}_r \tag{4.3}$$

Replacing the acceleration, we obtain for the temporal average of $\vec{S}$:

$$\bar{\vec{S}}(\vec{r}) = \frac{d_0^2}{32\pi^2\varepsilon_0}\frac{\omega^4}{c^3}\frac{\sin^2\theta}{r^2}\,\vec{u}_r \tag{4.4}$$

which shows that $\bar{\vec{S}}$ is parallel to the vector $\vec{u}_r$ and varies in $1/r^2$. The Poynting vector flux through a surface corresponds to the energy flux leaving the dipole. This energy flux per solid angle gives the differential cross section:

$$\frac{d\mathcal{P}}{d\Omega}(t) = r^2\vec{S}\cdot\vec{u}_r = \frac{2q^2}{32\pi^2\varepsilon_0}\frac{a^2}{c^3}\sin^2\theta \tag{4.5}$$

and it can be represented by displaying the norm of $\bar{\vec{S}}$ versus θ and ϕ (Fig. 4.12). The radiation is emitted in all directions except for the dipole direction $(\theta = 0)$.

Let us consider a charged particle moving at a relativistic speed over a circular orbit. The acceleration is perpendicular to the speed $(\vec{\beta} \perp \vec{a})$. Let us also place the origin at the instantaneous position of the particle and choose the orientation of the axes so that the acceleration is following Ox and the instantaneous speed following Oz. From the expression of the Poynting vector, and taking into account the delay effects,

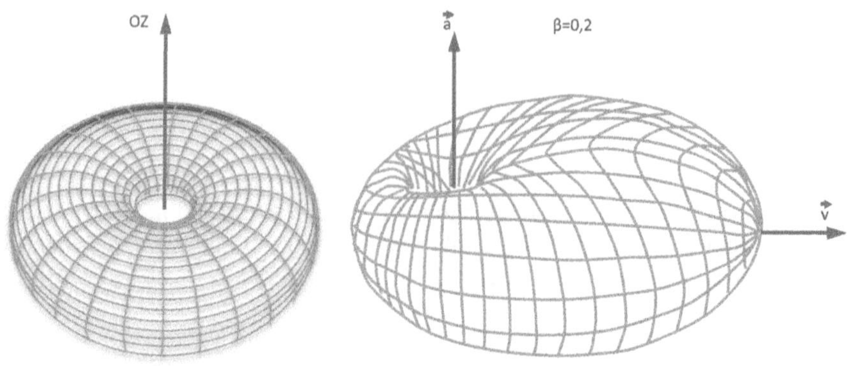

Figure 4.12 *Left: Radiation emitted by an electric dipole oscillating along* Oz. *The emitted radiation is zero in the dipole direction and maximum in the perpendicular plane. Right: Radiation emitted by an accelerated particle along a curved trajectory, travelling at a speed* $v = \beta c$ *(β =0.2). For an ultra-relativistic particle ($\beta \approx 1$), the radiation is in a very narrow solid angle centered on the speed direction.*

the differential radiated power is:

$$\begin{aligned}\frac{d\mathcal{P}(t)}{d\Omega} &= \frac{2q^2a^2}{32\pi^2\varepsilon_0 c^3}\frac{(1-\beta\cos\theta)^2-(1-\beta^2)\sin^2\theta\cos^2\varphi}{(1-\beta\cos\theta)^5}\\ &= \frac{2q^2}{32\pi^2\varepsilon_0 c^3}\frac{a^2}{(1-\beta\cos\theta)^3}\left(1-\frac{\sin^2\theta\cos^2\varphi}{\gamma^2(1-\beta\cos\theta)^2}\right)\end{aligned} \tag{4.6}$$

The corresponding emitted radiation for $\beta = 0.2$ is presented in figure 4.12. It presents two lobes, the more intense one being in the speed direction. In the plan $\varphi = 0$, the expression of the radiated power is simplified in:

$$\frac{d\mathcal{P}(t)}{d\Omega} = \frac{2q^2a^2}{32\pi^2\varepsilon_0 c^3}\frac{(\beta-\cos\theta)^2}{(1-\beta\cos\theta)^5} \tag{4.7}$$

The emitted radiation vanishes for $\theta_c = \arccos\beta$ so that for an ultra-relativistic particle, the radiation is emitted in a narrow cone of angular width $\theta_c \approx 1/\gamma$ where $\gamma = 1/\sqrt{1-\beta^2}$ is a very large number when compared to 1. This is precisely one of the characteristics of synchrotron radiation: the radiation is focused in a very narrow solid angle centered on the speed of the accelerated electrons.

Synchrotrons are particle accelerators that exploit precisely this property of relativistic particles. They are used as intense sources of electromagnetic radiation over a wide frequency spectrum. Non-relativistic particles on a circular trajectory do not have this property, as they radiate at the frequency of the movement. In the synchrotron radiation, the forward directionality and the Doppler effect allow us to understand the emitted spectrum by a curvature element of a synchrotron. The Doppler effect is the variation of the frequency when the radiation source moves with respect to the observer. This is the case in a synchrotron since electrons that radiate in the curvature elements move (at an ultra-relativistic speed) with respect to the observer. The frequency of the light from a moving source as a function of the emission angle θ with respect to the speed direction is:

$$\omega = \omega_0 \frac{1}{\gamma(1-\beta\cos\theta)} = \omega_0 \frac{\sqrt{1-\beta^2}}{1-\beta\cos\theta} \tag{4.8}$$

It can be observed that the frequency decreases if the source moves away (red shift, $\cos\theta = -1$), while it increases if the source gets closer to the observer ($\cos\theta = 1$). The radiation in the axis of the velocity is not the natural frequency of the particle on its orbit ($\omega_0 = v/R$). Taking into account the Doppler effect:

$$\omega = \omega_0 \frac{\sqrt{1-\beta^2}}{1-\beta} = \omega_0 \frac{\sqrt{1+\beta}}{\sqrt{1-\beta}} \tag{4.9}$$

and in the ultra-relativistic domain, β is close to 1 and γ is very large:

$$\beta = \sqrt{1-\frac{1}{\gamma^2}} \approx 1 - \frac{1}{2\gamma^2} \tag{4.10}$$

so the radiation frequency is:

$$\omega \approx 2\gamma\,\omega_0 \tag{4.11}$$

For example, at the European Synchrotron Radiation Facility (ESRF), electrons are accelerated to an energy $E = \gamma m_0 c^2$ = 6 GeV and $\gamma \simeq 1.2\ 10^4$ ($\beta = 0.999999996!$). The relativistic Doppler effect therefore allows us to understand the large increase in the emitted frequency with respect to the natural frequency of the electron movement inside the ring.

The polychromatic nature of the synchrotron radiation (or white spectrum) is a consequence of the high directionality of this radiation. Indeed, for ultra-relativistic particles, we have seen that electromagnetic radiation is emitted in a small opening cone ($\theta \approx 1/\gamma$) centered at the speed vector. An observer close to the synchrotron sees the radiation over a very short delay ΔT at each turn of the electron in its orbit. This pulse is so short that it is smaller than the radiation period. Therefore, according to the properties of the Fourier transform, this wave-packet of spatial length $c\Delta T$ is associated to the frequency $\omega_c \sim 1/\Delta T$ corresponding to the maximum intensity of the emitted radiation. It can be shown that the pulse duration as seen by the observer is:

$$\Delta T \approx \frac{R}{2c\gamma^3}$$

where R is the curvature radius of the trajectory. The maximum frequency is then:

$$\omega_{max} \sim \left(\frac{c}{R}\right)\gamma^3 \sim \omega_0\gamma^3 = \omega_0\left(\frac{E}{m_0c^2}\right)^3 \tag{4.12}$$

where $\omega_0 = v/R$ is the rotation frequency of the particle in its orbit and E is its total energy ($E = \gamma m_0 c^2$). We obtain thus a spectrum which can range up to several tens of keV (*i.e.*, hard X-rays).

Electron synchrotrons are remarkable, very intense photon sources with a very wide emission spectrum that depends on the energy of the electrons. This continuous spectrum allows you to choose the energy of the photons and to perform experiments impossible to achieve with conventional sources (UV lamps or X-ray tubes). The emitted radiation is polarized: it is linear in the plane of the orbit and elliptical outside it. Moreover, this radiation is well collimated, and its temporal structure due to the electron beam distribution inside the ring allows you to carry out time-resolved experiments. The pulses of photons are of a duration of 100 ns to 1ps, or even 0.1 ps in certain facilities, with frequencies ranging between a few hundreds of kHz and a few hundreds of MHz.

This exceptional tool began to be used in the 1960s and it has its origin in the particle physics field. Indeed, to study high-energy collisions, it is necessary to accelerate charged particles. This led to the development of linear accelerators, where charges are accelerated at high speeds by an oscillating electric field, and of synchrotrons, where particles are confined on a closed path by magnetic fields.

The curvature of the trajectory and the associated centripetal acceleration leads to the emission of radiation, which was observed for the first time in 1946 [14]. In particle physics, this parasitic radiation limited, for example, the maximum energy of the betatrons [15]. In 1956 Tomboulian and Hartman realized the utility of synchrotron radiation [16]. From then on, solid-state physicists began to use synchrotron radiation in accelerators designed and optimized for particle physics. In particular, synchrotron accelerators were of pulsed character, so the beam intensity decreased rapidly, this being inconvenient for solid state experiments. In 1966, the first storage ring, specially dedicated to radiation production, was built at the University of Wisconsin-Madison. Tantalus was the first "second" generation machine, where electrons were introduced into a storage ring with a longer lifetime. In this type of synchrotron, radiation was produced at the bending magnets that curved the beam trajectory. Eastman and Lepeyre made there the first photoemission experiments. The instrumentation was limited and CMA detectors were used, despite the difficulties in aligning them. Hemispherical analyzers were spread afterwards and subsequently two-dimensional detection was included in them. In parallel with these developments, third generation synchrotrons appeared, with a decrease in the size and divergence of the source, and with the generalization of the insertion elements. The first third-generation synchrotron was the European synchrotron installed in Grenoble (ESRF, 1994).

After this historical parenthesis, we will make a brief description of a synchrotron. We have chosen as an example the SOLEIL synchrotron in France. It consists of

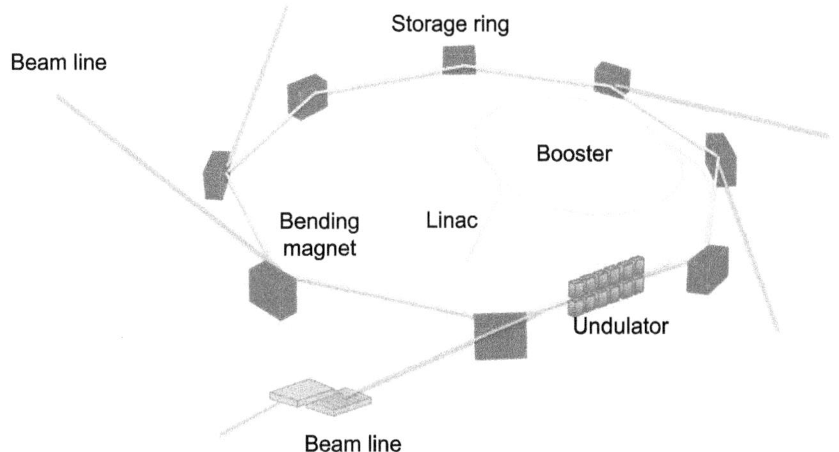

Figure 4.13 *Schematic diagram of a synchrotron with LINAC, booster, storage ring, bending magnets and undulators.*

four elements: a linear accelerator (LINAC), a circular accelerator (Booster), a storage ring and lines of light (Fig. 4.13). The particles rotating in a synchrotron are either electrons (produced by a heated filament in an electron gun) or positrons (obtained by beta decay). The production of positrons is less efficient than that of electrons, but when stored in the ring, the amount of ionized particles produced is lower, which increases the average beam lifetime. However, the improvement of the vacuum techniques nowadays makes it possible to work with electrons under excellent conditions. It is, for instance, the case of SOLEIL. At the output of the electron gun, the particles have an energy of the order of 100 keV. The electrons are then accelerated in a linear accelerator (LINAC) up to an energy of 100 MeV and grouped into packets before being sent to the booster (a ring 157 m of perimeter). The booster is a small synchrotron that allows to reach the nominal energy of the storage ring (2.75 GeV for SOLEIL synchrotron). The electron packets are then injected into the storage ring (354 m of perimeter for SOLEIL). The intensity of the stored current (a few hundred mA) decreases with time because of collisions with residual gas molecules and with electrons of the same packet. To maintain the beam intensity within adequate limits, particles are injected "continuously" to keep the intensity constant (top-up operating mode).

In the ring, several optical elements (quadrupoles, sextupoles or even octupoles) act on the beam to optimize its trajectory, its dimensions and its lifetime. Moreover, two types of elements allow to produce synchrotron radiation in the curved parts and in the linear sections of the ring. The curvature of the trajectory is obtained by means of magnets (bending magnets), which lead to the production of synchrotron radiation (due to the centripetal acceleration). The insertion elements (undulators and wigglers) are inserted in the straight segments of the ring. They are made up of magnetic elements that make the beam “wiggle” along the straight

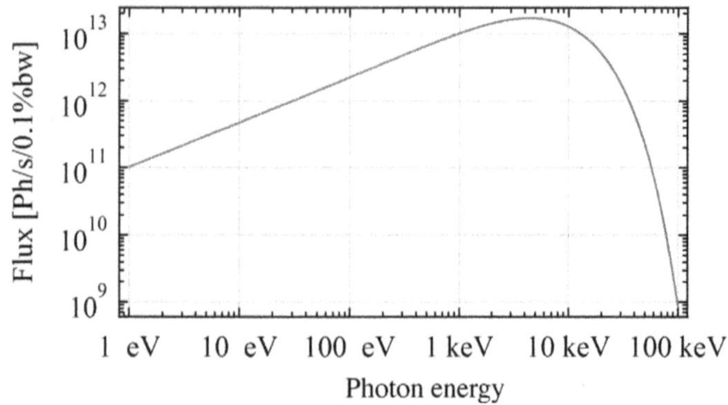

Figure 4.14 *Flux of SOLEIL bending magnets (curvature radius of 5.36 m, magnetic field 1.7 T) through a 4 mm (vertical) aperture × 6 mm (horizontal) located 10 m from the center of the magnet, as a function of the photon energy. Courtesy of Fabien Briquez, SOLEIL synchrotron.*

section, triggering the photon emission. We have explained above the characteristics of the radiation in a bending magnet. The total power emitted by all the particles is [17]:

$$P \propto i\gamma^4/R \propto iE^4/R \tag{4.13}$$

where i is the intensity in the ring, E is the energy of the particles, and R is the curvature radius of the bent part, which does not necessarily coincide with the average radius of the ring. R is a function of the magnetic field which curves the trajectory:

$$R = \gamma\frac{mc}{qB} \tag{4.14}$$

At constant energy, the power emitted decreases according to the particle mass power four. This is the reason why electrons or positrons are used to obtain synchrotron radiation. A proton in a synchrotron does not radiate much! Finally, as we have seen previously, the emitted spectrum by a bending magnet is polychromatic as illustrated in Fig. 4.14.

Another characteristic of synchrotron radiation is its well-defined polarization. In the plane of the orbit, which corresponds to the maximum intensity, the radiation is linearly polarized (in-plane polarization). When deviating from the orbit plane, a perpendicular contribution appears so that the radiation becomes elliptically polarized.

Let us now consider the insertion devices. These devices are installed in an astute, intelligent manner in the straight sections of the synchrotron and they allow you to obtain more intense radiation than that obtained in the bending magnets. The insertion devices, *i.e.*, wigglers and undulators, operate both on the same principle and are distinguished by the brightness of the emitted radiation as will be discussed below. They consist of magnets of alternating polarity (figure 4.15) that force the electrons to oscillate along the otherwise linear trajectory with the period (λ_u) of the magnets. We will explain the basic principle of the radiation production by these insertion devices.

Assuming that the magnetic field in the undulator is sinusoidal ($\vec{B} = B_0 \sin\left(\frac{2\pi x}{\lambda_u}\right)\vec{u}_z$), it can be shown that the displacement of the particle perpendicularly to the undulator axis is also sinusoidal. The relativistic laws for changes of frame of reference allow us then to qualitatively describe the emitted radiation. Indeed, if we place ourselves in the Galilean frame of reference moving at the average (ultra-relativistic) velocity of the particle along the undulator axis, there is an electric field along Oy due to the relativistic transformation of the magnetic field in the fixed frame of reference (that of the laboratory). This field is:

$$E'_y = \gamma\beta c B_z = \gamma v B_0 \sin\left(\frac{2\pi x}{\lambda_u}\right) \tag{4.15}$$

which can be expressed with the coordinates (x', t') of the frame of reference in motion:

$$E'_y = \gamma v B_0 \sin\left[\frac{2\pi\gamma(x' + vt')}{\lambda_u}\right] = \gamma v B_0 \sin\left(\frac{2\pi\gamma vt'}{\lambda_u}\right) \tag{4.16}$$

If the origin of the frame of reference is selected on the mean position of the charge ($x' = 0$), we are dealing with an oscillating classical dipole (and low velocities due to small amplitude oscillations) since the charge feels an electric field $\vec{E} = \vec{E}_0 \sin\omega_0 t'$ with

$$\vec{E}_0 = \gamma v B_0 \vec{u}_y$$

and:

$$\omega_0 = \frac{2\pi\gamma v}{\lambda_u} = \gamma\,\omega_u \quad \text{with} \quad \omega_u = \frac{2\pi\, v}{\lambda_u} \tag{4.17}$$

It will be noted that, owing to the relativistic contraction of the lengths, λ_u/γ is the period of the undulator in the moving frame of reference. The field radiated by a particle in an undulator is therefore that of an oscillating dipole moving with an ultra-relativistic velocity in a direction perpendicular to the oscillation. The emitted spectrum can easily be calculated in the moving frame of reference, as it is that of an oscillating dipole as presented above. The spectrum in the fixed frame of reference can be found just by making a change of reference.

In the moving frame of reference, the radiation is that of a classical oscillating dipole (equation (4.4) and figure 4.12). In this frame of reference, the electron radiates in all directions at the natural frequency of the dipole (ω_0). In the frame of reference of the laboratory, the Doppler effect and the angle modification by a Lorentz transformation allow to understand the synchrotron radiation emitted by the undulators. Consider a wave emitted in the direction of the electron motion. The frequency measured by the observer (ω) is shifted to higher frequencies (Doppler effect), according to equation (4.8):

Figure 4.15 *Diagram of an undulator: the alternating polarities of the magnets create a sinusoidal magnetic field in the Oyz plane. Electrons travelling through the undulator along Ox oscillate then in the Oxy plane. The electrons emit radiation in an angular width $1/\gamma$ along their trajectory.*

$$\omega = \omega_0 \frac{\sqrt{1-\beta^2}}{1-\beta} \approx 2\gamma\omega_0 \tag{4.18}$$

since for ultra-relativistic particles $\beta \approx 1 - \frac{1}{2\gamma^2}$. The frequency of the light emitted along the undulator axis is strongly shifted by the Doppler effect and it is:

$$\omega = 2\gamma^2\omega_u \tag{4.19}$$

Similarly, the wavelength of the light emitted in the laboratory frame of reference is related to the period of the undulator:

$$\lambda = \frac{\lambda_u}{2\gamma^2} \tag{4.20}$$

At the European Synchrotron Radiation Facility (ESRF), the machine runs at $E = \gamma m_0 c^2 = 6$ GeV ($\gamma \simeq 1.2\ 10^4$), the wavelength[1] emitted by an undulator with period $\lambda_u = 3$ cm is $\lambda \simeq 10^{-8}$ cm!

As seen above, the ultra-relativistic character of the particle motion leads to the very large directionality of the emission. The forward emission is contained in a cone of angular opening $\theta = 1/\gamma$. The Doppler effect then modulates the energy of the emitted light in this emission cone. In the direction ϑ with respect to the axis, when $\beta \simeq 1$, we have $\cos\vartheta \simeq 1 - \vartheta^2/2$ and from the equation (4.8):

[1] In the moving frame of reference, the wavelength emitted by the oscillating dipole is $\lambda' = \lambda_u/\gamma \simeq 2\ 10^{-4}$ cm. It is therefore the Doppler effect which allows to obtain a small wavelength radiation in the laboratory frame of reference.

$$\omega(\vartheta) \simeq \frac{\omega_0}{\gamma(1-\beta+\vartheta^2/2)} \tag{4.21}$$

The term in the denominator gives:

$$1-\beta+\frac{\vartheta^2}{2} = \frac{1}{2\gamma^2}+\frac{\vartheta^2}{2} = \frac{1}{2\gamma^2}\left(1+\gamma^2\vartheta^2\right)$$

and then:

$$\omega(\vartheta) \simeq \frac{2\gamma\omega_0}{1+\gamma^2\vartheta^2} = \frac{2\gamma^2\omega_u}{1+\gamma^2\vartheta^2} \tag{4.22}$$

There is therefore a significant variation since the frequency varies by a factor of two in the emission cone. The wavelength reads:

$$\lambda(\vartheta) = \frac{\lambda_u}{2\gamma^2}(1+\gamma^2\vartheta^2)$$

which is an approximate result. Indeed, depending on the applied field, the oscillation amplitude is more or less important and is characterized by the parameter K defined by:

$$K = \frac{qB_0\lambda_u}{2\pi\, m_0 c} \tag{4.23}$$

and the correct expression for the wavelength is:

$$\lambda(\vartheta) = \frac{\lambda_u}{2\gamma^2}\left[1+\frac{K^2}{2}+\gamma^2\vartheta^2\right] \tag{4.24}$$

The wiggler introduces a magnetic field and a sufficiently strong amplitude of oscillation ($K > 3$–4) so that the emission angle is greater than the photon divergence (Fig. 4.18). An observer close to the axis sees, for N magnetic periods, a set of $2N$ independent sources (half-periods) whose intensities add up[2]. The spectral emission of a wiggler induces a shift towards the high energies with a white spectrum of flux proportional to B_0 and N.

Moreover, for an undulator the magnetic field is small ($K < 1$) so that the sources are coherent and they lead to interference phenomena (Fig. 4.18), giving rise to an emission spectrum that gets narrower with the increasing number of periods. Not only the wavelength of the fundamental frequency is obtained but also wavelengths associated to its higher harmonics. For the nth harmonic we have:

$$\lambda_n(\vartheta) = \frac{\lambda_u}{2\gamma^2 n}\left[1+\frac{K^2}{2}+\gamma^2\vartheta^2\right] \tag{4.25}$$

[2] Strictly speaking there is on one hand the addition of the amplitudes for the particles radiating in the angular range $1/\gamma$ towards the observer and, on the other, the addition of the intensities for the other parts of the angular range.

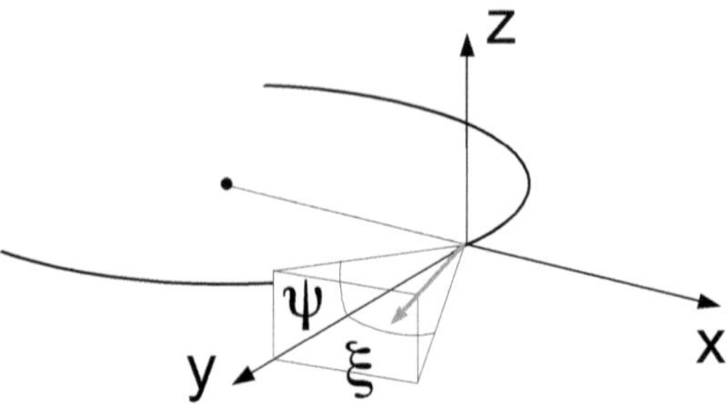

Figure 4.16 *Definition of the frame of reference and the angles that characterize the radiation emission.*

The dependence as a function of B_0 allows to settle the emission peak at the desired photon energy. Further control on the emission light is possible by using asymmetric or helical insertion elements to control the polarization of the radiation.

In addition to the photon energy, the intensity of the radiation and the number of emitted photons are important properties. For synchrotron users, a fundamental parameter is the photon number at a given energy arriving to the sample. To quantify these parameters, several magnitudes are introduced: flux, brightness and power that we will define below. The radiated power by an element, for instance a bending magnet, is the energy radiated per unit time in all directions and for all wavelengths. It depends on the electron energy power four and the intensity of the ring current. The flux is the number of emitted photons per unit of time for a horizontal angular aperture of 1 mrad, integrated in the vertical plane and for a spectral width $\Delta\lambda/\lambda = 10^{-3}$. Finally, for experiments that focus the beam on the sample, the relevant magnitude is the spectral brightness, which is the number of photons emitted around the energy $h\nu$ with a spectral width $\Delta\lambda/\lambda = 10^{-3}$, in a solid angle $d\Omega = d\xi\, d\psi$, being $d\xi$ the horizontal angle and $d\psi$ the vertical angle (Fig. 4.16). The spectral brightness is normalized by the surface unit of the source, the time unit and the source intensity i:

$$b = \frac{dN_{0.1\%}}{idSd\Omega dt} \tag{4.26}$$

It is expressed as the number of photons per mrad square, per square mm and it allows a comparison among the various synchrotrons (Fig. 4.17) or the different sources within the same ring: bending magnets, wigglers and undulators) (Fig. 4.18(c)).

The radiation emitted by the charges decreases their energy. Radiofrequency cavities installed on straight sections of the ring restore the lost energy. They create an electric field that accelerates the particles in the right direction at each turn. This electric field must be synchronized with the T_0 employed by particles in completing a complete revolution. In order to act synchronously at each turn, the electric field frequency must then be a multiple of the frequency of the electrons in the ring:

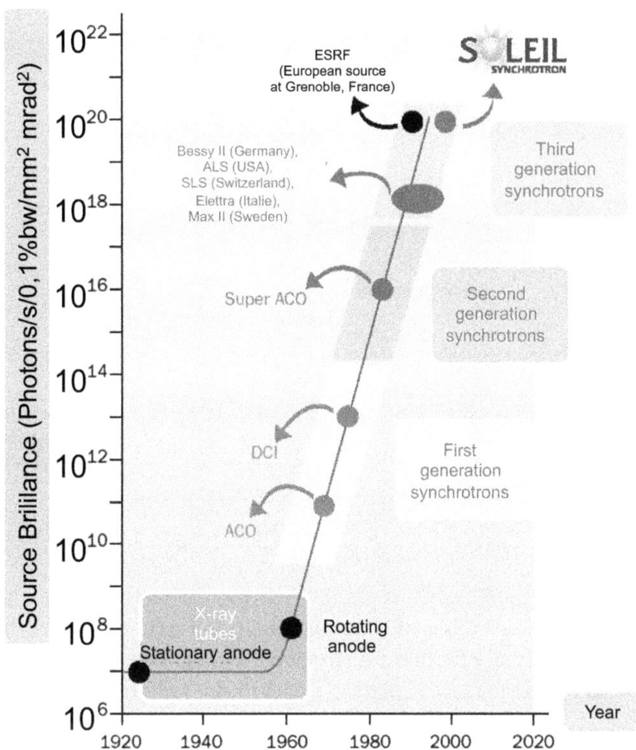

Figure 4.17 *Evolution of the source brightness, from X-ray tubes to synchrotrons of 3rd generation (RS - synchrotron radiation). ©SOLEIL synchrotron.*

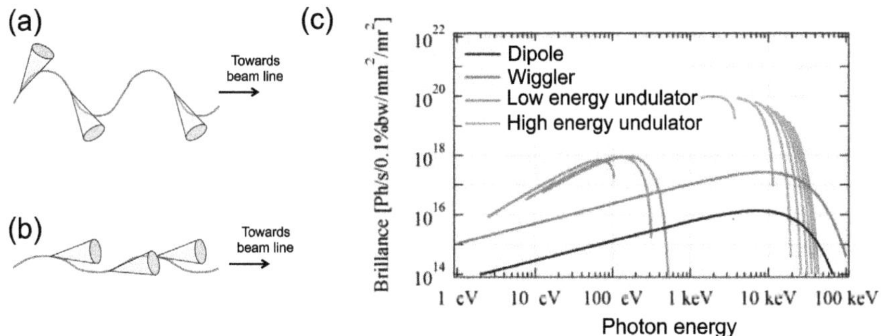

Figure 4.18 *a) Schematic diagram of a wiggler. b) Schematic diagram of an undulator. c) Comparison of the emitted intensity of a bending magnet, a wiggler, an undulator (courtesy of Olivier Marcouillé, SOLEIL synchrotron).*

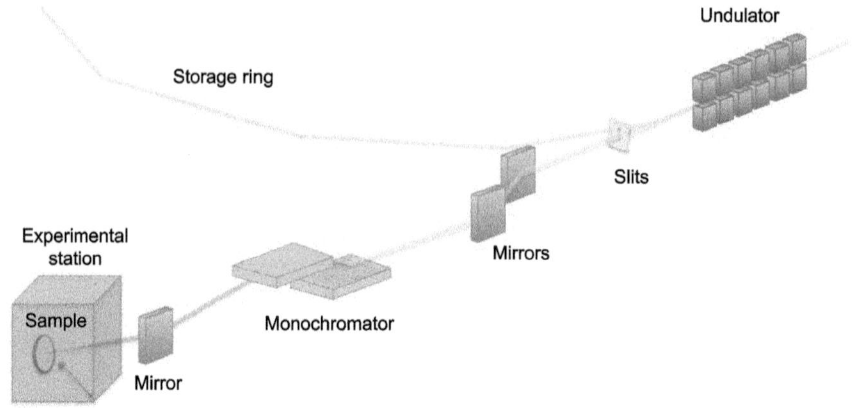

Figure 4.19 *Diagram of a beam line. The light emitted by the undulator, the wiggler or the bending magnet is monochromatized and focused on the sample.*

$$\omega_{rf} = n_b \frac{2\pi}{T_0} \tag{4.27}$$

being n_b the harmonic number of the ring, which is a natural integer that gives the maximum number of packets that can receive the ring. Because of the value of T_0, the electric field frequency ω_{rf} is precisely within the radiofrequency spectrum.

4.3.4.1 Beam lines

We have just seen how undulators, wigglers and bending magnets emit radiation. The main objective of the beam lines is to select the desired energy (monochromatization) for the experiment and to focus it on the sample. Figure 4.19 shows the principle of a beam line with the main elements: diaphragms and slits for collimating the beam, mirrors, monochromator and the experimental station at the end. The monochromator is a very important element of the beam line, as it selects the photon energy according to Bragg's law and can also focus the beam on the sample if the crystals or the arrays are curved, although the focusing is essentially performed by mirrors, which also eliminate harmonics of the main energy. The collimators limit the beam size to improve the resolution in wave-vector. All these manipulations of the beam, monochromatization and focusing, are accompanied by a decrease in the number of photons.

These optical elements for UV or X-rays undergo severe stress because of the very high intensity of the radiation. In particular, the mirrors, especially those before the monochromator and which receive the white beam, must be designed to dissipate the heat produced by the absorption.

Their reflectivity must be preserved as a function of time and to this end it is necessary to avoid the carbon contamination of the mirrors. Indeed, it has been found that

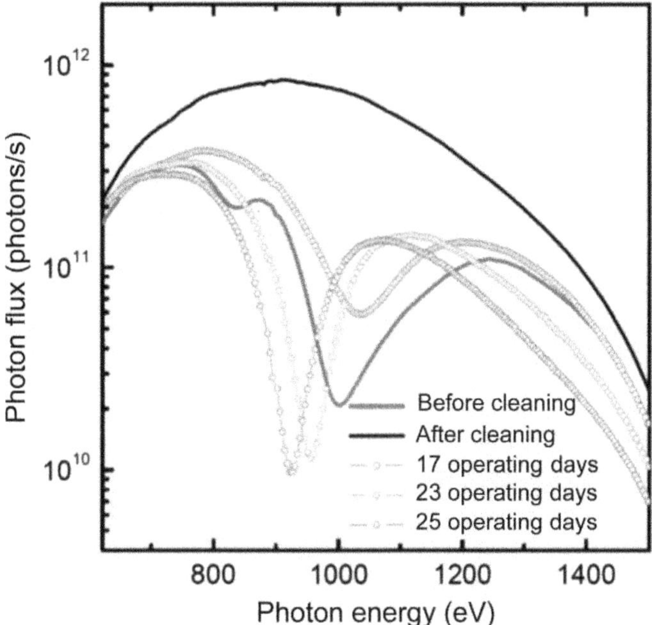

Figure 4.20 *Photon flux as a function of time at the TEMPO beam line of the SOLEIL synchrotron when analyzing the carbon contamination problems of the mirrors. Carbon forms a thin layer affecting the reflectivity of the mirror (from [18]). Reproduced with permission from the International Union of Crystallography.*

the residual CO in the chamber (present even in ultra-high vacuum) decomposes and leads to a carbon deposition on the mirrors. Figure 4.20 shows the effect of the reflectivity change of the mirrors because of this C deposition [18]. Such a degradation of the performance makes difficult to carry out experiments at high photon energy and renders practically impossible to work with photon energies close to the carbon K absorption threshold (280 eV). Methods for the mirror regeneration have been developed and consist on the exposure of the mirror to a small amount of oxygen which removes the C film [19].

Finally, the key element of the line is the monochromator which selects photon energy by diffraction.[3] The principle of these monochromators is based on Bragg's law:

$$2d \sin \theta = m\lambda \tag{4.28}$$

[3] Monochromators are also used with conventional sources, such as He lamps or X-ray tubes. The spectral width of the emitted lines by the UV lamps is of the order of 1 meV when satellites are well resolved and of the order of 1 eV for the X-ray tubes. The monochromator can eliminate the UV emission satellites and refine the spectral line in the case of X-rays.

where θ is the angle between the incident and diffracted beams, m is the diffraction order, λ is the selected wavelength and d is the period in the diffracting element. It is therefore sufficient to choose the right angle after the diffracting element to select the desired wavelength λ. Since d must be of the same order of magnitude than λ, the nature of the diffractive element depends on the working energy. For UV photoemission, grids of holographically etched lines are used. For hard x-rays, the diffracting element is often a quartz, Si or Ge crystal. Behind the monochromator, it is often necessary to refocus the radiation on the sample using refocusing mirrors.

4.3.5 *Free electron lasers*

Another radiation source is the free electron laser (FEL), which relies on the stimulated emission by relativistic electrons in the presence of a periodic magnetic field, *i.e.*, in an undulator or a wiggler. The advantage of FEL is that the active medium is the electron beam of the storage ring. Unlike the discrete levels of atoms or molecules, the spectrum is continuous, which allows to obtain simulated emission with adjustable energy. The mechanism was proposed for the first time in 1971 by J.M.J. Madey [20–22], then developed six years later [23]. FEL provides radiation in a range from microwaves to UV. Its temporal structure can vary between quasi-continuous operation and picosecond fraction pulses. Its average power is of the order of kW, with peaks at GW. Its flux for certain energies may be of an order of magnitude greater than that of the present synchrotrons. For all these characteristics, they might become the fourth generation of synchrotron. More information can be found, for example, in the following references [24–27].

In a free electron laser, the relativistic electrons travel through a wiggler (Fig. 4.21). If the emitted radiation is confined in an optical cavity by mirrors, stimulated emission may appear. However, it is difficult to obtain effective mirrors which work at normal incidence for certain wavelengths, which limits the maximum energy of this configuration. In spite of this constraint, it is possible to obtain self-amplified spontaneous emission (SASE) amplification in long wigglers with very bright beams [28].

For the FEL with optical cavity or for SASE, the stimulated emission principle is quite similar. For a FEL with optical cavity, the radiation produced by the beam is confined, *i.e.*, a stationary electromagnetic field with nodes and anti-nodes is established. After a few passages of the beam by the wiggler, the anisotropy of the stationary field will eventually group the electrons into packets. The magnetic field of the wiggler and the velocity of the electrons are chosen so that after a period of the wiggler all the charges of a given velocity are *exactly* in the same situation as in the preceding period. Charges with lower speeds are not in phase and lose energy. Electrons are structured in spatially separated packets according to the spatial period of the wiggler and they all emit radiation in phase. Before the separation into packets takes place, there is no stimulated emission. As time goes on, radiation begins to be strong enough to trap the electrons in its field, triggering the stimulated emission. As in lasers, the energy lost by radiation makes the necessary to pump. In the FEL, it is the radiofrequency cavity that pumps the system.

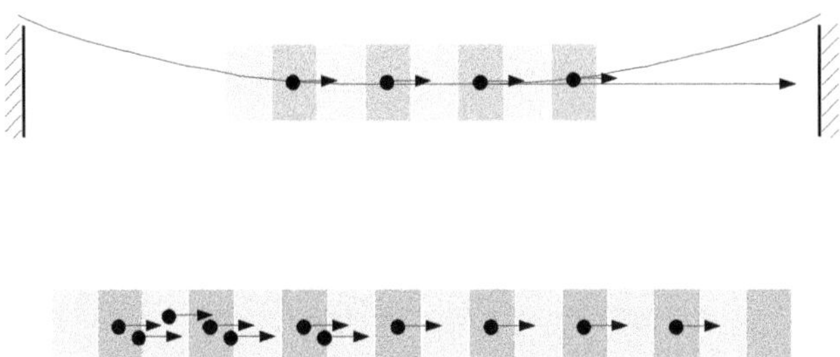

Figure 4.21 *Schematic diagram of a FEL. Above, FEL with optical cavity. Bottom, SASE principle.*

4.4 Electron analyzers

Electron detectors or analyzers play a fundamental role in photoemission, and technical advances in detection have been crucial for the recent advances in photoemission spectroscopy. An electron detector selects electrons in energy and often in emission angle. Being charged particles, electrons are easily manipulated with electrostatic lenses and their detection is relatively simple. However, their strong interaction with matter imposes the vacuum as the working environment (*cf.* section 4.1). The parameters that characterize analyzers are transmission, brightness, resolution and work function. The transmission corresponds to the solid angle of the analyzer acceptance. Brightness is the product of the transmission by the surface of the sample detected by the spectrometer. The relative energy resolution is the ratio between the absolute resolution (obtained for example through the FWHM of a peak) and the mean energy. Sometimes the resolving power ρ is used, which is the inverse of the relative resolution:

$$\rho = \frac{E_0}{\Delta E} \tag{4.29}$$

Angular detection can be obtained in modern two-dimensional detectors (*cf.* section 4.4.2). It is controlled by slits placed at the entrance of the analyzer which determine the emission angles. The sample or the analyzer must therefore be moved in order to explore the entire angular range. Sometimes, high angular acceptance analyzers are used to minimize the diffraction effects of photoelectrons in the measurements of core levels.

The work function is an important quantity because it determines the energy range which can be explored with the same photon energy. Indeed, the detected kinetic energy is $E^d_{kin} = h\nu - E_B - \Phi_a$ where E_B is the binding energy and Φ_d is the

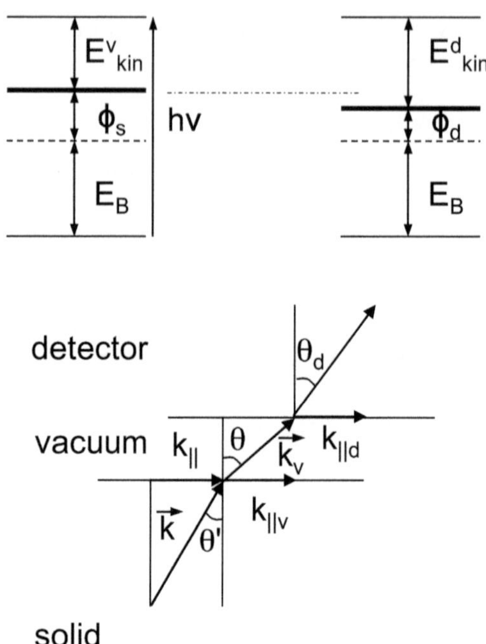

Figure 4.22 *Schematic of the detection process in an analyzer. The sample and the analyzer generally do not have the same work function but the Fermi levels are aligned at equilibrium. Therefore, the electron has a different kinetic energy E^d_{kin} in the analyzer than in the vacuum, at the output of the solid surface E^V_{kin}. The work function difference also introduces a new refraction of the electron at the input of the analyzer, which is analogous to the refraction when passing through a solid surface. Here we see the case where $\Phi_d > \Phi_S$.*

work function of the analyzer. The maximum kinetic energy corresponds to the last occupied states, *i.e.* the Fermi level in a metal which is the reference for binding energies ($E_B = 0$) and is $h\nu - \Phi_d$.

The minimization of the analyzer's work function is therefore critical when working at low photon energy (for example, laser photoemission is typically performed with a photon energy of 6 eV whereas the work function is on the order of 3-5 eV). Finally, it should be noted that it is the work function of the analyzer and not that of the sample the relevant one in the formula of the parallel moment conservation (equation 3.35). Figure 4.22 shows a diagram of the photoemission process taking into account the analyzer. The work function of sample Φ_s may be larger or smaller than that of analyzer Φ_d. Since the sample is in electrical contact with the analyzer (same Fermi levels), an electric field accelerates or slows down the photoelectrons between the sample surface and the analyzer entrance. Assuming a planar symmetry at the scale of the electronic paths, the difference of the two work functions constitutes a one-dimensional potential that the electron must overcome. Its wave vector is

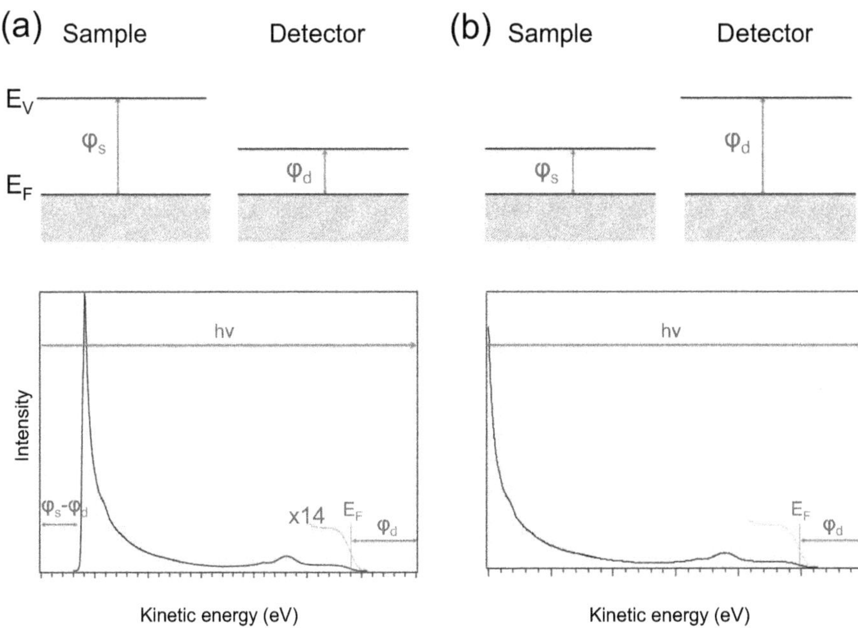

Figure 4.23 *Diagram of the determination of the work functions of the sample and of the analyzer. The electrons that leave the solid with the lowest kinetic energy (zero energy) have just the energy to overcome the sample work function. Their energy in the solid is therefore Φ_S. The difference between $h\nu$ and the kinetic energy of the Fermi level E_F is the analyzer work function Φ_d.*

controlled by a Descartes refraction that conserves the parallel component of the wave vector:

$$\vec{k}_{||} = \vec{k}_{||v} = \vec{k}_{||d} \tag{4.30}$$

$$k_{||} = \sqrt{\frac{2m}{\hbar^2}(h\nu - E_B - \phi_a)} \sin\theta_d \tag{4.31}$$

Knowing the angle and the detector work function allows therefore to determine the $\vec{k}_{||}$ of the initial state in the solid. Figure 4.23 shows how the analyzer work function is experimentally determined. To do this, it is sufficient to measure the kinetic energy of the electrons at the Fermi level of a sample. The difference between $h\nu$ and this energy is the work function of the detector Φ_d. If the analyzer work function is weaker than the work function of the sample Φ_s, it is also possible to determine Φ_s from the threshold energy of the secondary electrons, which corresponds to $\Phi_s - \Phi_d$ (Fig. 4.23a). On the other hand, if $\Phi_s < \Phi_d$, electrons exiting the sample with zero kinetic energy cannot enter the analyzer, unless the sample is biased so that $\Phi_s + eV > \Phi_d$.

After these general notions, we will describe in more detail the most commonly used detectors. These are the hemispherical, toroidal, cylindrical mirror and time-of-flight analyzers. The cylindrical mirror analyzer has been primarily used in the past because it detects the electron energies sequentially and not in parallel. Another disadvantage is its low sensitivity, due to the reduced angular acceptance, as well as the difficulty of adapting to its entrance a set of high-performance delay lenses. In return, it has a high energy resolution and can be used in a wide range of kinetic energies. The hemispherical detector is now very widespread, because the improvement of electronic optics has allowed us to change the entrance aperture by an entrance slit, which allows to detect simultaneously on a two-dimensional detector a whole range of kinetic energies and an entire angular range defined by the entrance slit. The time-of-flight detector has a resolution limited by the time resolution of the electronics and by the time of flight of the electrons. On the other hand, its principle is extremely simple, it simultaneously detects a large angular range and it is perfectly adapted to time-resolved experiments. It requires to illuminate the sample with a pulsed source (laser or synchrotron). Finally, the toroidal detector consists of a much more complicated electronics than that of the time-of-flight detector, but it has a high luminosity and it allows parallel detection of large regions of the reciprocal space. In the following, we will describe each of these detectors.

4.4.1 *Cylindrical Mirror Analyzer*

The Cylindrical Mirror Analyzer (CMA) is a simple and inexpensive detector, consisting of two coaxial concentric cylinders (Fig. 4.24). The outer cylinder is at a negative potential $-V$ and the inner cylinder is earthed. The sample is placed in the axis of the two cylinders. The photoelectrons pass through an opening in the inner cylinder, and are then deflected by the electric field depending on their initial kinetic energy. To reach the detector, placed again in the axis of two cylinders, they must cross a second opening in the inner cylinder. Only those with the desired energy E travel through the entire detector and are detected by the electron multiplier. The relation between E and the applied potential V is [29]:

$$\frac{E}{eV} = \frac{K}{\ln (R_2/R_1)} \tag{4.32}$$

with R_1 and R_2 the radii of the inner and outer cylinders respectively, and K a constant which is 1.3099 in the commonly used geometry [30]. A spectrum can therefore be obtained by systematically varying V.

The resolution of a CMA when the transmission angle $\delta\alpha$ is small is [31]

$$\frac{\Delta E}{E_0} = 2.75(\delta\alpha)^3 \tag{4.33}$$

Equation 4.33 shows that the energy resolution $\Delta E/E$ is constant: it is generally below 1%. Moreover, the transmission is high and this detector is generally used for Auger spectroscopy.

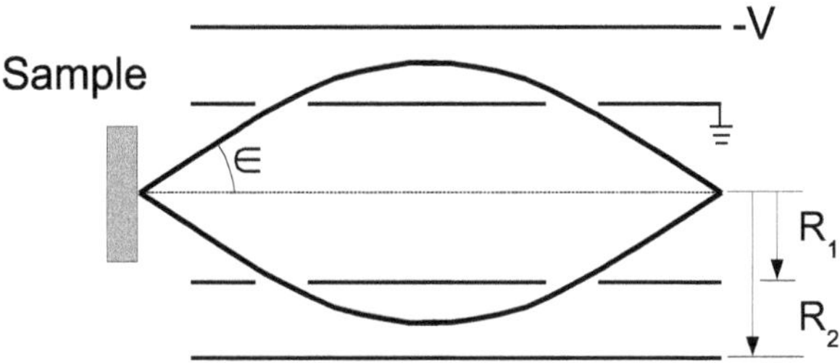

Figure 4.24 *Diagram of a cylindrical analyzer.*

4.4.2 *Hemispherical Analyzer*

The Concentric Hemispherical Analyzer (CHA) is the most widely used analyzer for photoemission, although it is more complicated than others. Its main advantage is the possibility of obtaining a good resolution compatible with multi-detection. The first hemispherical detectors counted all the electrons in their entire angular acceptance. Today, two-dimensional analyzers that detect a whole range of angles in a given direction are widespread, with a lens system focusing on different channels of a multichannel detector (Fig. 4.25). This parallel counting appeared around 1995 [32] and was a breakthrough for photoemission due to the high energy (~3 meV) and angular (~0.1°) resolutions. Their energy resolution is less than that of the time-of-flight detector (*cf.* subsection 4.4.5), but it can be kept constant over a whole spectrum.

These hemispherical analyzers consist of input lenses, hemispheres and electron multipliers. Electrostatic lenses are the first element of an analyzer. Their role is to select the solid angle of detection, to focus the electrons at the entrance slit and to control the energy of the electrons arriving at the entrance of the hemispherical detector. The lens system for UPS is simpler than that for the XPS, because at high energy, the potentials to be applied are more important, and the studied surface varies with the energy. These problems can be solved with a system of four lenses to introduce a delay potential and to form an amplified and inverted image in the equatorial plane of the analyzer. Today, there are usually several delay lenses to operate the analyzer in two modes: the Constant Retarding Ratio (CRR) and the Fixed Analyzer Transmission (FAT) modes. The CRR mode preserves the $\Delta E/E$ factor, which is preferred for Auger spectroscopy measurements. The FAT mode is preferred for XPS and stabilizes the energy of the electrons after the entrance lenses to the pass energy E_0. The pass energy controls the energy resolution, together with the size of the entrance slits. In a first approximation, the resolution is proportionate to the energy of the electrons

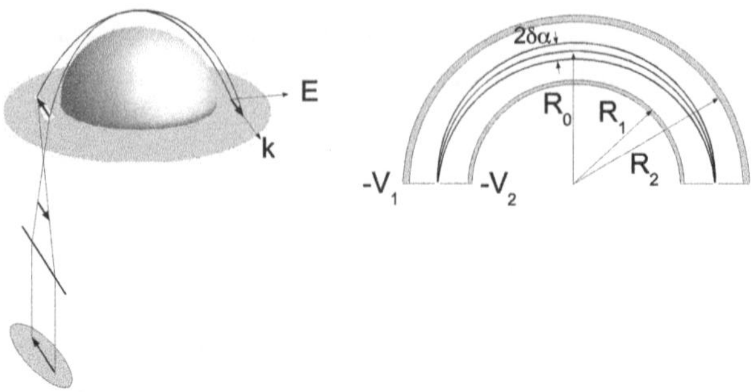

Figure 4.25 *Hemispherical detector. A lens system before the hemispheres focuses electrons at the entrance slit and settles the energy of the photoelectrons to the pass energy. The electrostatic field between the two hemispheres defines a circular trajectory for the electrons up to the detector.*

in the analyzer ($\Delta E = KE$). Since the transmission energy is constant, ΔE is constant. The lower the pass energy, the better the resolution, but this improvement in resolution is concomitant with a decrease in the measured signal.

After the delay grids, which focus the electrons at the hemisphere entrance, electrons go through the entrance slit of width w_1. An electron beam of angle α in the analyzer plane and of angle β in the orthogonal direction is transmitted. The two hemispheres, with radii R_1 and R_2, are subjected to a potential difference U_k. This potential is chosen so that only electrons having an energy equal to the pass energy arrive at the electron counter. The pass energy depends on the potential between the two hemispheres:

$$E_0 = \frac{U_k}{\left(\frac{R_2}{R_1} - \frac{R_1}{R_2}\right)} \tag{4.34}$$

which shows that a precise control of the voltage is fundamental to obtain a good energy resolution.

The hemispheres focus the beam on the w_2 width slit. Because of the spherical symmetry of the detector, electrons that differ only from the β angle are all focused in the same way, allowing to make longer entrance slits in the direction orthogonal to the plane of the analyzer. The compromise for simultaneously having good energy resolution and good intensity consists in choosing entrance and exit slits of the same size $w_1 = w_2 = w$. The energy resolution is then approximately [33]:

$$\frac{\Delta E}{E_0} = \frac{w}{2\overline{r}} + \frac{\alpha_m^2}{4} \tag{4.35}$$

where α_m is the maximum acceptance angle to the analyzer, $\overline{r}$ is the average radius between the two hemispheres and E_0 is the pass energy. The resolution in energy depends thus on the pass energy and the entrance slits of the detector. The dependence of the resolution as a function of the angle is proportional to the square of the angular tolerance, unlike the CMA. For typical values, the energy resolution is typically of the order of 1%.

After passing through the exit slits, the electrons reach the electron amplifier/counter. The signal can be amplified with dynodes, which are multipliers of discrete steps, or with a so-called 'channeltron' which consists of a continuous dynode subjected to an electric field. In two-dimensional analyzers, there are sets of 'channeltrons' called 'channel-plates'. The signal is thus amplified, with a gain of the order of 10^7. It is transformed into a current that excites a fluorescent screen or a CCD camera. This camera gives images in which the axes depend on the operation mode of the detector. Indeed, they can operate in spatial resolution mode or in angular resolution mode. In spatial mode, the electrons coming from the same point of the sample arrive at the same point of the detector, allowing to spatially map the sample. The two dimensions of the detector correspond here to the kinetic energy and the position. In angular mode, the electrons with the same emission angle arrive at the same channel of the detector. This time, the axes of the two-dimensional detector are the kinetic energy and the emission angle. This representation gives a very direct picture of the band structure of the systems.

4.4.3 *Toroidal detectors*

Toroidal detectors appeared in the 1980s [34–36]. The main interest of these analyzer type is their high luminosity, favorable for dynamic studies or for the detection of electrons with high kinetic energies, whose photoionization cross-section is low. The toroidal detector for electron spectroscopy was developed by Riley and Leckey [35]. It is based on the use of toroidal electrodes to analyze the electron energies. Because of its geometry, it measures in parallel the kinetic energy of the electrons at different polar angles in a fixed azimuthal direction, in a faster measurement with respect to a single-channel detector. The different azimuthal angles are explored by rotating the sample around its normal.

The operation of this analyzer is therefore similar to that of the hemispherical analyzer. At the entrance of the detector there is a set of electrostatic lenses. Their role is to focus electrons from the sample on the entrance slit of the analyzer, as well as to accelerate or decelerate the electrons to the pass energy. The emission angle and the kinetic energy of the electrons are simultaneously analyzed by the toroidal lenses.

As in the hemispherical analyzer, electrons whose energy differs from the pass energy do not reach the detector. The latter can be two-dimensional and analyze simultaneously an entire energy range, further reducing the detection time. This detector can also be improved with the introduction of a second set of toroidal lenses after the first one, which focuses electrons in a plane instead of focusing them in a cone (Fig. 4.26 (a)). This dual toroidal analyzer was developed in the late 1990s [38]. The result of a measurement with this type of analyzer is shown in 4.26 (b) for Cu_3Au.

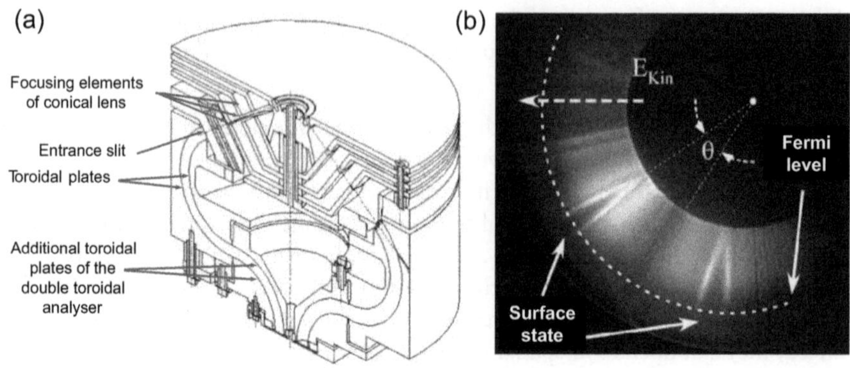

Figure 4.26 *Toroidal detector. (a) Operation diagram. The input lenses fix the pass energy through the toroidal lens. Reprinted from C. Miron et al., Rev. Sci. Instrum. 68, 3728 (1997), with the permission of AIP Publishing. (b) Image formed on the detector during the emission of electrons close to the Fermi level from a Cu_3Au sample. The parabolic surface state is observed on two different Brillouin zones (according to [37]). Reprinted from J. Electron. Spectrosc. Rel. Phenom. vol. 144, L. Brockman, A. Tadich, E. Huwald, J. Riley, R. Leckey, T. Seyller, K. Emtsev, L. Ley, "First results from a second generation toroidal electron spectrometer", ©2005, with permission from Elsevier.*

4.4.4 *High pressure electron detection*

High pressure photoemission was developed to study systems under operating conditions. It is therefore necessary to succeed in detecting electrons at high pressures approaching atmospheric pressure. Since photons and especially electrons are strongly absorbed by gases, it is necessary to minimize the paths of electrons and photons under high pressure conditions. A tube closed by a window of about 100 nm, often of Si_3N_4, leads the photons to the vicinity of the sample (Fig. 4.27). The electrons must be collected before colliding with the gas molecules within the chamber. The detector entrance must therefore be at a distance comparable to the mean free path of the electrons at the working pressure. This mean free path depends on the gas, the pressure and the energy of the electron. For 400 eV with a pressure of 1 mbar, the mean free path is 4 mm [39]. This type of analyzer allows the study of chemical reactions in the vapor phase.

4.4.5 *Time-of-flight detector*

The time-of-flight (TOF) analyzer is based on a very simple principle. The time to travel a given distance depends on the speed and therefore on the kinetic energy of the electrons. This type of detector is well-suited to experiments resolved in time. The optics are less complicated than in the hemispherical analyzer and require fewer

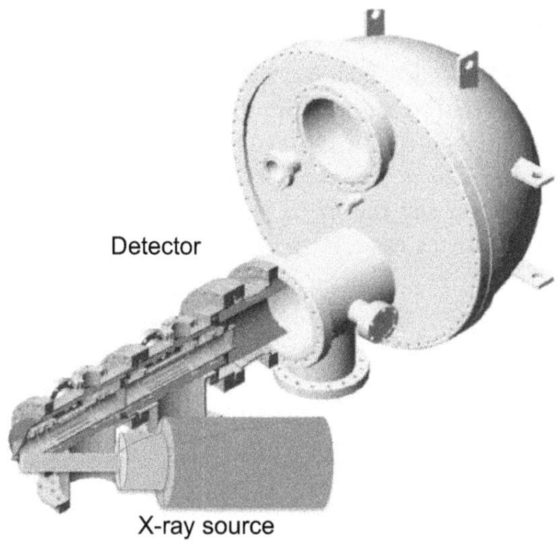

Figure 4.27 *Experimental configuration showing the proximity of the X-ray source and the detector entrance to the sample. The detector for high-pressure photoemission has a differential pumping between the sample and the analyzer. At the input of the detector, electrostatic lenses are placed to capture electrons of diverging angles and increase the detected signal (from [39]). Reprinted from Surf. Sci. Rep. Vol. 63, M. Salmeron, R. Schl., "Ambient pressure photoelectron spectroscopy: A new tool for surface science and nanotechnology", page 169, ©2008, with permission from Elsevier.*

corrections of aberrations. However, its resolution at high kinetic energy is limited and of course it needs time-resolved sources such as a pulsed laser or a synchrotron operating with a single electron packet or with a low number of packets.

Flight time detectors collect electrons in a free-field region (*drift region*) where electrons have only kinetic energy. Electrons travel a time τ along the tube length L from the entrance given by:

$$\tau = \frac{L}{\sqrt{2E_{kin}/m}} \tag{4.36}$$

The energies of the electrons are then determined according to their moment of arrival at the detector. The energy resolution is

$$\frac{dE}{d\tau} = -\frac{2}{L}\sqrt{\frac{2E_{kin}^3}{m}} \tag{4.37}$$

and it can be several orders of magnitude better than that of a hemispherical detector, depending on the length L.

4.4.6 *Spin analyzers*

We have seen that conventional analyzers allow to obtain the energy and the angle of the electrons, *i.e.*, the band structure. With the detection of the spin, the initial state of the electrons in the solid is fully characterized. However, the magnetic moment of the electron $e\hbar/mc$ can never be observed directly because of the uncertainty principle on the position and the speed [40]. Let us recall Mott's argument. The magnetic field created by the electron spin decreases in R^3:

$$B_s \sim \frac{\mu_0}{4\pi}\frac{M_s}{R^3} \tag{4.38}$$

where M_S is the electron spin moment ($M_s = \frac{q\hbar}{2m}$). The magnetic field due to the charge of the electron in motion is:

$$B_q = \frac{\mu_0}{4\pi}\frac{qv}{R^2} \tag{4.39}$$

To have access to the magnetic moment associated with the spin, B_s must be greater than ΔB_q due to the uncertainty on the electron velocity:

$$\Delta B_q = \frac{\mu_0}{4\pi}\frac{q\Delta v}{R^2}$$

The condition $B_s > \Delta B_q$ leads to:

$$\frac{M_s}{R^3} > \frac{q\Delta v}{R^2} \quad \text{d'où} \quad \frac{\hbar}{2} > R\, m\Delta v$$

The uncertainty relationship for position and speed ($\Delta R\, m\Delta v > \hbar/2$) shows that if $B_s > \Delta B_q$,

$$\Delta R > R$$

where R is typically the macroscopic distance between the electron and the detector. This reasoning demonstrates that it is impossible to have direct access to the magnetic spin moment of the electron. Similarly, the Pauli uncertainty principle prevents the Stern-Gerlach experiments from measuring the spin of the electrons [40]. For this reason Mott was interested in an alternative method based on the spin dependence of the electron scattering by a potential (today known as Mott diffusion). Mott diffusion is based on the spin dependent diffusion of relativistic electrons by heavy atoms for which the spin-orbit coupling is important. This diffusion can be explained in a formalism beyond the Rutherford diffusion (a non-relativistic Coulomb diffusion which is only sensitive to the charge) and taking into account the relativistic effects and especially the magnetic effects related to the electron spin.

The mechanism of the Mott diffusion can be qualitatively understood by considering the diffusion of a relativistic electron by an atom of charge Zq. In its frame of reference, the electron detects a nucleus moving at high speed and its associated

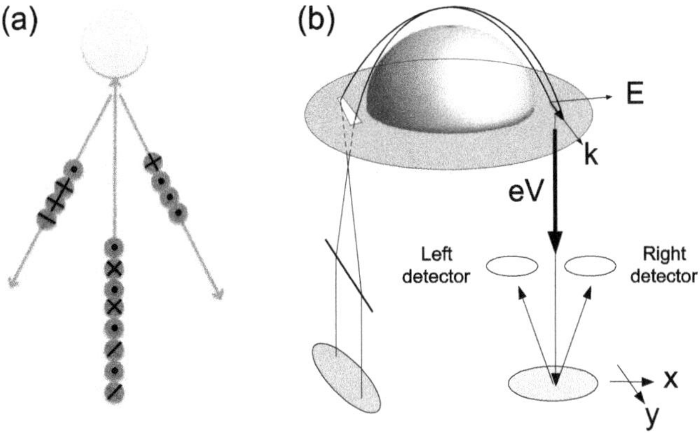

Figure 4.28 *Mott detector. (a) Principle of operation. An electron beam impinges a target of heavy atoms after being accelerated. Electron back-scattering is asymmetric for up and down spins. (b) Experimental realization of a Mott detector. The Mott detector is placed after a conventional analyzer that filters the energy of the incident electrons. The electrons are afterwards accelerated towards the heavy-atom target and measured on two detectors so that the polarization on the axis perpendicular to the detectors can be determined.*

current creates a magnetic field. The gradient of this field induces a force on the moment of the electron spin:

$$\vec{F} = (\vec{M}_S \cdot \vec{\nabla})\vec{B}$$

The force changes sign when the electron spin is reversed, leading to a spin-dependent asymmetry in the effective scattering section (Fig. 4.28a). This mechanism is exploited in the Mott detectors allowing to measure the photoelectron spin. A Mott detector, installed behind a conventional photoemission energy analyzer (Fig. 4.28(b)) usually consists of a very narrow sheet of gold (which is a heavy and chemically inert element) and two detectors (right - R and left - L) located symmetrically with respect to the normal. Electrons after the conventional analyzer are accelerated in the normal direction to energies of the order of tens of keV and are back-scattered towards the two detectors (in the direction Ox). If an unpolarized electron beam is sent, the same number of electrons will be counted in each detector, although electrons with up (down) spin will be mainly diffused towards the left or right detector as illustrated in Fig. 4.28(a). Moreover, if the incident beam is polarized, the number of detected electrons will be different in the two detectors. The asymmetry along Oy (A_y) of the number of electrons N_G and N_D detected in the detectors is defined by:

$$A_y = \frac{N_{Lx} - N_{Rx}}{N_{Lx} + N_{Rx}} \tag{4.40}$$

This asymmetry depends on the scattering angle and an asymmetry maximum is obtained for angles of $\pm 120°$ at relatively low energies. Beyond 200 keV, this angle increases. This parameter allows to trace the polarization of the electron beam (only the polarization component normal to the diffusion plane can be detected), but it is necessary to know the differential cross-section of scattering. It can be calculated theoretically for a gold atom:

$$A_y = S(\theta)P_y \tag{4.41}$$

where $S(\theta)$ is called the Sherman function and P_y is the polarization of the incident beam along Oy. The Sherman function depends on the Z of the detector atoms and on the energy. For $\theta = 120°$, it is of the order of 0.4, although experimentally it is weaker because of additional mechanisms such as multiple and inelastic scattering in the gold target [41, 42]. An effective Sherman factor S_{eff} is usually associated with a given experimental set-up (typically between 0.1 and 0.2) which must be perfectly known in order to determine the spin polarization of electrons according to:

$$P_y = A_y / S_{eff} \tag{4.42}$$

and if N_{tot} is the total number of detected electrons, the number of electrons with each spin along to x is:

$$N_\uparrow = \frac{N_{tot}}{2}(1 + P_y) \qquad N_\downarrow = \frac{N_{tot}}{2}(1 - P_y) \tag{4.43}$$

with a statistical error for the y polarization:

$$\Delta P_y = 1/\sqrt{\epsilon S_{eff}^2} \tag{4.44}$$

being N_0 the total number of electrons reaching the detector. The diffusion efficiency is $\epsilon \equiv (N_G + N_D)/N_0$ and for a Mott detector, this efficiency is $\sim 10^{-2}$. The overall efficiency Q of the Mott detector is [43]:

$$Q = \frac{N_L + N_R}{N_0} S^2 \tag{4.45}$$

which is rather small, of the order of $10^{-3} - 10^{-4}$, which imposes large acquisition times. Even if the efficiency of the Mott detectors is quite low, with the high flux of synchrotrons it is possible to carry out photoemission experiments resolved in angle *and* in spin (for example, the Cophee experimental setup of the Swiss Light Source proposes 70 meV of resolution in energy and 0.5° of angular resolution).

To improve the efficiency of the spin detection, it is necessary to abandon the Mott diffusion and move to a new detection principle. Some detectors exploit the diffraction of very slow electrons polarized by a magnetic surface. This type of detector was introduced for the first time in 1989 [44] and it is based on the difference in reflectivity of the two spins by a ferromagnetic surface. It allows to increase the detection efficiency by an order of magnitude and eventually by 4 if multichannel detectors are used for the detection [45], which makes it possible to carry out measurements with resolution better than 10 meV and 0.2° of angular resolution [46].

The magnetic surface (often Fe (001)) must here be prepared according to surface physics standards with a passivation to increase the optimal operation time of the detector[47].

4.4.7 Photoemission microscope

Photoemission microscopy covers the place between optical microscopy and local microscopies such as the tunneling microscope. The concept of performing microscopy with photoelectrons was born in the 1930s, with the development of electrostatic lenses [48–51]. Nevertheless, the true rise of this microscopy came after the development of low-energy electron microscopy (LEEM) between the decades of 1980 and 1990 and the construction of third-generation synchrotrons. Although the scanning microscopy approach with a focused photon beam has been developed, parallel imaging has been so successful that it is virtually synonymous to photoemission microscopy (PEEM - *PhotoEmission Electron Microscopy*) (Fig. 4.29).

The principle of PEEM is simple, even if the technical difficulties to achieve it are significant. A large region of the sample is illuminated by a beam of photons. The photoemitted electron beam coming from a photon-illuminated region is then enlarged by lenses to form a magnified image. In the early stages of the technique, the electron beam was expanded by a magnetic divergent lens [52]. Nowadays the lenses are electrostatic, accelerating electrons to high electrical potentials. Thanks to these lenses, the PEEM reaches its present spatial resolution of approximately 20 nm, with a theoretical limit of 2 nm. The absence of a focused photon beam prevents the surface damage, and the intensity of the current sources allows to obtain images in a reasonable time.

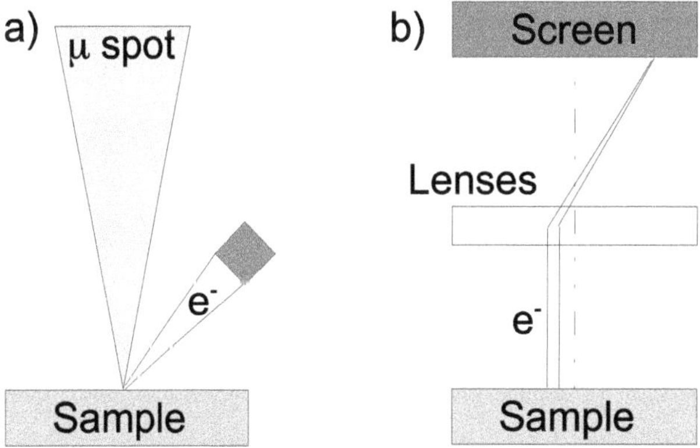

Figure 4.29 *Experimental configurations for photoemission microscopes (a) scanning photoemission microscope (b) PEEM.*

Depending on the type of the detected electrons, the contrast of the image reveals different information. Secondary electrons, for example, are used to obtain the surface topography. These electrons reflect the bulk properties of the material and they are also sensitive to the work function and to the screening effects due to the different emission angles ... Therefore different surface structures have a distinct contrast with secondary electrons. A chemical contrast can also be obtained from the core levels when the microscope is equipped with an analyzer capable of detecting different kinetic energies. In microscopes without an electron analyzer, the chemical contrast can be obtained from the absorption edges, as they can be measured just by varying the energy of the incoming photons. Moreover, the study of the absorption edges with circular polarizations allows to obtain a magnetic contrast through the dichroic signal.

Experimentally, the sample is placed on a piezoelectric motor. There is one objective lens with a variable aperture that collects photoelectrons into the PEEM column. A strong electrostatic field on the sample accelerates electrons up to 10-15 keV to decrease the chromatic aberration of the objective lens and increase the resolution. Another consequence of this strong field is that the image is formed from a very large solid angle. The other elements of the column are the octupolar stigmator, the projection lenses, and adjustable pinholes at the first image plane to define the field of view. Finally, electrons are sent to a channel-plate detector and a fluorescent screen. The PEEM system has sometimes an associated LEEM to take advantage of the analogy of the electron detection, in order to make the instrument even more versatile. Another fairly frequent change is to associate it with time-of-flight sensors to obtain dynamic information.

Bibliography

[1] D. Turner and M. Al-Joboury, J. Chem. Phys. **37**, 3007 (1962).
[2] J. Asmussen, J. Vac. Sci. Technol. A 7, 883 (1989).
[3] P. Baltzer, L. Karlsson, M. Lundqvist, and B. Wannberg, Rev. Sci. Instr. **8**, 2179 (1993).
[4] J. Samson, Rev. Sci. Instr. **40**, 1174 (1969).
[5] J. Schweppe, R. Deslattes, T. Mooney, and C. Powell, J. Electron Spectrosc. **67**, 463 (1994).
[6] C. Powell, Surf. Int. Analysis **25**, 777 (1997).
[7] J. Moulder, W. Stickle, P. Sobol, and K. Bomben, *Handbook of X-ray Photoelectron Spectroscopy* (Perkin Elmer Corporation, Minnesota, 1992).
[8] J. Faure *et al.*, Review of Scientific Instruments **83**, 043109 (2012).
[9] R. Haight and D. Peale, Rev. Sci. Instrum. **65**, 1853 (1994).
[10] P. B. Corkum, Phys. Rev. Lett. **71**, 1994 (1993).
[11] G. Dakovski, Y. Li, T. Durakiewicz, and G. Rodriguez, Rev. Sci. Instrum. **81**, 073108 (2010).
[12] B. Christensen, M. Raarup, and P. Balling, Nuclear Instruments and Methods in Physics Research Section A: Accelerators, Spectrometers, Detectors and Associated Equipment **615**, 114 (2010).

[13] K. Oguri *et al.*, Japanese Journal of Applied Physics **51**, 072401 (2012).
[14] F. Elder, A. Gurewitsch, R. Langmuir, and H. Pollock, Phys. Rev. **71**, 829 (1947).
[15] D. Ivanenko and J. Pomeranchuk, Phys. Rev. **65**, 343 (1944).
[16] D. Tomboulian and P. Hartmann, Phys. Rev. **102**, 1423 (1956).
[17] G. Margaritondo, *Introduction to Synchrotron Radiation* (Oxford University Press, New York, 1988).
[18] C. Chauvet *et al.*, J. Synchrotron Rad. **18**, 761 (2011).
[19] P. Risterucci *et al.*, J. Synchrotron Rad. **570**, 19 (2012).
[20] H. Motz, W. Thon, and R. N. Whitehurst, J. Appl. Phys. **24**, 826 (1953).
[21] V. L. Granatstein *et al.*, Appl. Phys. Lett. **30**, 384 (1977).
[22] J. M. J. Madey, J. Appl. Phys. **42**, 1906 (1971).
[23] D. A. G. Deacon *et al.*, Phys. Rev. Lett. **38**, 892 (1977).
[24] G. Dattoli and A. Renieri, *Experimental and theoretical aspects of Free-Electron Lasers, Laser Handbook* (North Holland, Amsterdam, 1984), Vol. 4.
[25] J. C. Marshall, *Free Electron Lasers* (Mac Millan Publishing Company, New York, 1985).
[26] P.Luchini and H. Motz, *Undulators and Free Electron Lasers* (Claredon Press, Oxford, 1990).
[27] *Free-Electron Lasers*, edited by C. Brau (Academic Press, Oxford, 1990).
[28] P. O'Shea and H. Freund, Science **292**, 1853 (2001).
[29] *Electron Scattering and Related Spectroscopies*, edited by M. de Crescenzi and M. Piancastelli (World Scientific, Singapore, 1996).
[30] V. Zashkvara, M. Korsunskii, and O. Kosmachev, Sov. Phys. Tech. Phys. **11**, 96 (1966).
[31] H. Sar-El, Rev. Sci. Instrum. **41**, 561 (1970).
[32] N. Martensson *et al.*, J. Electron Spectrosc. Relat. Phenom. **70**, 117 (1994).
[33] R. Matzdorf, Surf. Sci. Rep. **30**, 153 (1998).
[34] H. A. Englehardt, W. Back, and D. Menzel, Review of Scientific Instruments **52**, 835 (1981).
[35] F. Toffoletto, R. C. G. Leckey, and J. D. Riley, Nucl. Instrum. Methods B **12**, 282 (1985).
[36] G. J. A. Hellings *et al.*, Surf. Sci. **162**, 913 (1985).
[37] L. Broekman *et al.*, Journal of Electron Spectroscopy and Related Phenomena **144**, 1001 (2005).
[38] C. Miron, M. Simon, N. Leclercq, and P. Morin, Review of Scientific Instruments **68**, 3728 (1997).
[39] M. Salmeron and R. Schlögl, Surf. Sci. Rep. **63**, 169 (2008).
[40] N. Mott, Proc. Roy. Soc. A (London) **124**, 425 (1929).
[41] S. Qiao *et al.*, Rev. Sci. Instrum. **68**, 4390 (1997).
[42] S. Qiao and A. Kakizaki, Rev. Sci. Instrum. **68**, 4017 (1997).
[43] J. Kessler, *Polarized electrons* (Springer, Berlin, 1985).
[44] D. Tillman, R. Thiel, and E. Kisker, Z. Phys. B **77**, 1 (1989).
[45] M. Kolbe *et al.*, Phys. Rev. Lett. **107**, 207601 (2011).
[46] T. Okuda *et al.*, Rev. Sci. Instrum. **82**, 103302 (2011).
[47] R. Bertacco *et al.*, Rev. Sci. Instrum. **73**, 3867 (2002).
[48] E. Brüche and H. Johanson, Physik. Zeitschr. **33**, 898 (1932).
[49] E. Brüche, Z. Phys. **86**, 448 (1933).
[50] J. Pohl, Z. Tech. Phys. **15**, 579 (1934).
[51] H. Mahl and J. Pohl, Z. Tech. Phys. **16**, 219 (1935).
[52] G. Beamson, H. Porter, and D. Turner, Nature **290**, 556 (1981).

Part II

Applications

Transitions from localized states

As we saw in the introductory chapter, core levels can be studied by photoemission with sufficiently energetic photons (generally in the X-ray domain). Let us recall briefly the characteristics of core level electrons. They are very localized so their wave functions and their energies are very close to those of an isolated atom. This property is the origin of the core level spectroscopy since the binding energies constitute a signature of the chemical nature of the emitter. Indeed, two different atoms, even neighbors in the periodic table, have very different atomic levels. For example, the 3s levels of potassium and calcium respectively have binding energies of 35 eV and 44 eV. This sensitivity lies at the origin of core level spectroscopies, such as X-ray Photoelectron Spectroscopy (XPS) or Auger spectroscopy. More elaborated information can also be obtained thanks to the intensity and the energy resolution of the synchrotron radiation. In this way it is possible to observe the small energy variations of the same element placed in different environments (chemical shift). Indeed, even if the energy of a photoelectron from a core level depends essentially on the potential of the emitter, it can be modified slightly by the potential due to neighboring atoms. The precise determination of this energy allows us to distinguish, for example, the different oxidation states of the same atom. Core level spectroscopy can also follow ongoing chemical reactions. An example is given in figure 5.1(a) viz., a study of the dehydrogenation of methyl on a platinum surface. The image presents - in false colors - the variation of the photoemission intensity of the 1s carbon peak as a function of the temperature. A single peak at 282.6 eV is observed up to 220 K corresponding to the CH_3 group, whereas above 250 K, the peak decreases in favor of a second peak at 283.6 eV, corresponding to the CH group. This behavior is a signature of methyl dehydrogenation [1]. Other interesting applications result from

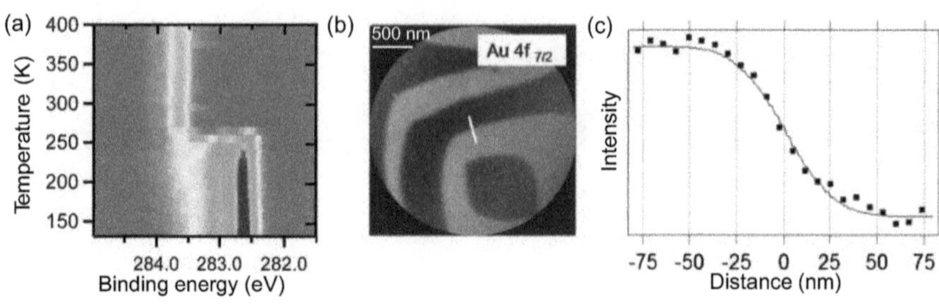

Figure 5.1 *(a) Study of the dehydrogenation of a fraction of monolayer of methyl on a (111) oriented platinum surface. Intensity of the carbon spectrum 1s as a function of temperature (from [1]). Reprinted with permission from C. Papp et al., J. Phys. Chem. C 111, 2177 ©2007 American Chemical Society. (b) Photoemission microscopy image of an oxidized Au-Pd surface (2.7 μm $\times$2.7 μm) detecting the Au 4f level at $h\nu = 200$ eV. (c) Profile on the image. This experiment has 50 nm of spatial resolution, estimated from fitting the experimental data to a step convoluted with a Gaussian (Reprinted from "Photoemission electron microscopy with chemical sensitivity: SPELEEM methods and applications", A. Locatelli* et al., *Surface & Interface Analysis, 38, 1554). ©2006 John Wiley & Sons, Ltd.*

combining core level spectroscopy and photoemission microscopy, giving rise to chemical maps with a good spatial resolution. Figure 5.1(b) shows a photoelectron microscopy image of an oxidized Au-Pd surface. The gray levels are proportional to the photoemission signal at the energy of the $4f$ states of Au. There are dark and light regions corresponding respectively to non-oxidized (Au-rich) and oxidized (Au-poor) zones. The acquisition time of this type of images is a few seconds and their resolution in energy can reach 300 meV, which allows us to perform precise spectroscopic analysis with spatial resolution. Other examples of the versatility of core-level spectroscopy will be presented in this chapter.

5.1 Spectral shape of core level transitions

As we have seen, electrons detected in a photoemission analyzer can originate via a transition induced by the photoelectric effect but can also be due to desexcitation processes (Auger electrons). In the latter case, their energy does not depend on the energy of the incoming photons. In the following section we will explain the shape of different spectral structures appearing in a photoemission spectrum.

5.1.1 *Auger transitions*

In photoemission spectra, features with a kinetic energy independent of the photon energy correspond to Auger transitions. The mechanism of Auger transitions was first explained by Pierre Auger in 1925 [2, 3]. The emission of an Auger electron corresponds to the atomic relaxation after an ionization process with photons or high-energy electrons. Indeed, a core hole is an unstable excited state. An electron of higher energy will lose its energy to fill the core hole. The energy of the relaxation can lead to the emission of an electron (Auger process) or a photon (fluorescence). The total balance of the Auger process is the creation of two holes and the emission of two electrons (the primary plus the Auger electron). The energies of the Auger electrons are directly related to the energy differences between the atomic states involved in the atomic transitions and are thus characteristic of each element. As

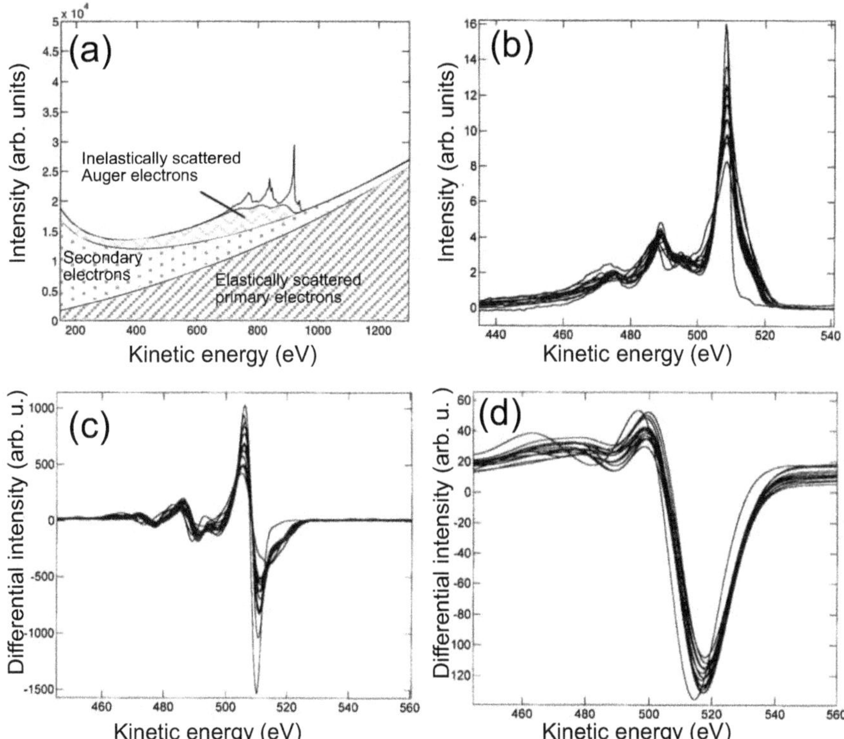

Figure 5.2 *(a) Auger spectrum of copper after excitation with an electron gun. The signal is superimposed on the continuous background due to both the inelastic electrons coming from the transition and those coming from primary electrons. When the background is too intense, its subtraction helps to analyze the Auger peak. (b) I(E) Auger spectra of oxygen for a series of oxides. (c) dI(E)/dE Auger spectra obtained from the derivative of the signal in (b) as a function of energy. (d) Differentiated and widened spectra with a Gaussian of 20 eV. Reprinted from "Quantitative AES. V. Practical analysis of intensities with detailed examples of metals and their oxides". M. P. Seah et al. Surface & Interface Analysis, 26, 701. ©1998 John Wiley & Sons, Ltd.*

these energies are known and tabulated, the composition of the materials can be obtained, with a sensitivity of 1%. Auger transitions are also sensitive to chemical shifts in the elements. The indexing nomenclature of these transitions has been described in section 3.4.1. Auger transitions are often used to determine qualitatively the chemical composition of a material. It is also possible to obtain a quantitative analysis by comparing the relative intensity between the different transitions, often corrected by factors associated to the experimental sensitivity of each transition. In the Auger spectra compilations, the derivative of the intensity is shown on the spectra to eliminate the background, very high when electrons are used to excite the sample as many backscattered electrons are created (Fig. 5.2(a)). The Auger intensity is generally defined as the peak-to-peak height of the differentiated spectrum. This quantification of the peak intensity has some disadvantages. Indeed, the spectral shape depends on the experimental resolution, the disorientation of the sample with respect to the detector, the chemical state and the differentiation algorithm [4]. Figure 5.2(b) shows the Auger spectra of oxygen for a series of oxides (MgO, Al_2O_3, SiO_2, Co_3O_4...), all of which have been normalized to have the same integral. The intensity of the most intense peak can vary almost by a factor of 2 from one oxide to another. The method of measuring the peak intensity is possibly inaccurate. Moreover, the peak-to-peak height in the differentiated spectrum (Fig. 5.2(c)) gives also a factor 3 between the different systems. The reason of this disagreement is the variation in width of the peaks. For this reason, some studies prefer to measure the area under the peak of the transition. The main problem here is the determination of the background to be subtracted in order to obtain the intrinsic intensity of the transition, as well as the range of the energy integration, so there is also some uncertainty in this method [5]. A more elaborate procedure - which nevertheless remains fairly simple - has been developed to overcome these problems. It consists of widening the spectrum with a Gaussian to deliberately degrade the resolution and eliminate the width dependence of the differentiated spectrum. In some systems, the peak-to-peak height of the differential mode gives errors of only 5% [6]. Once the intensity associated with the Auger transition has been determined, it is possible to carry out quantitative analysis such as those that will be presented hereafter.

5.1.2 *Photoemission transitions*

We have already seen that X-ray photoemission spectroscopy provides information on the chemical analysis of materials. Further information can be obtained by analyzing the energy and the shape of the photoemission peak. The line shape of photoemission core levels is much simpler than Auger transitions because the final state has only one hole, whereas there are at least two holes in the final state of an Auger transition. This facilitates quantitative studies, often involving the analysis of the number of components in the core level, their shape and their binding energy. As a first approximation, the binding energy of each component comes from atoms that can be differentiated by their electronic structure or by their structural environment. The chemical shifts for bulk atoms can reach several eV, although they are generally

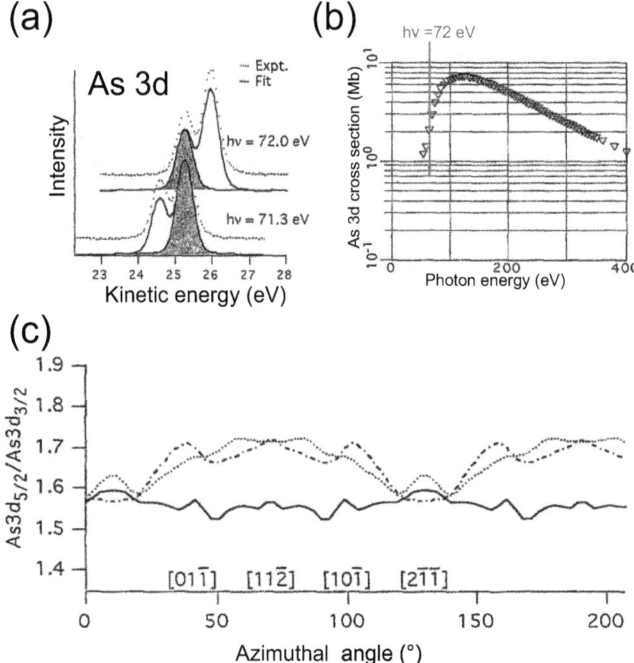

Figure 5.3 *As/Si(111). (a) As 3d core level measured at $h\nu = 72.0$ and 71.3 eV (for emission angles at 45° from the surface normal (the [11$\bar{2}$] direction). (b) Cross section of the As 3d core level as a function of the photon energy. (c) Intensity ratios of the $3d_{5/2}$ and $3d_{3/2}$ lines of As for $h\nu = 72.0$ eV (dotted line) and 71.3 eV (dashed line) as a function of the azimuth angle (from [7]). Reprinted from Surf. Sci. vol. 352-354, E. L. Bullock, R. Gunnella, C. R. Natoli, H. W. Yeom, S. Kono, L. Patthey, R. I. G. Uhrberg, L. S. O. Johansson, "Angle dependence of the spin-orbit branching ratio", page 352, ©1996, with permission from Elsevier.*

smaller for surface atoms. In some cases, the energy and the shape of photoemission lines may be affected by the so-called 'final state' effects, resulting from the dynamics of the relaxation processes at the core hole.

In essence, the sometimes complex shape of a photoemission line has multiple origins. There may be several components associated with the different inequivalent sites in the solid but also with an intrinsically complex shape related to the coupling between the core hole created by the photoemission process and the other electrons of the system. Computation modelling is often necessary to reproduce the profile of the photoemission lines. As we have already introduced the theoretical concepts for describing the photoemission lines in Chapter 3, we will focus here on the decomposition of a core level into several components. We have seen that the simplest form of a photoemission line is a Lorentzian:

$$L(E) = \frac{1}{1 + [\frac{E - E_0}{\Gamma}]^2} \tag{5.1}$$

centered at E_0 and with width Γ. This simple form, which frequently appears in semiconductors and insulators, is due to the finite lifetime of the core hole of the final state, as $\Gamma\tau = \hbar$. On the other hand, in the case of metals, the hole screening by the conduction electrons leads to a more complex line form described by the Doniach-Šunjić:

$$I_{DS}(E) = \Gamma \frac{\cos\left[\pi \frac{\alpha}{2} + (1 - \alpha)\arctan\left(\frac{E - E_0}{\Gamma}\right)\right]}{\left[(E - E_0)^2 + \Gamma^2\right]^{\frac{1-\alpha}{2}}}. \tag{5.2}$$

We saw in Chapter 3 that the asymmetry parameter α is associated to the presence of electron-hole pairs close to the Fermi level, that screen the photoemission hole. The higher the density of states at Fermi level, the greater the α and the more asymmetric the line. At the limit $\alpha = 0$ we find the Lorentzian form (symmetric line).

These lines (Lorentzian or Doniach-Šunjić) are the intrinsic forms. They are modified and expanded due to the finite experimental resolution. To simulate the effect of the experimental resolution, the intrinsic form is to be convolved by a Gaussian of ΔE width. ΔE is the half-height width (FWHM) of the total Gaussian that takes into account the spectral width of the X-ray source and the analyzer resolution (Fig. 2.5).

It is possible, of course, to observe more complex line shapes with several components or satellites well separated from the main line, which are found especially when the valence states themselves are localized. These line shapes result from coupling between the angular momenta of the core hole and those of valence band electrons or because of particular screening processes. We will discuss these complex cases later and we will focus now on the simplest intrinsic components (Lorentzian form or Doniach-Šunjić).

Experimentally, one doublet per component is often observed. The reason for this splitting is the spin-orbit coupling, which gives rise to two peaks of different intensities whose ratio (*branching ratio*), is in principle given by the degeneracy associated with the total angular momentum. This relativistic interaction couples the orbital angular momentum of the core hole $\vec{\ell}$ with its spin momentum $\vec{s}$ and it is written $\lambda\vec{\ell} \cdot \vec{s}$ (a hole in a complete atomic subshell carries the same angular momentum as an electron). The coupling constant λ increases with higher speeds of electrons in the subshell, i.e., for deeper core levels. The eigenstates are the states of the total angular momentum $\vec{j} = \vec{\ell} + \vec{s}$. Since $s = 1/2$, we obtain two levels $j = \ell \pm 1/2$ of degeneracy $2j + 1$.

However, deviations in the branching ratio from the theoretical value may appear under certain conditions, such as, for instance, when the photoionization cross section is varying rapidly. Differences in the radial wave function may also be responsible for the modification of the intensity ratio. Finally, photoelectron diffraction can affect the branching ratio because of a different de Broglie wavelength for each of the components of the doublet. Accordingly, the branching ratio can be considered an adjustable parameter to some extent. An example of the need to vary this parameter is illustrated in 5.3(a). This figure shows the analysis of the 3d core level of As for a monolayer of As on Si (111)-(1×1). Two effects are observed. First, the intensity ratio of the two spin-orbit structures ($j = 2 \pm 1/2$) as a function of the azimuthal angle is always greater than the maximum expected value of 1.5 because of the cross-section variation of the low binding energy component (Fig. 5.3 (b)), due to its larger cross-section. In addition, the intensity ratio varies with the emission angle of the photoelectron. The photoelectron diffraction is at the origin of this behavior as suggested by the figure 5.3(c)) which shows that the emission of electrons of the $3d_{5/2}$ and $3d_{3/2}$ levels has the same behavior as a function of the azimuthal angle.

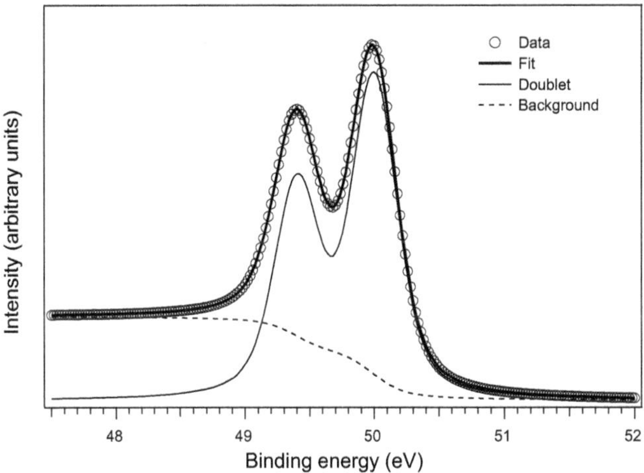

Figure 5.4 *Theoretical core level with $\ell \neq 0$, simulated by a spin-orbit doublet (continuous line). A Shirley background, proportional to the integral of the signal from the higher binding energies to the considered energy, is shown in dashed line.*

The description of the elastic contributions of a photoemission spectrum is not enough to accurately describe its spectral features, as it is also necessary to reproduce the background from the inelastic electrons. There are many approaches to describe the background. The most common approach was proposed by Shirley [8]. The determination of the background in this approach is based on the following argument: at a given kinetic energy, the inelastic signal is proportional to the integral of the elastic signal for all the larger kinetic energies. Indeed, the background electrons having lost part of their energy must come from all the elastic transitions with higher kinetic energy. Considering that the energy loss probability is constant, Shirley's background at each energy is simply calculated from the integral of the spectrum to a given energy (Fig. 5.4).

We also cite the more elaborate procedure proposed by Tougaard [9–11] which models the probability of an electron losing its energy and contributing therefore to the background. This background is calculated from the experimental spectrum $I(E)$ and with the hypothesis that the dispersion cross-section is the same for all electrons in the spectrum. The background here is described by [12]:

$$F_T(E) = \int_E^{\infty} F(E' - E) I(E') dE' \tag{5.3}$$

where $F(E)$ is the energy loss probability obtained from the dielectric theory, which is often modeled for a number of materials from:

$$F(E) = \frac{BE}{(C + E^2)^2} \tag{5.4}$$

and which can be further modified with a third parameter to increase its generality [13] (readers interested by the quantitative analysis of the quality of an adjustment will find relevant information in the Appendix B).

5.1.3 Spectra from complex core levels

We have seen above how the number of components of a spectral line can be determined when the form of these components is simple (Lorentzian or Doniach-Šunjić). Here, we will briefly discuss more complex spectral shapes composed of several structures. These forms result from the interaction between the core hole in the final state and the valence electrons or the vibrational degrees of freedom.

We have seen in Chapter 3 that multiplets may appear when the photoexcited atom has an external shell that is only partially filled. This is the case, for example, of the ionic compounds of transition metals or of rare earths, whether they be ionic or metallic, because of the quasi-atomic character of the $4f$ electrons. Let us take the example of the $4d$ line of holmium ($4d^{10}4f^{10}$). The coupling of the $4d$ hole and the $4f$ layer in the final state is particularly important since the two subshells have the same principal quantum number. The multiplet states extend over more than 30 eV in this case. The multiplet energy is well described from the Slater integrals F_k and G_k computed in the Hartree-Fock approach introduced in Chapter 3, although they are to be reduced by a factor of 0.7-0.8. This reduction with respect to the atomic values is due to the screening of interactions within the solid. However, the experimental spectra cannot be reproduced by widening the calculated spectra by a Lorentzian as for a single line. This means that the notion of the hole lifetime loses its meaning in the case of large multiplet effects i.e., in the presence of a coupling between the core hole and the incomplete layer. We have seen in Chapter 3 that the widening of spectral structures depends on the imaginary part of the self-energy and therefore on the relaxation processes of the multiplet state $3d^9 4f^{10}$. The

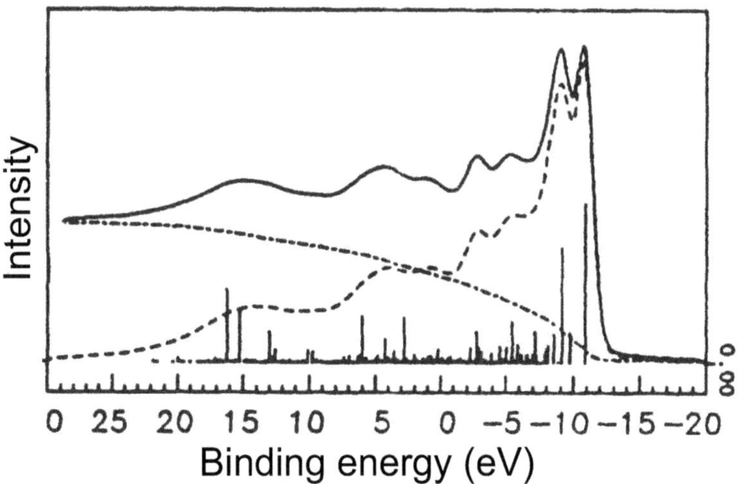

Figure 5.5 *Multiplet effects: the experimental spectrum of the 4d line of Ho is compared with a calculation taking into account the lifetime effects on the multiplet structure [14]. Figure reprinted with permission from H. Ogasawara, A. Kotani, and B. T. Thole, Phys. Rev. B 50, 12332 (1994). ©1994 by the American Physical Society.*

main mechanism involved in the lifetime calculation is the Auger relaxation of type $N_{4,5}N_{6,7}N_{6,7}$ super-Coster-Krönig ($4d$, $4f$, $4f$). This process involves the same interactions that determine the multiplet structure. The calculation shows that the broadening of the multiplet states depends strongly on their energy and it increases with the binding energy. The calculation of the $4d$ Ho photoemission spectrum is in excellent agreement with the calculated spectrum when taking into account the lifetime, as seen in Fig. 5.5.

Multiplet effects are only one of the possible effects giving rise to complex spectra. In localized electron systems, mixtures of electronic configurations can lead to satellites such as in rare earth oxides or in transition metal oxides. This is the case, for example, of CeO_2, where each of the $Ce3d_{3/2}$ and $Ce3d_{5/2}$ components exhibits three peaks. The Anderson impurity model presented in Chapter 3 allows you to understand this effect, due to the Coulomb interaction between the core hole and the d or f localized electrons and the hybridization between these states and the $2p$ band of oxygen. Let us consider the simple case of La_2O_3. In a purely ionic approach, the ground state corresponds to a filled $2p$ band of oxygen and to a $4f$ empty band of La (Fig. 5.6(a)). The first excited states located at the energy Δ, called charge transfer energy, are associated with an electronic structure $f^1\underline{v}$ where $\underline{v}$ designates a hole in the $2p$ band (transfer of a $2p$ electron of oxygen to the $4f$ states of La). These excited states extend over the width of the $2p$ band (Fig. 5.6(b)). However, hybridization, characterized by the parameter V_{eff}, must be introduced between the $2p$ valence band and the $4f$ states. In the case of La, V_{eff} ($\sim$ 2.8 eV) is very small compared with Δ ($\sim$ 12.5 eV). This term will slightly separate the energy of the ionic states as schematized on the left part of the figure and mix the states very weakly. In the final state, the deep $3d$ hole strongly interacts with the $4f$ states, leading to an energy decrease U_{fc} ($\sim$ 12.7 eV) of the f^1 configuration while the $4f^0$ configuration is not affected. Since Δ and U_{fc} are of the same order of magnitude, the $4f^1\underline{v}\underline{c}$ and $4f^0\underline{c}$ final states, where $\underline{c}$ designates the $3d$ hole, are close in energy ($U_{fc} - \Delta \sim 0.2$ eV) with the $4f^1\underline{v}\underline{c}$ state lower than the $4f^0\underline{c}$. The hybridization effects, energy shift and mixing of the two states (covalence), will then be important. The lowest-energy final state ($4f^1\underline{v}\underline{c}$ in the figure) is actually a combination of the type:

$$\beta\,(4f^1\underline{v}\underline{c}) + \alpha\,(4f^0\underline{c}) \text{ with } \alpha < \beta$$

so that the weight of the corresponding spectral structure, given by the coupling to the $4f^0$ ground state, is $|\alpha|^2$. We thus expect two structures in the $3d$ photoemission spectrum, whose separation is close to $U_{fc} - \Delta$ and whose intensities reflect the mixture of configurations induced by the hybridization. The photoemission thus allows to determine these electronic parameters. The intensity evolution of the two structures observed in the R_2O_3 oxides directly reflects the decrease of the hybridization parameter due to the contraction of the spatial extension of the $4f$ wave functions in the rare earth series (Fig. 5.7b-f).

The situation for dioxides, in particular for CeO_2, is different. The ground configuration is $4f^0$ and the first excited state is $f^1\underline{v}$, very close in energy ($\Delta \sim 1.6$ eV). We therefore expect very important hybridization effects since $V_{eff} > \Delta$ ($V_{eff} = 2.8$ eV). This leads to a very strong mixture of the two configurations and a number of f electrons, at the ground state, close to 0.5[15]. To describe the photoemission process, it is necessary to take into account the $f^2\underline{v}^2$ configuration located at the energy $2\Delta + U_{ff}$ ($U_{ff} \sim 10.5$ eV). In the final state of photoemission from $3d$ level, the energies of $4f^1\underline{v}\underline{c}$ and of $4f^2\underline{v}^2\underline{c}$ are lowered respectively by U_{fc} and $2U_{fc}$ as illustrated in 5.6(b). These energies are very close and can therefore mix. On the other hand, the $f^0\underline{c}$ state is very far from the other final states and it is associated with the observed peak at higher binding energy. We thus expect three peaks for each spin-orbit contribution (Fig. 5.6(b)).

The photoemission process may be accompanied by the creation of collective excitations (plasmons, phonons, etc.) which will be manifested by the presence of additional structures (satellites) in the spectrum. Plasmon-type excitations are easily observed in metallic systems because of the order of magnitude of their energy (around ten eV). These plasmons result from the response of the system to the creation of a core hole and constitute a screening mechanism of the hole. Part of the kinetic energy is spent in the plasmon creation. The plasmon satellite is thus located at lower kinetic energy (and higher binding energy) when compared to the main photoemission line of the core level, corresponding to the transition without plasmon creation. Figure 5.8 shows the photoemission $1s$ spectrum of oxygen for an aluminum film with less than one atomic layer of oxygen at the surface. It consists of the main line at 532 eV and a weak intense satellite displaced from $\hbar\omega_p = 11$ eV

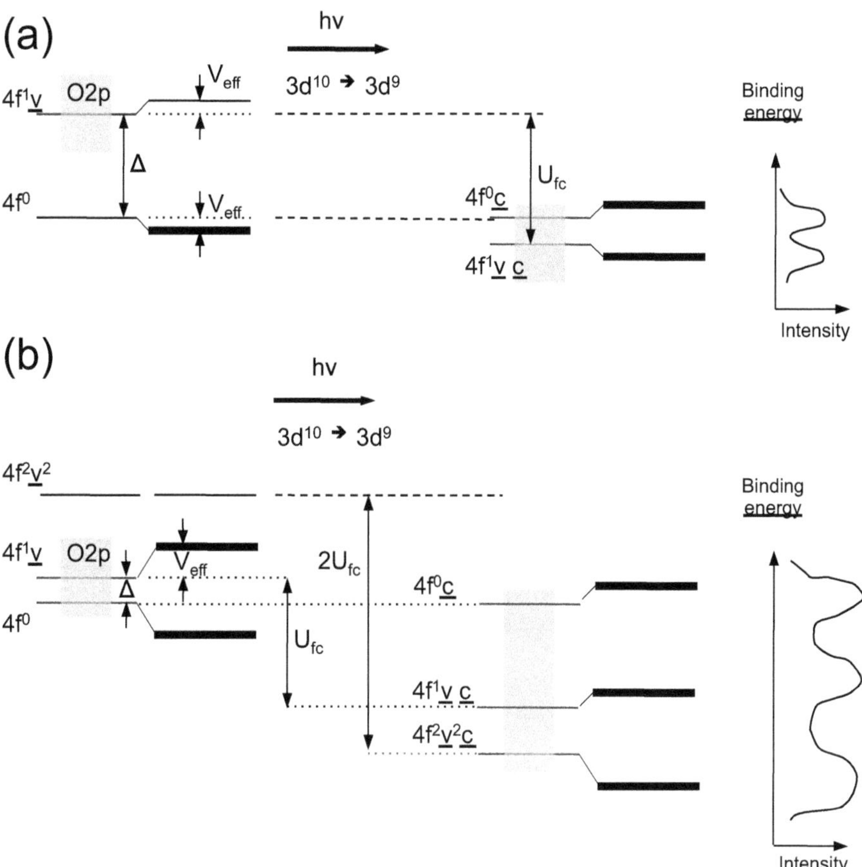

Figure 5.6 *Rare-earth energy level diagram for the initial state and for a 3d final state, in the case of (a) large and (b) low charge transfer energy between the valence band and the f states.*

corresponding to a transition with plasmon creation. Sometimes a second satellite at $2\hbar\omega_p$ can be observed for a transition with two-plasmon creation, etc. The satellite intensity depends on the electron-plasmon coupling and for creations of several plasmons, it decreases according to a Poisson distribution [17, 18]. The plasmons created during the photoemission process are called intrinsic plasmons. However, they can also be created by inelastic scattering processes during the photoelectron propagation in the solid, the so-called extrinsic plasmons. Finally, the photoemission can lead to the creation of surface plasmons of $\hbar\omega_s$ energy when crossing the surface. Their energy is lower than the energy of the bulk plasmons.

The mechanism for creating vibrational excitations (i.e., phonons in crystals) is very similar to that of creating plasmons, except that the characteristic energies for phonons are much smaller. During the emission of a photoelectron, a hole appears

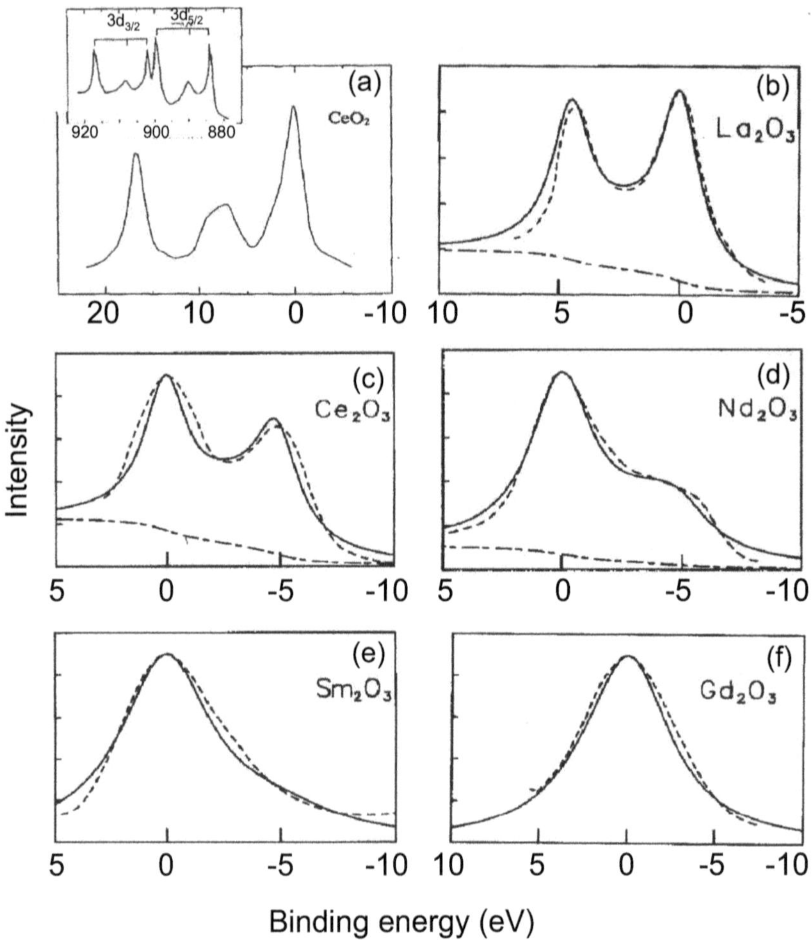

Figure 5.7 *3d core levels of rare earth oxides (- theory, - - experiment, .- background). (a) CeO_2, (b) La_2O_3, (c) Ce_2O_3, (d) Nd_2O_3, (e) Sm_2O_3 and (f) Gd_2O_3 (according to [15]). Reprinted from J. Electron. Spectrosc. Rel. Phenom. vol. 60, A. Kotani and H. Ogasawara, "Theory of core-level spectroscopy of rare-earth oxides", page 257, ©1992, with permission from Elsevier.*

which, due to the coupling to the vibration modes, can be accompanied by several vibrational quanta. This mechanism is the object of Franck-Condon model whose principle we shall describe. Let us consider the case where the equilibrium geometry in the final state is the same as that in the initial state. If the degrees of freedom of the atomic motion are represented by a harmonic oscillator, the potential minima correspond to the same value (Fig. 5.9(a)). In this situation, the ground state of the harmonic oscillator is identified with the ground state of the excited state. Therefore only this ground state is coupled to the initial state. On the other hand, if the harmonic oscillator corresponding to the final state is shifted with respect to that of the

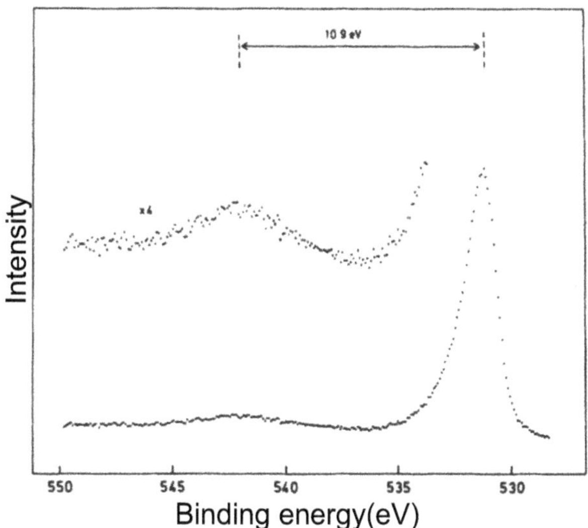

Figure 5.8 *(a) O 1s of a Al (111) surface coated with 0.17 monolayers of oxygen, with the main line and the plasmon at 542 eV (from [16]). Reprinted figure with permission from A. M. Bradshaw, W. Domcke, and L. S. Cederbaum, Phys. Rev. B 16, 1480 (1977). ©1977 by the American Physical Society.*

initial state (i.e., there is a different equilibrium geometry), several vibrational states can be coupled to the ground state. Indeed, within the Born-Oppenheimer approximation it is considered that the electronic characteristic times are much shorter than the atomic characteristic times, so atoms do not have time to move during the photoemission process. The atomic positions immediately after the emission of the electron are therefore unchanged and the vibrational state is the fundamental state of the initial state $|\phi^i_{\nu=0}\rangle$. We must therefore decompose this state on the eigenstates of the oscillator ($|\psi^f_\nu\rangle$) displaced from the excited state:

$$|\phi^i_{\nu=0}\rangle = \sum_{\nu'} c_{\nu'}|\psi^f_{\nu'}\rangle \quad \text{with} \quad c_{\nu'} = \langle\psi^f_{\nu'}|\phi^i_{\nu=0}\rangle \tag{5.5}$$

The system can therefore be found in several vibrational states $|\psi^f_{\nu'}\rangle$ with probability $|\langle\psi^f_{\nu'}|\phi^i_{\nu=0}\rangle|^2$, each final state $|\psi^f_{\nu'}\rangle$ corresponding to the creation of ν' phonons of energy $\hbar\omega_{ph}$ accompanying the photoemission process (Fig. 5.9(b). The kinetic energy of the photoelectron is reduced by the energy necessary to the phonon creation, so there is a spectral line corresponding to zero phonons and as many satellites as states in equation 5.5. The intensity of each satellite is proportionate to the probability of observing it in the ground state, i.e., $\propto |c_{\nu'}|^2$.

In molecules, the energy of the vibration modes is generally high and the different transitions can be solved as illustrated in figure 5.10. This figure shows the core level

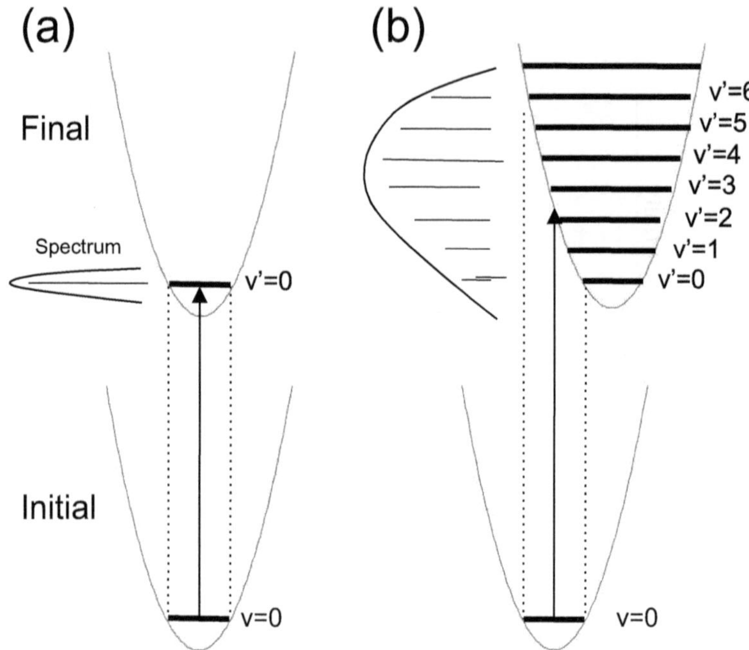

Figure 5.9 *Franck-Condon principle. Diagrams of the so-called 'vibronic' potential for initial and final states. When the potential is the same, there is only one possible transition to the ground state (a), whereas when the final-state potential is strongly displaced, several states are coupled, leading to satellites in the photoemission spectrum (b).*

of the O1s in CO_2. There are several peaks corresponding to transitions associated to the creation of an increasing number of vibration quanta [19]. In solids, the phonon energy is lower (of the order of the tenth of eV) and the vibrational modes form bands. The mechanism of Franck-Condon here does not lead to clearly resolved satellites but rather to an asymmetric widening of the spectral lines.

5.2 Core level spectroscopy applications

5.2.1 *Quantitative chemical analysis*

One of the main applications of photoelectron spectroscopy is the composition analysis of materials. The qualitative chemical analysis by photoemission is relatively simple as chemical elements are identified from the characteristic energy of the core levels present in the spectrum. Subsequently, it is possible to determine the composition of a material (at least on its surface) from the relative intensities of the core levels and the tabulated cross-sections. It is also possible to demonstrate

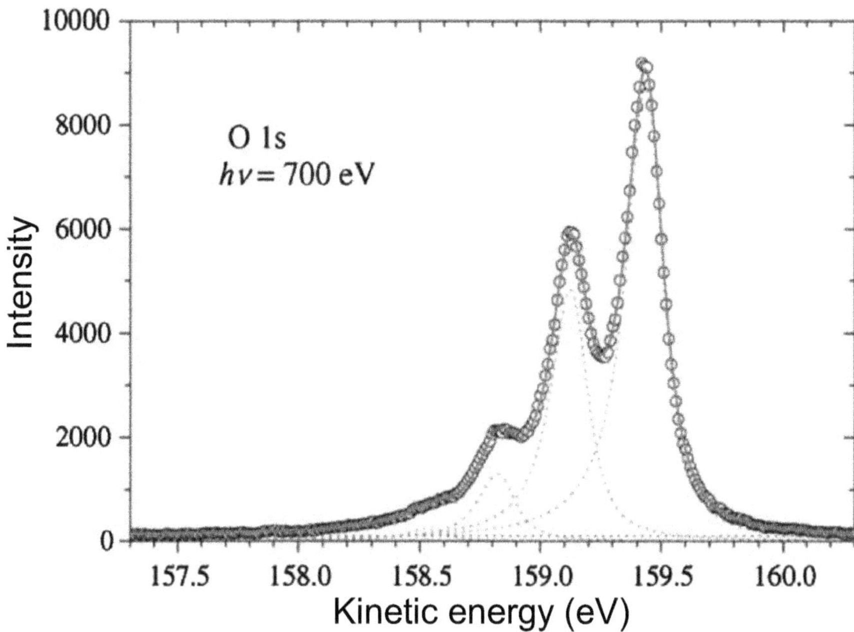

Figure 5.10 *Spectrum of the O1s core level of the CO_2 molecule. Reprinted from J. Electron. Spectrosc. Rel. Phenom. vol. 155, T. Hatamoto, M. Matsumoto, X.-J. Liu, K. Ueda, M. Hoshino, K. Nakagawa, T. Tanaka, H. Tanaka, M. Ehara, R. Tamaki, H. Nakatsuji, "Vibrationally resolved C and O 1s photoelectron spectra of carbon dioxide", page 54, ©2007, with permission from Elsevier.*

the presence of a surface contaminant, if the spectrum shows the characteristic lines of C and O revealing an oxidation of the surface and/or the adsorption of CO_2 or H_2O. One of the most powerful applications is possibly the combination of the chemical sensitivity and the spatial resolution in a XPEEM microscope. For example, the dynamics of chemical reactions can be studied. Figure 5.11 studies the water production in the catalytic reaction $O_2 + 2\ H_2 \rightarrow H_2O$ on a Rh(110) surface with small quantities of gold. Each image is recorded at the energy of the core level of the species involved in the reaction, so that the contrast reveals the spatial distribution of the species. In particular, these images show that the reaction forms alternating stripes of Au + Pd and O islands.

The quantitative determination of the concentration can be obtained with a sensitivity of the order of 5 to 10%, when all the factors are carefully considered. The intensity ratio of the photoemission peaks of two different elements depends essentially on the ratio between the number of their respective emitters and their cross sections. Let us analyze the situation more precisely. The intensity of a photoemission peak is proportional to the photon flux per unit time F, the integration time t of the spectrum and the number of emitters. It is also necessary to consider the volume

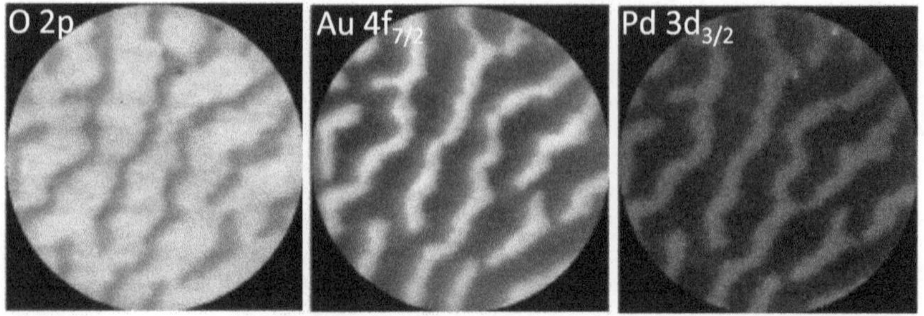

Figure 5.11 *XPEEM images showing the formation of phase separation (from [20]). Reprinted with permission from A. Locatelli et al., J. Phys. Chem. B 110, 19108 ©2006 American Chemical Society.*

v explored by the beam and the photo-ionization cross section σ. Therefore, the intensity ratio between two photoemission peaks of elements 1 and 2 is:

$$\frac{I_1}{I_2} = \frac{\sigma_1}{\sigma_2}\frac{v_1}{v_2}\frac{F_1 t_1}{F_2 t_2}\frac{n_1}{n_2} \tag{5.6}$$

which makes it possible to obtain the ratio n_1/n_2. By integrating the same time with the same flux and when the same volume is explored:

$$\frac{I_1}{I_2} = \frac{\sigma_1}{\sigma_2}\frac{n_1}{n_2} \tag{5.7}$$

5.2.2 *Chemical shifts*

The core level spectroscopy allows to determine whether the same element appears in the material under different chemical states or different atomic environments. For this, it is necessary to analyze the different components within a core line of the element. To illustrate this application, we present, in Fig. 5.12, the line Si $2p_{3/2}$ for a thin layer of SiO_2 on Si(001) and Si(111). The components are associated with the different oxidation states between the Si^0 of the substrate (which is the highest binding energy, taken as the energy reference in the figure) and the Si^{4+} of the SiO_2 (the lowest binding energy).

The separation between components is important, of the order of the eV. If the different sites are distinguished not by their degree of oxidation but only by a different atomic environment, the energy shift is much smaller. Fig. 5.13 shows the example of the stepped Rh surface (553). This surface has steps of width corresponding to five atomic distances. There are three types of surface atoms: 3/5 are atoms on the terraces with coordination 9 (T), 1/5 are atoms on the step edges with coordination 7 S) and 1/5 are atoms under the step edges with coordination 11 (U). The spectrum

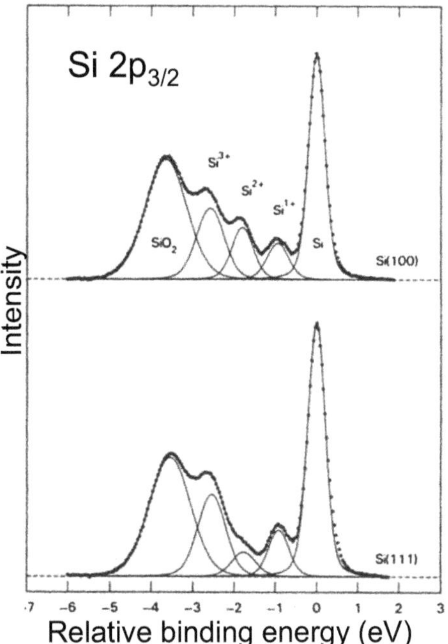

Figure 5.12 *Si 2p core level of thin films of SiO_2 on Si(001) and Si(111) (from [21]). Figure reprinted with permission from F. J. Himpsel, F. R. McFeely, A. Taleb-Ibrahimi, J. A. Yarmoff, and G. Hollinger, Phys. Rev. B 38, 6084 (1988). ©1988 by the American Physical Society.*

of Rh 3d shows shoulders that indicate the presence of several components. It can be decomposed with the Doniach-Šunjić line shape convoluted with a gaussian.

As a first approximation, the difference in binding energy between the components can be explained by the coordination. Indeed, in a tight binding model, the surface shift Δ (*surface core level shift*) is given by [22]:

$$\Delta \sim 1 - \left(\frac{Z_s}{Z_b}\right)^{1/2} \tag{5.8}$$

where Z_s and Z_b are respectively the coordination numbers of the surface and bulk atoms.

This type of analysis can confirm the validity of relatively complex surface structure models. This is the case, for example, of the interface with 1/3 of monolayer of Sn on Ge (111) (Fig. 5.14). In this system, Sn atoms appear with three different heights, corresponding to the three visible color levels in the scanning tunneling microscopy images, in occupied or unoccupied states. This difference in height is best seen on the profiles shown below the images. The unequivalence between the three Sn atoms is reflected on the Sn 4d core level spectrum. There are three main components

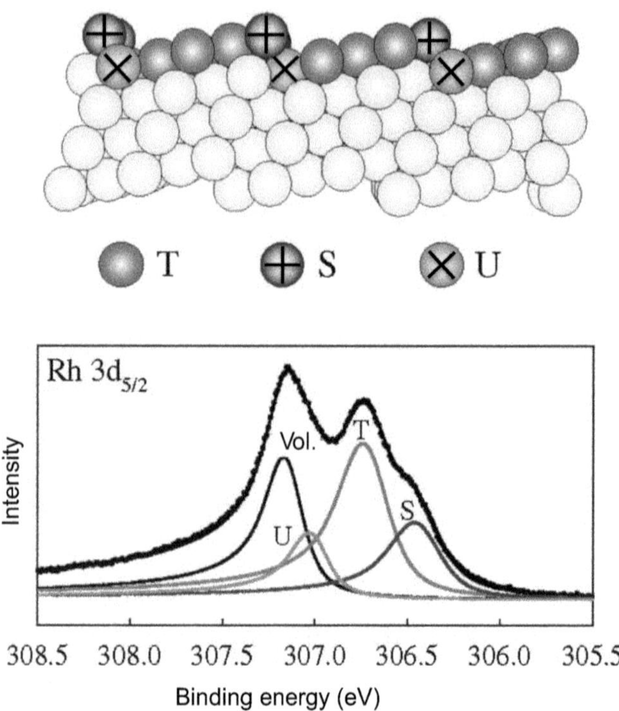

Figure 5.13 *(Top) A lateral view of a Rh(553) surface, showing three types of atoms with different coordination, i.e., terrace atoms (T), step-edge atoms (S), atoms under step edges (U) and bulk atoms (Vol.). (Bottom) Rh $3d_{5/2}$ spectrum and its decomposition into components associated with the U, S, T and Vol. atoms (from [23]). Reprinted figure with permission from J. Gustafson, M. Borg, A. Mikkelsen, S. Gorovikov, E. Lundgren, and J. N. Andersen, Phys. Rev. Lett. 91, 056102 (2003). ©2003 by the American Physical Society.*

associated with each type of atom. There is also a fourth component, whose origin cannot be identified from the STM measurements. The core level can finally be explained with four contributions duplicated by the spin orbit ($4d_{5/2}$ and $4d_{3/2}$). The decomposition remains valid for two different photon energies, the spectral shape being modified by photoelectron diffraction (cf. Chapter 6).

In addition to these fundamental studies, this type of analysis can be used to study catalytic reactions. Figure 5.15 shows a model reaction for car catalysis, that of NO reduction and CO oxidation on Rh (111). To understand the reaction mechanism, it is necessary to know what the adsorption sites of the molecules and their adsorption energies are. The experiment consists in saturating the surface with CO molecules, which occupy the two adsorption sites *top* and *hollow* of the surface. The progressive exposure to NO replaces the CO of the *hollow* sites. The substitution of the COs of the *top* sites requires a much higher temperature.

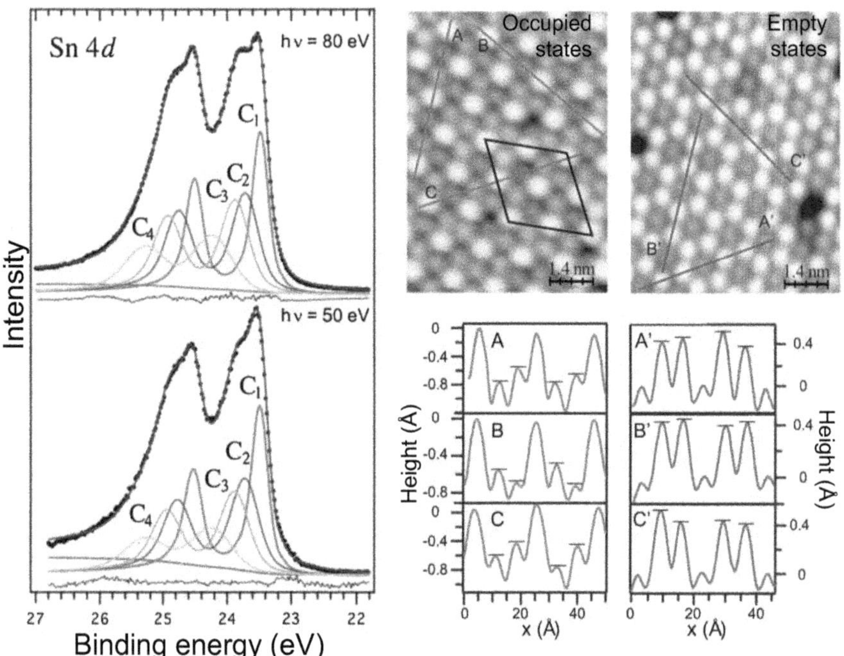

Figure 5.14 *(Left panel) Sn 4d spectrum decomposed with four doublets. The three main components correspond to the inequivalent Sn atoms of the unit cell, which have different heights and different charges. The different heights are observed in scanning tunneling microscopy images and their profiles (right panel) (from [24]). Reprinted figure with permission from A. Tejeda, R. Cortés, J. Lobo-Checa, C. Didiot, B. Kierren, D. Malterre, E. G. Michel, and A. Mascaraque, Phys. Rev. Lett. 100, 026103 (2008). ©2008 by the American Physical Society.*

To finish this part, let us point out a technical difficulty that can appear in insulating systems, viz. that to determine the binding energy, it is necessary to unambiguously figure out the energy reference which is the Fermi level. If the latter is perfectly defined for a metal, its experimental determination for semiconductors and insulators is far from trivial. In addition, insulators and semiconductors can be charged leading to a surface photovoltage effect that moves electronic levels. The photoelectron kinetic energy changes as the accumulation of holes at the surface (photo-holes) creates a potential that the photoelectron has to overcome. Sometimes the effect is so strong in large gap insulators that it is not possible to reach a steady state without a charge compensation system. This is obtained by neutralizing photo-holes by electrons using an electron gun. The surface charge has the effect of broadening the photoemission spectra (which exhibit an asymmetry towards the low binding energies) and the effect of shifting the spectral lines towards larger binding energies (Fig. 5.16). A charged sample has an apparent binding energy:

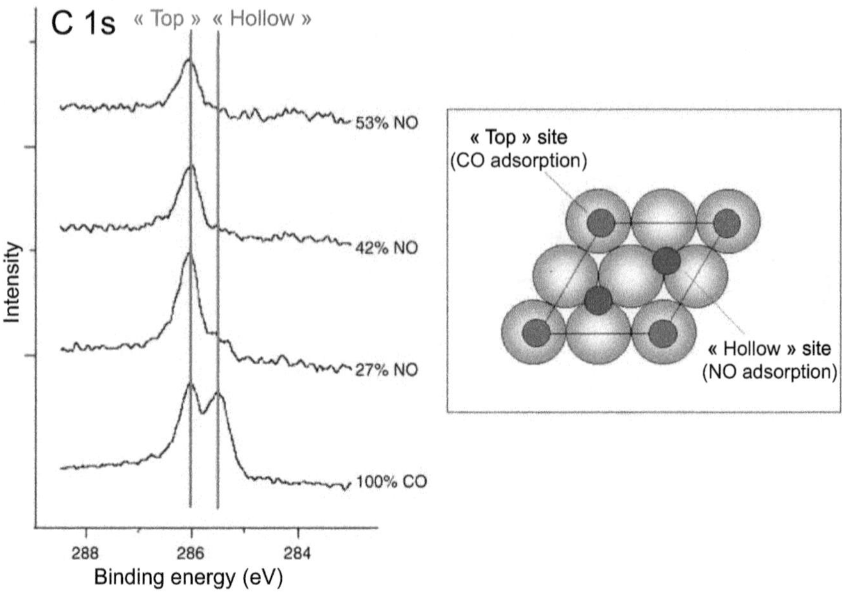

Figure 5.15 *C1s spectrum for CO + NO on Rh (111). The surface is initially coated with CO which occupies both types of adsorption sites (*top and hollow*). Exposure to NO leads to the desorption of the site 'hollow' (from [25]). Reprinted from Surf. Sci. Rep. vol. 63, M. Salmeron, R. Schlögl, "Ambient pressure photoelectron spectroscopy: A new tool for surface science and nanotechnology", page 169, ©2008, with permission from Elsevier.*

$$E_b^* = h\nu - E_k - E_{ch} \tag{5.9}$$

where E_k is the observed kinetic energy and E_{ch} is the change in kinetic energy due to the accumulated charges.

5.2.3 Composition profile under the surface

Estimation of the depth at which an element is located can be done in a destructive manner by eroding the surface by ion bombardment. It is sometimes possible to obtain this information in a non-destructive manner by photoemission. However, some assumptions are necessary. The most common are the homogeneity of the sample and the absence of significant photoelectron diffraction effects in single crystals. With these assumptions, the intensity I_n of a plane n, located at a distance $n \times d$ from the surface, decreases exponentially with the mean free path λ. For a transmission at the angle θ of the normal, we have:

$$I_n = I^0 \exp\left(-nd/\lambda\cos\theta\right) \tag{5.10}$$

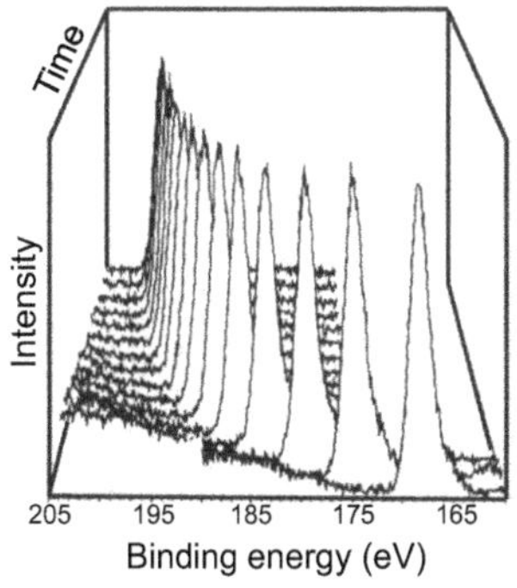

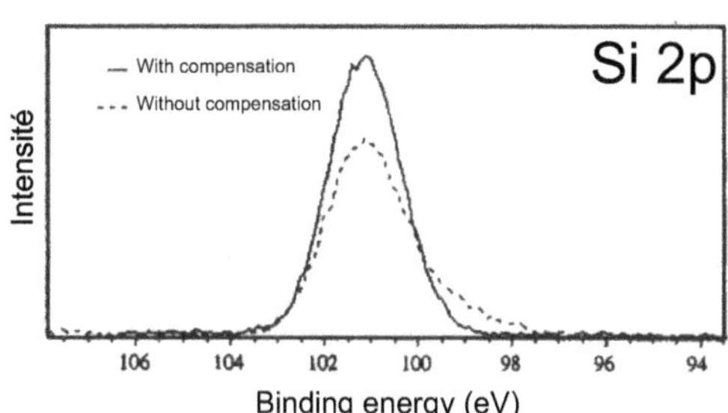

Figure 5.16 *(a) Si 2p spectrum on glass as a function of time. (b) Si 2p with and without charge compensation (continuous and discontinuous line, respectively). (Reprinted from "Differential Charging in XPS. Part I: Demonstration of Lateral Charging in a Bulk Insulator Using Imaging XPS", B. J. Tielsch and J. E. Fulghum, Surface & Interface Analysis, 24, 28). ©1996 John Wiley & Sons, Ltd.*

The prefactor I^0 is the intensity of a plane without attenuation, and it is equal to:

$$I^0 = K \times N \times S \times T(E) \tag{5.11}$$

where N is the concentration of the atoms detected in the plane, S is the atomic sensitivity factor in the photoionization section of Scofield [26], $T(E)$ is the transmission of analyzer according to the kinetic energy of the electrons and K is the constant of proportionality which includes the experimental factors such as the explored volume, the photon flux and the detection efficiency [27–29].

This expression allows you, knowing λ, to determine the depth nd of each emitting element. Moreover, in systems where the same element gives two photoemission peaks at very different kinetic energies (and hence different λ), the depth can be estimated from the intensity ratio between the two lines. Another possibility consists in varying the angle θ between normal emission and grazing emission.

5.2.4 Coverage estimation

To estimate the coverage of a thin film A, the intensity of the substrate B can be used as a reference. The intensity from B is the signal of an infinite substrate I_B^∞, attenuated in the thickness nd of the film:

$$I_B = I_B^\infty \exp[-nd/\lambda(E_B)\cos\theta] \tag{5.12}$$

The intensity of the film can be obtained from the intensity of an infinite film I_A^∞, by removing the contribution of the intensity beyond the thickness nd [30]:

$$I_A = I_A^{\infty}[1 - \exp(-nd/\lambda(E_A)\cos\theta)] \tag{5.13}$$

where $\lambda(E)$ is the inelastic mean free path (IMFP). Since I_A^{∞} and I_B^{∞} are not known *a priori*, it is necessary to use the tables to include the sensitivity factors S_A and S_B that are proportional to them. We then have:

$$\frac{I_A/S_A}{I_B/S_B} = \frac{1 - \exp(-nd/\lambda(E_A)\cos\theta)}{\exp(-nd/\lambda(E_B)\cos\theta)} \tag{5.14}$$

and if $E_A \sim E_B$, λ is the same for both peaks and finally we obtain:

$$nd = \lambda\cos\theta\ln\left(1 + \frac{I_A/S_A}{I_B\ S_B}\right) \tag{5.15}$$

However, the elastic effects can also contribute to the attenuation and lead to a non-exponential dependence of the substrate intensity. For this purpose, the effective attenuation length (EAL) is introduced, which acts similarly to the mean free path, but taking into account the elastic scattering. However, it is not an intrinsic property of the system because it depends on the experimental geometry. It allows in the end to give a better estimate of the coverage than the mean free path. In some cases, the difference between the inelastic mean free path and the effective attenuation length can reach up to 40%, depending on the atomic species of the film, the energy of the electrons and the experimental geometry. But generally it is estimated that for thicknesses less than 1.5 IMFP, the difference between the inelastic mean free path and the effective attenuation length is of the order of 10 % [31].

5.2.5 Growth kinetics

The growth mode of epitaxial thin films on a single crystal substrate depends on many thermodynamic and kinetic parameters. When the film grows layer by layer, the atoms complete the nth-layer before forming the n+1th layer. This situation is known as Frank-van der Merwe growth. A more complex growth mode is often observed, consisting of the formation of one or more complete layers, followed by the subsequent growth of three-dimensional islands. This growth mode is called Stranski-Krastanov. Finally, when islands form already in the early stages of growth, we are dealing with a Volmer-Weber growth. The evolution of the photoemission lines of the core levels or Auger transitions of the deposited substrate or film atoms can be used to determine the growth kinetics.

Let us look at what happens for layer-by-layer growth. In this situation the intensity associated with the film increases as the substrate signal decreases. Before the first layer is complete ($1 < x$), the film signal is simply $I_f^1 = I_f^0\ x$, with I_f^0 the intensity of a complete film layer and x the coverage. Between one and two monolayers, there are two contributions: a contribution associated with the second incomplete layer and

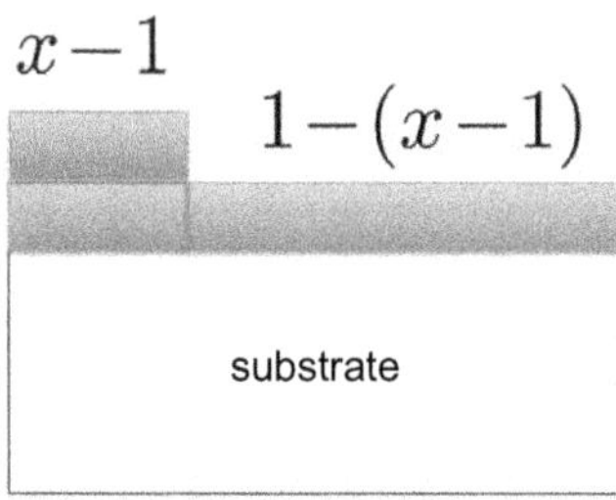

Figure 5.17 *Schematic diagram of a layer-by-layer growth with* $1 < x < 2$.

the uncovered part of the first layer, and a contribution associated with the covered part of the first layer, the signal of which is attenuated. The intensity for this coverage is therefore:

$$I_f(x) = I_f^0 \left[1 + (x-1) \exp\left(\frac{-d}{\lambda \cos\theta} \right) \right] \tag{5.16}$$

where d is the thickness of a monolayer, λ the mean free path and θ the emission angle. When generalizing to any number of layers, we obtain that the dependence with the coverage is always linear between 2 layers. The envelope of all the straight segments is the function:

$$I_f^{env} = I_f^{\infty} \left[1 - \exp\left(\frac{-x}{\lambda \cos\theta} \right) \right] \tag{5.17}$$

with I_f^{∞} the intensity of an infinitely thick layer. The signal from the substrate behaves symmetrically according to a decreasing exponential.

What happens in the case of Stranski-Krastanov growth? In the early stages of growth, there is no difference with the layer-by-layer growth. But when the islands appear, the substrate signal remains lower than that in the layer-by-layer growth, because some of the photoelectrons go through the islands and are therefore attenuated. Finally, in the Volmer-Weber growth, there will be no discontinuity in the intensity evolution of the core levels. The intensity will always be lower than for the other two growth modes.

Let us now examine a concrete case, that of the Bi growth on InAs(110) at ambient temperature (Fig. 5.18). The intensities of In 4d and Bi 5d have a linear dependence at the beginning, and then change their slope. Since the slope change is not very sharp, this suggests a deviation from an ideal layer-by-layer growth mode. This behavior may mean that the growth of a layer begins before the previous one has been completed. The intensity dependence of the surface component corroborates this hypothesis. This component is proportional to the uncovered fraction of InAs (110) and should disappear once the first Bi layer covers the entire surface. In a purely Franck-van der Merwe mode, the surface component should have a linear dependence and vanish strictly upon completion of a monolayer. The residual intensity beyond this thickness

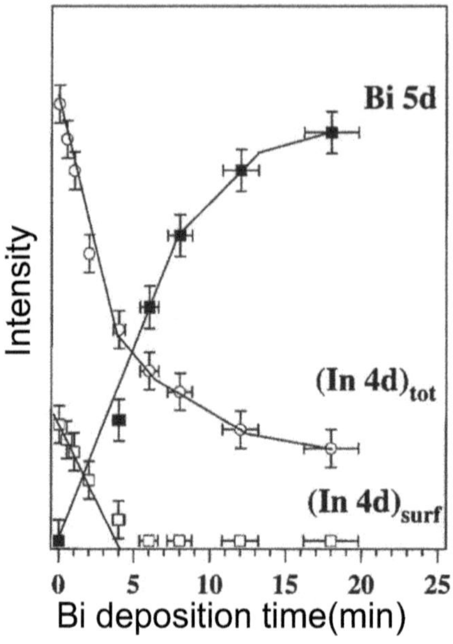

Figure 5.18 *In 4d and Bi 5d intensities as a function of the deposition time of Bi on InAs(110). The surface component is also shown (from [32]). Reprinted from V. De Renzi, M.G. Betti, V. Corradini, P. Fantini, V. Martinelli and C. Mariani, "A high-resolution spectroscopy study on bidimensional ordered structures: the (1 × 1) and (1 × 2) phases of Bi/InAs(110)", J. Phys.: Condens. Matter 11, 7447 (1999). ©IOP Publishing. Reproduced by permission of IOP Publishing. All rights reserved.*

proves that small regions of the substrate remain uncovered and thus indicates a deviation from the layer-by-layer growth mode.

5.2.6 *Dichroism in photoemission*

Dichroism is the different response of a system to the light polarization. This property is widely used in X-ray absorption spectroscopy for the study of magnetic properties. But dichroism can also provide specific information in photoemission. Chiral molecules in particular have a dichroic response. A chiral molecule is a molecule that is not the image of itself in a mirror. Two types of similar molecules exist (levogyre and dextrogyre), that are their respective mirror images and are called enantiomers. The right or left chirality of a molecule can indeed modify its chemical properties. Life on Earth, for instance, is organized around left chiral molecules, so that the right enantiomers are not assimilated by living organisms and may even be extremely toxic.

In core levels, it is possible to observe differences with the light polarization. This is the basis of Circular Dichroism in the Angular Distribution of the

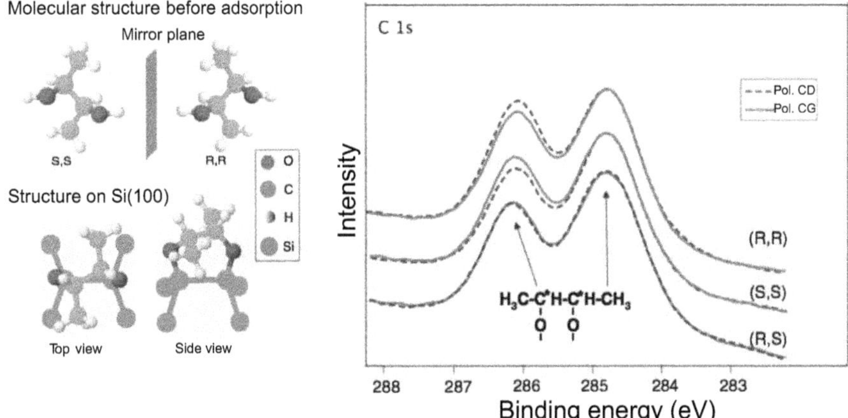

Figure 5.19 *(a) Structural model of the 2,3-butanediol enantiomers. (b) Adsorption of the S,S molecule on Si(001). (c) Carbon 1s lines of the enantiomers and the achiral stereoisomer with right and left circularly polarized light (from [33]). Reprinted figure with permission from J. W. Kim, M. Carbone, J. H. Dil, M. Tallarida, R. Flammini, P. Casaletto, K. Horn, and M. N. Piancastelli, Phys. Rev. Lett. 95, 107601 (2005). ©2005 by the American Physical Society.*

photoemission (CDAD). The CDAD intensity is defined as the normalized asymmetry of the intensity for each polarization:

$$I_{CDAD} = \frac{I_+ - I_-}{I_+ + I_-} \tag{5.18}$$

where I_+ (I_-) is the intensity for a circular polarization of right (left) helicity according to the convention used by physicists, i.e., the rotation direction of the electromagnetic field when the light goes away from the observer [34][1].

Let us look at a chiral system. Fig. 5.19 corresponds to the study of 2,3-butanediol molecules where it is possible to find chiral (S, S) and (R, R) molecules and an achiral (R, S) stereoisomer. In each molecule there are two types of carbon atoms which correspond to the chiral and achiral parts of the molecule and which are separated from 1.3 eV. With the appropriate experimental configuration, excitation with right and left helicities reveals differences in the C1s core level. The intensity of the carbons associated with the chiral part changes in the case of the enantiomers although no dichroism is present for the stereoisomer. This effect opens the possibility of selecting chiral molecules by photoemission.

[1] This convention is the opposite to that in optics, where circular polarization is defined from the rotation direction of the electric field when the radiation approaches the observer.

Bibliography

[1] C. Papp *et al.*, J. Phys. Chem. C **111**, 2177 (2007).
[2] P. Auger, J. Phys. Radium **6**, 205 (1925).
[3] P. Auger, Ann. Phys. **6**, 183 (1926).
[4] D. Briggs and M. Seah, *Practical Surface Analysis* (John Wiley & Sons, Chichester, 1983).
[5] M. Seah, Surf. Int. Anal. **24**, 830 (1996).
[6] M. Seah, I. Gilmore, H. Bishop, and G. Lorang, Surf. Int. Anal. **26**, 701 (1998).
[7] E. Bullock *et al.*, Surf. Sci. **352-354**, 352 (1996).
[8] D. A. Shirley, Phys. Rev. B **5**, 4709 (1972).
[9] S. Tougaard and P. Sigmund, Phys. Rev. B **25**, 4452 (1982).
[10] S. Tougaard and B. Jörgensen, Surf. Interface Anal. **7**, 17 (1985).
[11] S. Tougaard, W. Braun, E. Holub-Krappe, and H. Saalfeeld, Surf. Interface Anal. **13**, 225 (1988).
[12] S. Tougaard, Sol. State. Commun. **61**, 547 (1987).
[13] S. Tougaard, Surf. Interface Anal. **25**, 137 (1997).
[14] H. Ogasawara, A. Kotani, and B. Thole, Phys. Rev. B **50**, 12332 (1994).
[15] A. Kotani and H. Ogasawara, J. El. Spec. Rel. Phenom. **60**, 257 (1992).
[16] A. Bradshaw, W. Domcke, and L. Cederbaum, Phys. Rev. B **16**, 1480 (1977).
[17] P. Steiner, H. Höchst, and S. Hüfner, Z. Phys. **B30**, 129 (1978).
[18] S. Hüfner, *Photoelectron spectroscopy* (Springer, Heidelberg, 1995).
[19] T. Hatamoto *et al.*, J. El. Spec. Rel. Phenom. **155**, 54 (2007).
[20] A. Locatelli *et al.*, J. Phys. Chem. B **110**, 19108 (2006).
[21] F. Himpsel *et al.*, Phys. Rev. B **38**, 6084 (1988).
[22] J. V. der Veen, F. Himpsel, and D. Eastman, Phys. Rev. Lett. **44**, 189 (1980).
[23] J. Gustafson *et al.*, Phys. Rev. Lett. **91**, 056102 (2003).
[24] A. Tejeda *et al.*, Phys. Rev. Lett. **100**, 026103 (2008).
[25] M. Salmeron and R. Schlögl, Surf. Sci. Rep. **63**, 169 (2008).
[26] J. Scofield, J. Electron Spec. Rel. Phenom. **8**, 129 (1976).
[27] C. Fadley, Prog. Solid State Chem. **11**, 265 (1976).
[28] C. Fadley, J. Electron Spectrosc. **5**, 725 (1974).
[29] M. Seah, Surf. Int. Anal. **2**, 222 (1980).
[30] P. Cumpson and M. Seah, Surf. Int. Anal. **25**, 430 (1997).
[31] S. Tougaard and A. Jablonski, Surf. Interface Anal. **25**, 404 (1997).
[32] V. de Renzi *et al.*, J. Phys.: Condens. Matter **11**, 7447 (1999).
[33] J. Kim *et al.*, Phys. Rev. Lett. **95**, 107601 (2005).
[34] N. Böwering *et al.*, Phys. Rev. Lett. **86**, 1187 (2001).

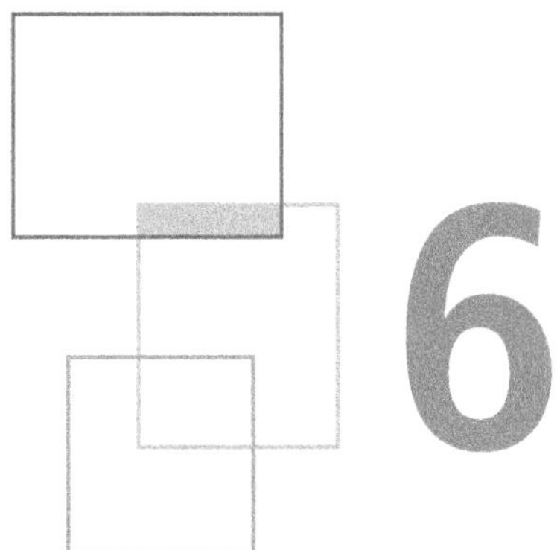

Photoelectron diffraction

The wave-particle duality of photoelectrons is evident in their diffraction by the atomic lattice. The photoelectron diffraction can be exploited to obtain structural information from the anisotropy of the photoemission signal intensity. This effect is especially interesting in the emission from localized states like core levels. For the latter, the emitter is well-defined and it is possible to identify around which type of atom (chemical element) the diffraction has taken place.

We have already introduced in Chapter 2 the principle of diffraction of photoelectrons. Here we shall focus on the practical aspects of data processing and analysis, as well as the information that this technique can provide.

6.1 Experimental techniques

Photoelectron diffraction data can be acquired in two ways. In the first method, the photon energy is fixed and the solid angle is varied; in the other method the opposite is performed. When fixing the photon energy, it is convenient to start the measurements by grazing emission in order to minimize the effects due to surface contamination. A uniform distribution of solid angles can be obtained by adjusting the azimuth angle variation according to the formula $\delta\phi = \delta\theta / \sin\theta$. In general, these data are then projected using the stereographic projection. The other experimental configuration consists in varying the photon energy at a fixed angle. The variation of $h\nu$ changes the path-length of the direct and scattered waves and therefore the relative phases, which gives rise to intensity variations. In these experiments, the initial state is always

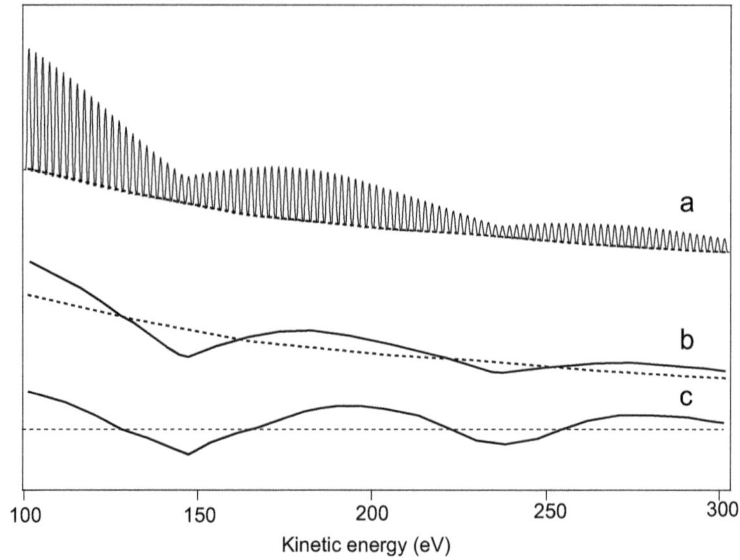

Figure 6.1 *(a) Schematic illustration of the intensity variation of a core level as a function of hν. Rapid oscillations are due to photoelectrons diffraction. (b) The total integral of each spectrum (continuous line) and its slow variation (dotted line). (c) After subtracting the slow variation, the anisotropy is obtained, which is compared with theoretical simulations.*

the same, despite the variation of $h\nu$, so they are experiments with Constant Initial State (CIS). This configuration is particularly sensitive to the binding distances of a localized adsorption geometry.

The first stage of the data processing involves transforming the experimental core levels as a function of photon energy or angle into a curve that contains only the information related to the photoelectron diffraction. It is therefore necessary to eliminate the background of the secondary electrons which does not contain diffraction information. As a second stage, the integral of the primary electrons is determined. This integral often gives a smooth variation as a function of θ or $h\nu$ which is superimposed on the rapid variation of the photoelectron diffraction signal. Schematically, figure 6.1 (a) illustrates the effect of the diffraction on the core level intensity as a function of photon energy. The smooth variation as a function of the energy (dotted line) can arise, for example, from the photoionization cross-section variation of the core level. In measurements as a function of the angle, when the angle between the radiation and the analyzer is constant, there is an experimental contribution to the intensity depending on the polar angle. Indeed, if the instrumental response is uncorrected, the background intensity decreases away from the normal emission because the diffraction effects exhibit a dependence of type $1/\cos\theta$ [1]. One way of obtaining this background experimentally - but not generally applicable- is to measure the polar dependence of the intensity in a polycrystalline sample of the same material. Another way to estimate it is to use the intensity at energies or angles very different from the considered peak, assuming that this signal is not correlated to the diffraction. Another strategy is to perform an additional measurement by integrating over an angle large enough to eliminate the diffraction effects. Finally, a phenomenological approach can be used to eliminate the slow variation of the curve, since the diffraction effects have a faster variation than the other experimental contributions. The slow variation can be obtained from the fit, or when a spherical sector has been measured, from the average of all measurements made at each polar angle that is subsequently smoothed. For a more detailed study of the angular dependence of theoretical and experimental backgrounds, we invite those readers interested to study the reference [2].

After this treatment, the signal which corresponds only to the diffraction, called anisotropy, is obtained according to:

$$\chi_{exp}(\theta, \phi, k) = \frac{I(\theta, \phi, k) - I_0(\theta, \phi, k)}{I_0(\theta, \phi, k)} \tag{6.1}$$

where $I(\theta, \phi, k)$ is the signal due to the diffraction and $I_0(\theta, \phi, k)$ is the signal coming from instrumental effects of slow variation. When comparing this anisotropy with simulations, it is relevant to apply also the Equation 6.1 to the simulations, in order to compare theoretical and experimental results which have undergone exactly the same treatment. Once the anisotropy curve has been determined, the amplitude of the diffraction oscillations can be calculated from:

$$(I_{max} - I_{min})/I_{max} \tag{6.2}$$

which can reach up to 50-70 % [3].

6.2 Methods for structural determination

There are two methods for determining atomic structures by photoelectron diffraction. The first consists of carrying out a large number of diffraction measurements to try to obtain the real space from a mathematical algorithm. This approach is called direct method. The other approach consists in simulating the photoelectron diffraction for different test structures and compare the simulation of the diffraction of these structures with the experiments. The atomic positions of the test structure are refined until a satisfactory agreement with the experiment is reached. We call this procedure the comparative method. We explain in the following the general lines of the two strategies.

6.2.1 *Direct methods*

There are three direct methods in diffraction of photoelectrons: the holographic method, the Fourier transform and the projection method. We will describe the basis of each.

The Fourier transform allows to obtain the real space from the diffraction patterns [4, 5]. The method allows us to identify the backscattering directions of the first neighbors, even if it does not provide information on the bond distances because it disregards the phase shifts [6]. This method does not work as satisfactorily as the Patterson function in X-ray diffraction analysis because of the importance in photoelectron diffraction of multiple scattering and the complexity of the electron-atom scattering.

The projection method [7, 8] is another direct method that can be applied to energy-resolved diffraction spectra originated from an adsorbate atom emission. This method uses the backscattering process. To compare the numerical value of the experimental spectra $\chi_{exp}(k)$ with the simulated spectra $\chi_{theo}(k)$, the projection $c(\vec{r})$ of one function on the other is determined:

$$c(\vec{r}) = \int \chi_{exp}(k)\chi_{theo}(k, \vec{r})dk \tag{6.3}$$

These projection integrals have maxima in the spatial regions corresponding to the most probable localizations of the first scattering neighbors in the backscattered configuration [9]. The positions of the emitter with respect to the surface can finally be determined with 0.2 Å precision.

The main difference with the Fourier transform method is that here the phase shifts are taken into account in the calculation of the single scattering. Therefore, this method does not have a disadvantage of the Fourier transform, that of leading to systematic errors on the atomic positions [6]. But it should be noted, in any case, that the projection method has some limits. For example, maxima sometimes appear at non-physical positions [9]. This problem results from the absence of experimental spectra sufficiently close to the backscattering directions of the first neighbors. Also, when there are different peaks arising from different scatterers, their superposition can displace the angular position of the maximum intensity, which may introduce an error in the determination of bonding angles. In fact, even in situations where the method works in a correct way, the information cannot be considered blindly [9]. However, the obtained 'images' of the real space may be the starting point for a comparative method.

The third direct method is holography. Szöke [11] and Barton [12] first proposed the analogy between photoelectrons and optical holography (Fig. 6.2) [13]. In the photoelectron diffraction process, the outgoing wave is divided into a wave impinging directly to the surface (equivalent to the reference wave of optical holography) and a second wave diffused by the atomic environment of the emitter before reaching the surface (equivalent to the wave diffused in the object). Interference between the two waves generates a diffraction pattern which contains the same information as an optical hologram. Therefore, the standard configuration of a photoelectron diffraction experiment is the one required for performing photoelectron holography. This is a difference with other holography techniques for which the experimental arrangement has to be modified. For example, in X-ray or neutron diffraction, there is no external reference beam and this is the cause of the phase loss [3].

Barton made the first theoretical demonstration that the Fourier transform of the photoelectron diffraction signal allows to obtain the three-dimensional environment of the emitter [12]. The first structural determinations from experiment were

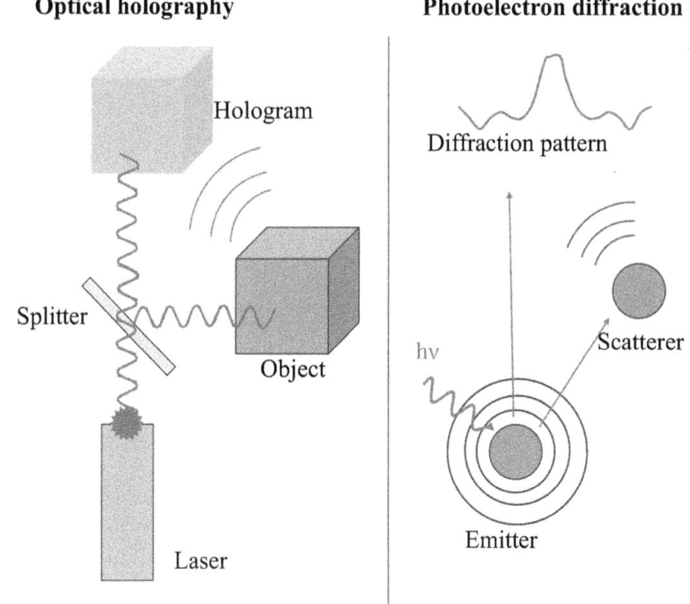

Figure 6.2 *Analogy between optical and photoelectron holographies.*

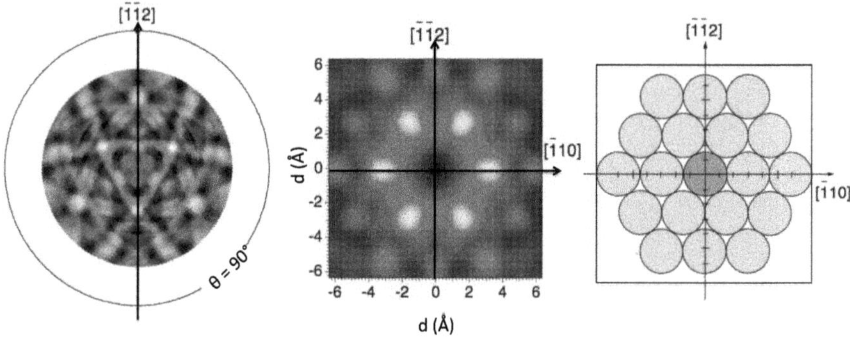

Figure 6.3 *(Left) Stereographic projection of Al 2s diffraction pattern (E_{kin} = 952 eV) for an Al (111) surface. (In the center) Real space obtained by holography in a plane parallel to the surface that contains the emitter. (Right) Expected scheme of the Al(111) plane (from [10]). Figure reprinted with permission from J. Wider, F. Baumberger, M. Sambi, R. Gotter, A. Verdini, F. Bruno, D. Cvetko, A. Morgante, T. Greber, and J. Osterwalder, Phys. Rev. Lett. 86, 2337 (2001). ©2001 by the American Physical Society.*

obtained in 1990 [14, 15]. Despite the success of the method, the structures were very elongated along the directions of forward scattering. This problem arises from the more complex nature of the electronic waves with respect to the optical waves [3]. The problem can be attenuated by performing the Fourier transform over a very large range of angles and energies. Because of these aberrations, it is difficult to accurately determine structures directly from the photoelectron holography [3]. Figure 6.3 shows the diffraction pattern of Al 2s of a surface of Al (111). It corresponds to a kinetic energy of the electrons of 952 eV for which forward scattering is predominant. The holographic method makes possible to reconstitute the real space from this type of measurements.

6.2.2 Comparative methods

Comparative methods are based on determining the structure from an iterative process of refinement of atomic positions. With each iteration, theory and experiment are compared in an objective way. This objectivity is reached by introducing what we call *reliability factors* or *R-factors*, which quantify the agreement between theory

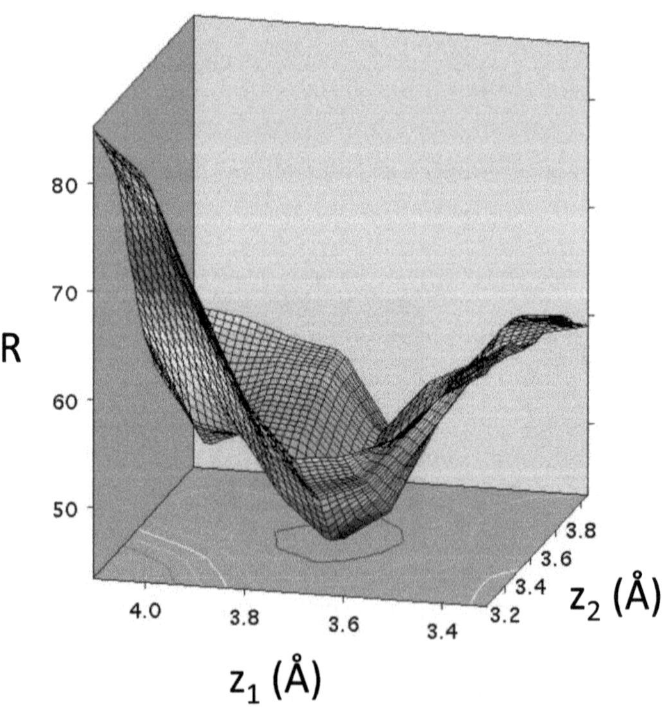

Figure 6.4 *R-factor optimization in the space of structural parameters. The global minimum corresponds to the real structure (from [29]).*

and experiment. These reliability factors must tolerate systematic errors in the data, so that this error type increases the value of the absolute minimum of the confidence factor, but the minimum position remains fixed in the structural parameter space. The structural refinement process consists in the minimization of the reliability factor as a function of the structural parameters (Fig. 6.4).

In contradistinction to X-ray diffraction, several types of confidence factors have been proposed for photoelectron diffraction (cf. Table 6.1), but there is no unanimous agreement on which is the best. In analogy to X-ray crystallography, a convenient and numerically robust reliability factor was proposed [16]. The factor that Pendry introduced for the LEED is also used in a modified version [30]. The most sophisticated of all is the confidence factor introduced by Zanazzi and Jona [28], which compares the number of maxima and minima, their positions and their shapes [31]. Its main disadvantage is the computation time and the difficulty of predicting its behavior in a conventional structural determination [32]. Other very different reliability factors have also been used, based on the multipolar expansion of the data [18, 33] or on combinations of several reliability factors [34–36]. For each reliability factor, the value that determines a reasonable agreement between theory and experience is obtained empirically from its use [28].

Van Hove has defined very robust reliability factors, R_1, R_2, R_3, R_4, R_5 (table 6.1), of reasonable calculation time and taking into account a large number of comparison criteria [24]. The average weight of these factors gives the same structural conclusions as the analysis of R_1 [26, 37], which has sometimes been used to simplify the analysis. These reliability factors were subsequently standardized by Saiki *et al.* [25] to account for the fact that the theoretical anisotropies are normally larger than the experimental ones by a factor 3-5 as in LEED [3]. This difference may be due to surface disorder, to emitters in non-ideal sites, defects or steps, or to failure to take into account anharmonic thermal vibrations by the Debye-Waller factors commonly used in the simulations of photoelectron diffraction.

Once the confidence factor has been introduced, the problem of the structural determination consists in locating a global minimum in a multidimensional hypersurface. The main difficulty of this process is the high number of parameters to consider, especially in large unit cells. The number of parameters can be reduced by using symmetry properties of the system and the existing information such as the chemical composition or the lateral surface periodicity and the presence of adsorbates.

In the simplest situation, the parameters are independent and can be optimized separately, but this decoupling seldom appears in Cartesian coordinates. In general, the parameters are coupled, which requires a simultaneous fitting. In addition to the number of considered parameters, the efficiency of the search depends on the number of grid points for each parameter, *i.e.*, the amplitude of the parameter variation and the step used. For a small step, the precision will in principle be higher, provided that the noise of the reliability factor topology is avoided. This noise starts to appear for intervals smaller than ~ 0.01 Å [38].

Table 6.1 *Reliability factors.*

Reliability factor	Formula	Optimal value
Cristallographic[1]	$R_2 = \sum_i [w_i(\chi_{exp,i} - c_{hk}\chi_{th,i})]^2$ where c_{hk} is the scaling factor: $c_{hk} = \frac{\sum_i \chi_{exp,i}}{\sum_i \chi_{th,i}}$	
Dippel[2]	$R_m = \frac{\sum(\chi_{th}-\chi_{exp})^2}{\sum(\chi_{th}^2+\chi_{exp}^2)}$	0.20–0.40[3]
Multipolar[4]	$R_{MP} = \sum_{l=0}^{l_{max}} \sum_{m=-l}^{l} \lvert a_{th,lm} - a_{exp,lm}\rvert$ with $a_{lm} = \frac{1}{4\pi}\int \chi(\theta,\phi) Y_{lm}^*(\theta,\phi)\, d\Omega$	0.30–0.50[5]
Pendry modified[6]	$R_P = \frac{\sum_i [\chi'_{exp,i} - \chi'_{th,i}]^2}{\sum [\chi'^2_{exp,i} + \chi'^2_{th,i}]}$	0.20–0.40[7]
R_{DE}[8]	$R_{DE} = \sum_g W_g \frac{\sum_i [\chi_{exp,i} - C_g \chi_{th,i}]}{\sum_i \chi_{exp,i}}$ $C_g = \sum \chi_{exp,i} / \sum \chi_{th,i}$ and $W_g = n_g / \sum n_g$	0.30[9]
R_N[10]	$R_N = \frac{1}{N}\sum_{k_i} [\chi_{th}(k_i) - \chi_{exp}(k_i)]^2$	0.005–0.010[11]
r_2[12]	$r_2 = \frac{1}{N}\sum_j \lvert \chi_{th,j} - \chi_{exp,j}\rvert$	0.10–0.15[13]
Van Hove[14]	$R_1 = \frac{\sum_n \lvert I^{\ddagger}_{exp}(n) - I^{\ddagger}_{th}(n)\rvert}{\sum_n \lvert I^{\ddagger}_{exp}(n)\rvert}$	0.025–0.050[15]
Van Hove[16]	$R_2 = \frac{\sum_n [I^{\ddagger}_{exp}(n) - I^{\ddagger}_{th}(n)]^2}{\sum_n [I^{\ddagger}_{exp}(n)]^2}$	0.1[17]
Van Hove[18]	R_3 = % of angles or slopes of $I^{\ddagger}_{exp}$ and $I^{\ddagger}_{th}$ have opposite signs	
Van Hove[19]	$R_4 = \frac{\sum_n \lvert I^{\ddagger\prime}_{exp}(n) - I^{\ddagger\prime}_{th}(n)\rvert}{\sum_n \lvert I^{\ddagger\prime}_{exp}(n)\rvert}$	
Van Hove[20]	$R_5 = \frac{\sum_n [I^{\ddagger\prime}_{exp}(n) - I^{\ddagger\prime}_{th}(n)]^2}{\sum_n [I^{\ddagger\prime}_{exp}(n)]^2}$	
Zanazzi and Jona[21]	$R_{ZJ} = \frac{A}{\delta E}\int \omega(E) \lvert c\chi_{th}' - \chi_{exp}'\rvert dE$ $c = \int \chi_{exp} dE / \int \chi_{th} dE$ and $\omega = \frac{\lvert c\chi_{th}'' - \chi_{exp}''\rvert}{\lvert \chi'_{exp}\rvert + \epsilon}$ $\epsilon = \lvert \chi'_{exp}\rvert_{max}$ and $A = \delta E / (0.027 \int \chi_{exp} dE)$	0.20–0.35[22]

Normally some parameters are refined by leaving other fixed, until they are all gradually released and refined, in the order of decreasing sensitivity to the reliability factor. It is convenient to perform the refinement in this way because if too many parameters are varied simultaneously, especially in the early stages of optimization, an initial error can result in large changes in the less sensitive parameters. These parameters can then reach values without a physical meaning to compensate for displacements of more sensitive atoms. In general, the coordinates perpendicular to the surface modify the reliability factor more than the parallel ones (due to the larger momentum transfer associated to the component perpendicular to the surface). Similarly, the reliability factor is more sensitive to the coordinates of the surface atoms than to the atoms of the deeper layers (due to the attenuation). The usual way of proceeding is therefore to first vary the perpendicular coordinates of the atoms depending on the increasing depth in the material.

In the refinement process, there are local minima that need to be avoided. The random minima of the reliability factor have often a value close to the average. On the contrary, the changes due to the diffraction coincidences are much deeper. Another way to identify local minima is to check if the lengths and the binding angles are physically meaningful. It is also possible to improve the description of the scattering potential with a greater number of phase shifts, which increases the difference between the local minima and the global minimum [39]. An alternative procedure is to use different reliability factors, all of which reach a minimum for the global minimum; otherwise, each reliability factor reaches a minimum for a different set of parameters [40].

The simplest method of searching the global minimum is the trial and error method. The problem of this method is the computation time, which varies exponentially with the number N of varied parameters. In order to reduce this computing time and achieve greater automation of the refinement process, mathematical algorithms have been developed which can be classified into global methods and refinement methods. In the early stages of the research, it is better to use global methods as the cooling algorithm[41], the genetic one[42], or mixed algorithms. Indeed, some refinement algorithms have been developed including that of Marquardt [43], the 'simplex' method [44, 45], the Hooke-Jeeves algorithm [46], which uses steepest descent procedures [45], the gradient algorithm [21, 46, 47], the Rosenbrook version[48] and the direction-set algorithm [44, 45, 49]. These refinement algorithms are not free of problems. Some are based on reliability factor derivatives, which are sensitive to experimental noise or numerical truncations. Sometimes, refinement strategies cannot distinguish between local minima and the global minimum, which requires starting the refinement process from a variety of structures with the hope of finding the overall minimum. In this situation, the calculation time is the same as that of the trial and error methods. In other methods as in the cooling algorithm, the probability of finding the global minimum depends on certain parameters introduced to reduce the search time. It is therefore necessary to have prior knowledge of the topography of the reliability factor hypersurface in order to be able to choose the appropriate searching parameters. All these aspects explain why the trial and error procedure is still used, in addition to the more refined methods.

Once the absolute minimum of the R-factor has been determined as a function of the structural parameters, the error associated with each parameter must be determined. Uncertainty in the structural determination is due to random and systematic errors. The error bars are only well defined for the main directions, which in general do not coincide with Cartesian coordinates. The error matrix is $\varepsilon = G^{-1}$, where the elements of the G matrix are:

$$G_{nm} = \frac{\partial^2 R}{\partial x_n \partial x_m}. \tag{6.4}$$

ε has non-diagonal terms because of the correlations between the parameters [50]. In X-ray diffraction, these correlations are often neglected, which allows to define the errors by the diagonal terms of the matrix without diagonalizing. Reference [51] proposes a study on this approximation. This approach may underestimate the standard deviations by a factor of two at least [16].

The complexity of the rigorous analysis of the reached precision has led to the search for other methods to estimate errors. Sometimes the accuracy has been estimated from the half-width of the global minimum. In experiments where the same initial state is studied for different photon energies, the accuracy can be determined from the R_P Pendry factor [30] or the standard deviation of Adams [50]. In the considerations of Adams and Pendry, the error bar for a coordinate is obtained from the $1/\beta$ curvature of the reliability factor minimum measured for this coordinate:

$$\sigma_A^2 = \frac{\beta R_0}{N} \tag{6.5}$$

$$\sigma_P^2 = \frac{\beta R_0}{\sqrt{N/8}} \tag{6.6}$$

with R_0 the minimum of the reliability factor and N the number of available points. Adams takes for N the number of Bragg peaks in the ΔE measurement regions and Pendry in turn estimates the number of peaks including multiple scattering effects and takes $N = \Delta E/4|V_{0i}|$, where V_{0i} is the imaginary part of the inner potential [16, 30, 53]. For the same structural determination, the error bar of Adams is generally optimistic and that of Pendry is conservative. There is often a factor 3 between them [32].

Other work defined the variance of the minimum of the R-factor, R_{min}, as [9]:

$$var(R_{min}) = \sigma^2 = R_{min}\left(\frac{2}{N}\right)^{1/2} \tag{6.7}$$

where N is the number of independent information objects, including the attenuation effects of the photoelectron. The use of different reliability factors has sometimes allowed to give a heuristic measure of the uncertainties. The error can be associated to the dispersion of the coordinates of the global minimum for each reliability factor. In all cases, it is possible to eliminate structures with larger reliability factors than $R_{min} + var(R_{min})$.

6.3 Photoelectron diffraction examples

Now that we have presented the principle of photoelectron diffraction, we shall deal with experimental considerations and those related to the data analysis, and we will discuss some examples. The difficulty of the low-energy regime analysis has promoted more studies exploiting the forward scattering in the high-energy regime. These studies include, for example, valuable metallurgical information on species segregation or understanding the growth mode of a film on a substrate. In the latter case, the absence of forward scattering peaks of the adsorbate when the first monolayer is completed indicates a layer-by-layer growth. Similarly, when a face-centered cubic structure (fcc) has a sufficiently large lattice parameter, a forward scattering peak cannot appear at normal emission. The presence of such a peak indicates stacking faults.

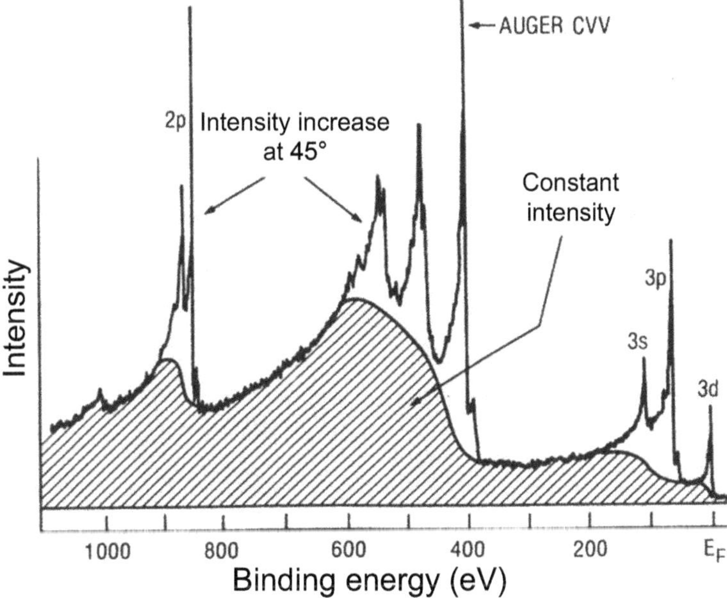

Figure 6.5 *A photoemission spectrum of Ni (001) which is decomposed into a background which does not depend on the emission angle (dashed part) and peaks whose intensity has a maximum at 45° when emission is along a bond direction (from [52]). Reprinted figure with permission from W. F. Egelhoff, Jr., Phys. Rev. B 30, 1052 (1984). ©1984 by the American Physical Society.*

Figure 6.5 illustrates the principle of photoelectron diffraction applications using forward scattering. In the figure we see a spectrum for Ni(001) which can be decomposed into a background (dashed line) and some peaks. In the directions connecting the emitter with a first neighbor (45° emission angle), there is an increase in the core level intensity due to forward scattering. A straightforward application of this fact is the qualitative comparison of diagrams giving immediate structural information. Figure 6.6 shows photoelectron diffraction of photoelectrons for Ni (110) and Nb (110). The difference between the two diffraction patterns indicates that the two

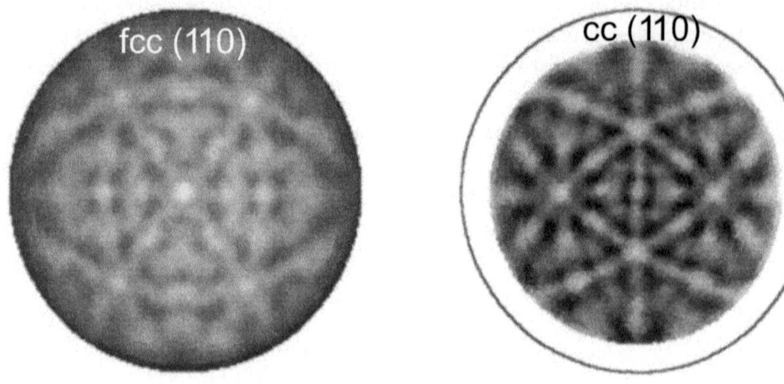

Figure 6.6 *Ni (110) and Nb (110). Emission of Ni 3p (left) and Nb 3d (right) excited with the K α line of Mg (from [54]). Reprinted from Surf. Sci. vol. 402-404, P. Aebi, R. Fasel, D. Naumović, J. Hayoz, Th. Pillo, M. Bovet, R. G. Agostino, L. Patthey, L. Schlapbach, F. P. Gil, H. Berger, T. J. Kreutz, J. Osterwalder, "Angle-scanned photoemission: Fermi surface mapping and structural determination", page 614, ©1998, with permission from Elsevier.*

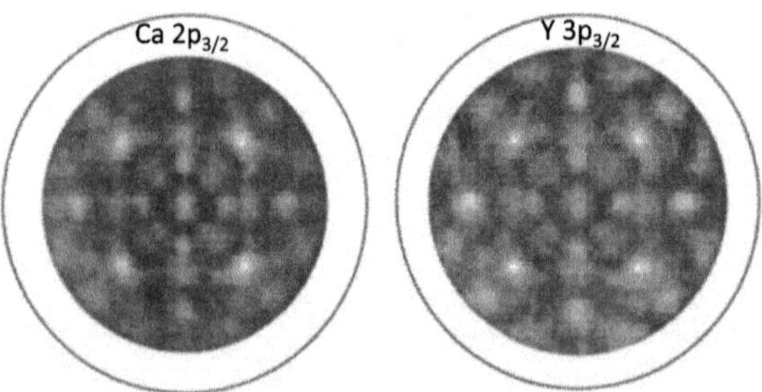

Figure 6.7 *Photoelectron diffraction of photoelectrons from $Bi_2Se_2CaCu_2O_{8+\delta}$ doped with Y. Since Y and Ca give the same diffraction signal, they occupy the same site (according to [54]). Reprinted from Surf. Sci. vol. 402-404, P. Aebi, R. Fasel, D. Naumović, J. Hayoz, Th. Pillo, M. Bovet, R.G. Agostino, L. Patthey, L. Schlapbach, F. P. Gil, H. Berger, T. J. Kreutz, J. Osterwalder, "Angle-scanned photoemission: Fermi surface mapping and structural determination", page 614, ©1998, with permission from Elsevier.*

structures are different. Indeed, the Ni has a fcc structure while the Nb is a centered cubic structure (cc).

Another application of photoelectron diffraction allows to know if one element replaces another in a crystal. Figure 6.7 shows the case of Bi2212 doped with Y. This system has a large unit cell, which makes difficult a detailed crystallographic study. However, the pattern shows that the photoelectron diffraction pattern of Y is identical to that of Ca, so it can be concluded that Y is substituting Ca in this material.

Beyond this qualitative information on the structure, photoelectron diffraction can also determine the lattice parameters of structures. An example of this application is the study of relaxations. Figure 6.8 shows forward scattering results for a thin film

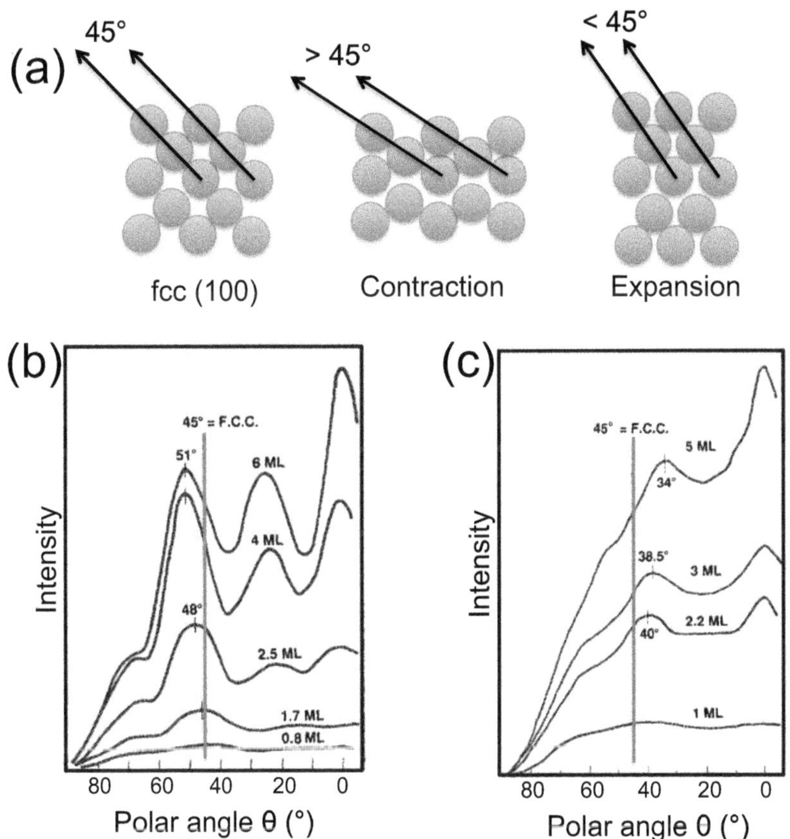

Figure 6.8 *(a) Use of forward scattering to determine dilatations and contractions of the lattice. (b) Mn/Ag(001) et (c) Mn/Cu(001) (from [55]). Reprinted with permission from W. F. Egelhoff Jr. et al., J. Vac. Sci. Technol. A 8, 1582 (1990). ©1990, American Vacuum Society.*

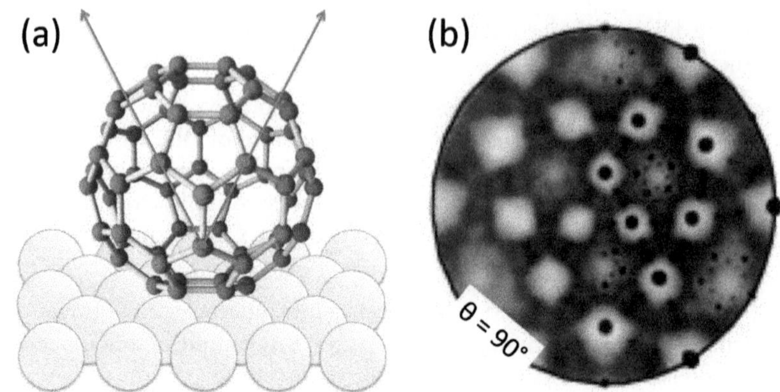

Figure 6.9 *(a) When the surface is illuminated, photoelectrons are emitted from each C atom and are diffused by the atoms in their neighborhood. (b) Diffraction pattern calculated for the structure shown on (a) (from [56]). Reprinted figure with permission from R. Fasel, P. Aebi, R. G. Agostino, D. Naumovic, J. Osterwalder, A. Santaniello, and L. Schlapbach, Phys. Rev. Lett. 76, 4733 (1996). ©1996 by the American Physical Society.*

of Mn on Ag (001). The forward scattering maxima appear at 48° and 51° and not at 45° as it would be the case for an fcc structure or at 54.7° for a cc structure, demonstrating a vertical contraction of the thin film and an intermediate structure between fcc and cc. This phenomenon is due to the pseudomorphic growth of Mn on the Ag substrate. In contradistinction, for a film of Cu on Ag (001), the situation is reversed. The film in this case is expanded vertically. Finally, in order to obtain a simple knowledge of the total volume of the unit cell, it is sufficient to combine this result with a Low Energy Electron Diffraction measurement which gives information on the in-plane lattice parameters.

Other applications involve atomic layers or molecules on surfaces. The analysis of the photoelectron diffraction is considerably simplified in these cases, where the chemical sensitivity allows to select a reduced number of emitters which are well-known. It is then possible to use forward scattering to determine the tilt angle of linear molecules on surfaces by analyzing the emission of one of their atoms. A more complex example, demonstrating the power of this approach, is the determination of the C_{60} molecule adsorption on noble metals. For this system, conventional diffraction techniques have failed to determine the structure, but photoelectron diffraction can benefit from its chemical sensitivity and the non-requirement of long-range order. Indeed, the forward scattering provides the projection of the atomic structure around the emitters and allows to differentiate if the C_{60} exposes a hexagon or a pentagon towards the surface. Figure 6.9 shows the calculation of the diffraction when a hexagon is in contact with the surface. The black points superimposed to the computation correspond to the experimental interatomic directions. The agreement allows us to conclude on the molecular orientation on the surface.

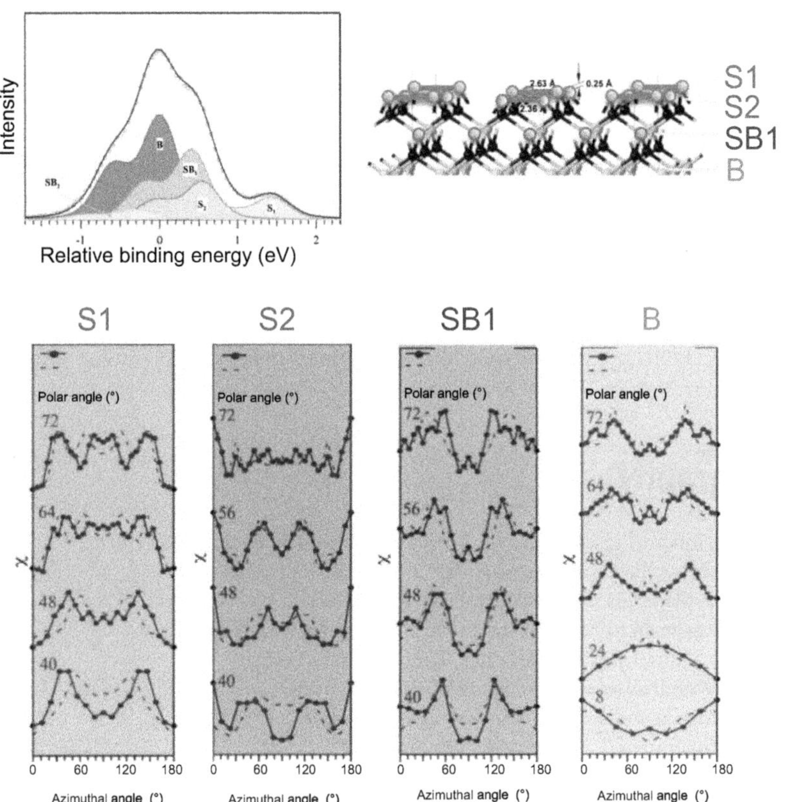

Figure 6.10 *Scheme of the reconstruction called AUDD of the surface c(4×2) of cubic silicon carbide. Comparison of the experimental diffraction of each of the core level components and the photoelectron diffraction simulation (from [57]). Figure reprinted with permission from A. Tejeda, E. Wimmer, P. Soukiassian, D. Dunham, E. Rotenberg, J. D. Denlinger, and E. G. Michel, Phys. Rev. B 75, 195315 (2007). ©2007 by the American Physical Society.*

These applications demonstrate that photoelectron diffraction provides rather straightforward information about the structure. Complete structural determinations and discrimination of structural models using photoelectron diffraction are less common. For surface reconstructions, it is necessary to use the low-energy regime to reach the required surface sensitivity. This energy range corresponds to the backscattering regime, where the multiple scattering effects are considerable and an intuitive interpretation impossible. This type of structural determination must therefore involve simulation and the comparison between theory and experience. Figure 6.10 presents the results of such an analysis for the $c(4 \times 2)$ reconstruction of cubic silicon carbide. The figure shows the experimental angular variation of the intensity for each of the core level components and the simulation of the theoretical photoelectron diffraction. Agreement between theory and experiment justifies the structural model for surface reconstruction and at the same time validates the interpretation of the core level decomposition by the good agreement between theory and experience. We thus see here a specificity of photoelectron diffraction which is the determination of the diffraction of every inequivalent atom.

Without being exhaustive, we have presented some examples of the application of photoelectron diffraction. The technique is particularly useful when the system does not have a long-range order because it is the only possible diffraction technique. Photoelectron diffraction allows us to study the structure around a given atom on a characteristic radius of the photoelectron mean free path. We have also seen the great advantage of chemical sensitivity, which simplifies problems by choosing specific emitters to facilitate data analysis. These properties make this technique unique.

Bibliography

[1] M. Seelmann-Eggebert and H. J. Richter, Phys. Rev. B **43**, 9578 (1991).
[2] C. Fadley, J. Electron Spectrosc. **5**, 725 (1974).
[3] C. S. Fadley, Surf. Sci. Rep. **19**, 231 (1993).
[4] V. Fritzsche and D. P. Woodruff, Phys. Rev. B **46**, 16128 (1992).
[5] K.-M. Schindler *et al.*, Phys. Rev. Lett. **71**, 2054 (1993).
[6] D. P. Woodruff *et al.*, Surf. Sci. **357-358**, 19 (1996).
[7] P. Hofmann and K.-M. Schindler, Phys. Rev. B **47**, 13941 (1993).
[8] P. Hofmann *et al.*, Nature **368**, 131 (1994).
[9] M. Kittel *et al.*, Surf. Sci. **470**, 311 (2001).
[10] J. Wider *et al.*, Phys. Rev. Lett. **86**, 2337 (2001).
[11] A. Szöke, *Short Wavelength Coherent Radiation: Generation Applications*, Vol. 147 of *AIP Conf. Proc.* (AIP, New York, 1986), edited by T. Attwood and J. Boker.
[12] J. Barton, Phys. Rev. Lett. **61**, 1356 (1988).
[13] D. Gabor, Nature **161**, 777 (1948).
[14] G. R. Harp, D. K. Saldin, and B. P. Tonner, Phys. Rev. Lett. **65**, 1012 (1990).
[15] G. R. Harp, D. K. Saldin, and B. P. Tonner, Phys. Rev. B **42**, 9199 (1990).
[16] D. L. Adams, V. Jensen, X. F. Sun, and J. H. Vollesen, Phys. Rev. B **38**, 7913 (1988).
[17] R. Dippel *et al.*, Chem. Phys. Lett. **199**, 625 (1992).
[18] R. Fasel *et al.*, Phys. Rev. B **50**, 14516 (1994).
[19] J. A. Martín-Gago *et al.*, Phys. Rev. B **55**, 12896 (1997).
[20] S. Bengió *et al.*, Phys. Rev. B **65**, 205326 (2002).
[21] G. Kleinle, W. Moritz, D. L. Adams, and G. Ertl, Surf. Sci. **219**, L637 (1989).
[22] A. E. S. von Wittenau *et al.*, Phys. Rev. B **45**, 13614 (1992).
[23] C. Rojas *et al.*, Phys. Rev. B **57**, 4493 (1998).
[24] M. A. V. Hove, S. Y. Tong, and M. H. Elconin, Surf. Sci. **64**, 85 (1977).
[25] R. S. Saiki *et al.*, Surf. Sci. **282**, 33 (1993).
[26] R. S. Saiki *et al.*, Surf. Sci. **279**, 305 (1992).
[27] Z. Huang *et al.*, Phys. Rev. B **48**, 1696 (1993).
[28] E. Zanazzi and F. Jona, Surf. Sci. **62**, 61 (1977).
[29] E. Michel *et al.*, J. Phys. IV France **132**, 49 (2006).
[30] J. B. Pendry, J. Phys. C: Solid St. Phys. **13**, 937 (1980).

[31] P. R. Watson, F. R. Sheperd, D. C. Frost, and K. A. R. Mitchell, Surf. Sci. **72**, 562 (1978).
[32] M. A. V. Hove *et al.*, Surf. Sci. Rep. **19**, 191 (1993).
[33] J. Osterwalder *et al.*, Surf. Sci. **331-333**, 1002 (1995).
[34] D. H. Rosenblatt *et al.*, Phys. Rev. B **26**, 3181 (1982).
[35] R. X. Ynzunza *et al.*, Surf. Sci. **459**, 69 (2000).
[36] R. X. Ynzunza *et al.*, Surf. Sci. **442**, 27 (1999).
[37] E. L. Bullock *et al.*, Phys. Rev. B **41**, 1703 (1990).
[38] P. J. Rous, Prog. Surf. Sci. **39**, 3 (1992).
[39] D. L. Adams, H. B. Nielsen, and M. A. V. Hove, Phys. Rev. B **20**, 4789 (1979).
[40] M. A. V. Hove and R. J. Koestner, *Determination of Surface Structure by LEED* (Plenum, New York, 1984), edited by M. Marcus and F. Jona.
[41] P. J. Rous, Surf. Sci. **296**, 358 (1993).
[42] R. Döll and M. A. V. Hove, Surf. Sci. **355**, L393 (1996).
[43] D. W. Marquardt, J. Soc. Ind. Appl. Math. **11**, 431 (1963).
[44] U. Starke *et al.*, Surf. Sci. **287**, 432 (1993).
[45] W. H. Press, B. P. Flannery, S. A. Teukolsky, and W. T. Vetterling, *Numerical Recipes* (Cambridge University Press, Cambridge, 1986).
[46] P. G. Cowell and V. E. de Carvalho, Surf. Sci. **187**, 175 (1987).
[47] G. Kleinle, W. Moritz, and G. Ertl, Surf. Sci. **238**, 119 (1990).
[48] H. H. Rosenbrook, Comput. J. **3**, 175 (1960).
[49] R. P. Brent, *Numerical Recipes: Algorithms for Minimization without Derivatives* (Prentice-Hall, Englewood Cliffs, NJ, 1973).
[50] H. B. Nielsen and D. L. Adams, J. Phys. C **15**, 615 (1982).
[51] E. Prince, *Mathematical Techniques in Crystallography and Materials Science* (Springer, Heidelberg, 1982).
[52] W. Egelhoff, Phys. Rev. B **30**, 1052 (1984).
[53] N. A. Booth *et al.*, Surf. Sci. **387**, 152 (1997).
[54] P. Aebi *et al.*, Surf. Sci. **402-404**, 614 (1998).
[55] W. F. Egelhoff, J. Vac. Sci. Technol. A **8**, 1582 (1990).
[56] R. Fasel *et al.*, Phys. Rev. Lett. **76**, 4733 (1996).
[57] A. Tejeda *et al.*, Phys. Rev. B **75**, 195315 (2007).

[31] [illegible] Mitchell, Surf. Sci. 72, [illegible]

[32] M.A. Van Hove et al., Surf. Sci. Rep. 19, 191 [illegible]

[33] J. Osterwalder et al., Surf. Sci. 331-333, 1002 (1995).

[34] [illegible]

[35] [illegible]

[36] [illegible]

[37] [illegible]

[38] [illegible] Surf. Sci. [illegible]

[39] [illegible] and M.A. Van Hove, Phys. Rev. [illegible]

[40] M.A. Van Hove [illegible]

[illegible]

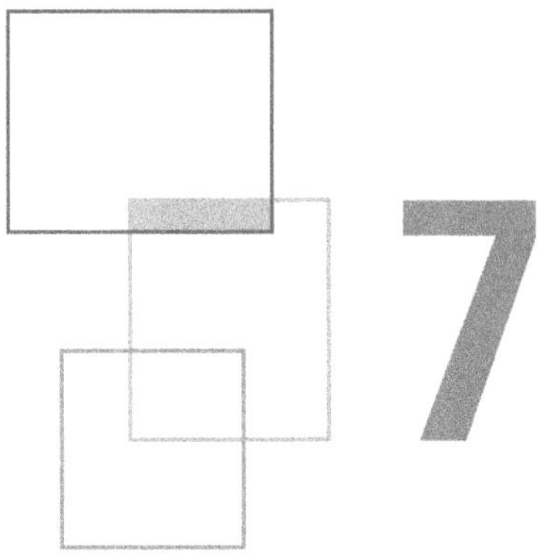

Dispersion relations

We saw previously that in a solid, there are two types of electrons. On the one hand there are the core electrons, strongly bonded to the core and characteristic of every atom. On the other, we have the valence electrons which are involved in the chemical bond due to the overlap of atomic orbitals. In crystals, the coherent superposition of these atomic orbitals forms collective electronic states (called Bloch states) that give rise to the band structure. The latter is characterized by the dispersion relations $(E(\vec{k}))$, *i.e.*, the relationship between the energy of the Bloch state of a given band and its wavevector. Photoemission spectroscopy measures in a straightforward way the energy dispersion and allows to determine experimentally the band structure. In figure 7.1, we present a map of the photoemission intensity as a function of the angle (*i.e.*, the wave vector) on silicon carbide. The dispersion of several bands of this system is clearly distinguishable.

However, angle-resolved photoemission provides more subtle information about electronic properties. Indeed, the mechanisms of the interaction between electrons and the collective excitations in a crystal (phonons, magnons, etc.) can modify the experimental dispersion close to the Fermi level. This modification appears in an energy domain characteristic of the considered excitation (e.g. Debye energy for electron-phonon coupling). Figure 7.1 b illustrates this behavior on the high temperature superconductor Bi2212. The dispersion shows a slope change (a *kink*), at the energy that separates two different regimes. For bond energies $E_b \gtrsim 75$ meV, the dispersion is that of a particle without interactions, whereas below this energy level, the dispersion reflects the coupling of the electrons with a collective excitation. The figure also shows that the dispersion relation does not change in the studied $h\nu$ range, whereas these spectra correspond to very different mean free paths. It also indicates that the best resolution is obtained with low energy photons (laser photoemission).

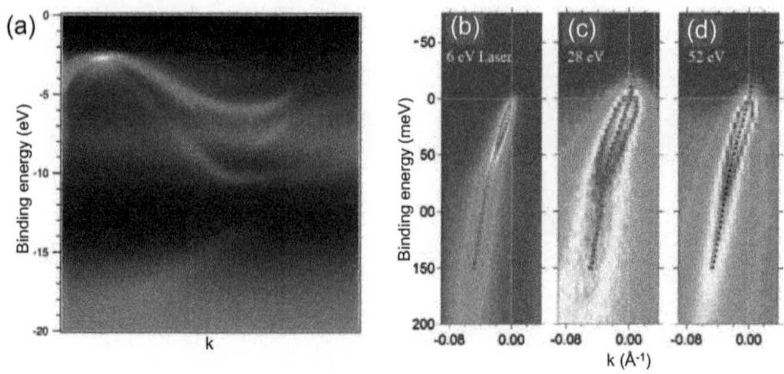

Figure 7.1 *(a) Image in false colors of the photoemission intensity as a function of the energy and the wavevector for a SiC surface. (b)-(d) Dispersion near the Fermi level of Bi2212 as measured with (b) a 6 eV laser source (T = 25 K) (c) a 28 eV synchrotron source (T = 26 K) (d) a 52 eV synchrotron source (T = 16 K) (after [1]). Figure reprinted with permission from J. D. Koralek, J. F. Douglas, N. C. Plumb, Z. Sun, A. V. Fedorov, M. M. Murnane, H. C. Kapteyn, S. T. Cundiff, Y. Aiura, K. Oka, H. Eisaki, and D. S. Dessau, Phys. Rev. Lett. 96, 017005 (2006). ©2006 by the American Physical Society*

Technical progress, particularly on the resolution in energy and angle (wave vector) allow to measure the intrinsic widths as seen in Fig. 7.2. It is represented the Ag(111) spectrum at the Γ point ($\vec{k} = 0$) with spectrometers of increasing resolution. Nowadays, low temperature measurements with the best resolution allow us to estimate the intrinsic width of the excitations and thus to figure out their life time. In the following, we will discuss the different ways of representing the photoemission data (*cf.* section 7.1) as well as the information that can be extracted (*cf.* section 7.2).

7.1 Experimental data representation

Several representations are commonly used to present the dispersion relations: a two-dimensional representation, consisting of an intensity map as a function of the energy and as a function of the angle or the wave vector, but also sets of spectra at constant energy or wave vector. It is also common to represent isoenergetic cuts of the band structure in a chosen plane of the reciprocal space, the zero-energy cut constitutes the Fermi surface.

7.1.1 *Two-dimensional representations of the dispersion*

In a two-dimensional representation, the grayscale or the color intensity map represents the measured photoemission signal as a function of the energy (vertical

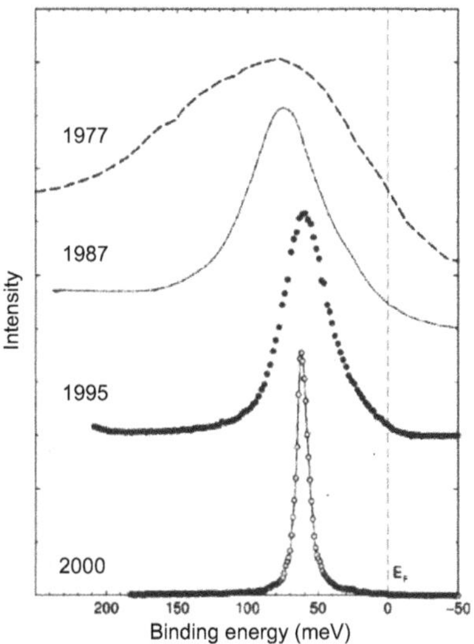

Figure 7.2 *Evolution of the performances of photoemission analyzers over time as illustrated by the measurement of the Ag(111) surface state. (a) Measurements at room temperature, integrated in angle, (b) measurements at ambient temperature and normal emission ($\Delta E \simeq 60$ meV, $\Delta\theta = 1°$), (c) measurements at 56K ($\Delta E \simeq 21$ meV, $\Delta\theta = 0.9°$) and (d) measurements at 30K ($\Delta E \simeq 3.5$ meV, $\Delta\theta = 0.3°$) (after [2]). Figure reprinted with permission from F. Reinert, G. Nicolay, S. Schmidt, D. Ehm, and S. Hüfner, Phys. Rev. B 63, 115415 (2001). ©2001 by the American Physical Society.*

axis) and the wave vector along a reciprocal space direction (horizontal axis). The generalization of Equation 3.35 allows us to transform angles into wave vector (Fig. 3.13):

$$k_x = \frac{\sqrt{2mE_{kin}}}{\hbar} \sin\theta \cos\phi \tag{7.1}$$

$$k_y = \frac{\sqrt{2mE_{kin}}}{\hbar} \sin\theta \sin\phi \tag{7.2}$$

or their numerical expression:

$$k_x(\text{Å}^{-1}) = 0.512\sqrt{E_{kin}(\text{eV})} \sin\theta \cos\phi \tag{7.3}$$

$$k_y(\text{Å}^{-1}) = 0.512\sqrt{E_{kin}(\text{eV})} \sin\theta \sin\phi \tag{7.4}$$

where the prefactor 0.512 is $\sqrt{2m}/\hbar$ and it allows to obtain k in Å^{-1} when the kinetic energy is expressed in eV. Equations 7.1-7.2 allow us to determine the angles (θ, ϕ) to vary $k_{||}$ along an arbitrary direction. For directions containing $\overline{\Gamma}$ (the projection of the Γ point in the first surface Brillouin zone), ϕ is made constant and the polar angle θ is varied. This procedure has been applied to obtain the

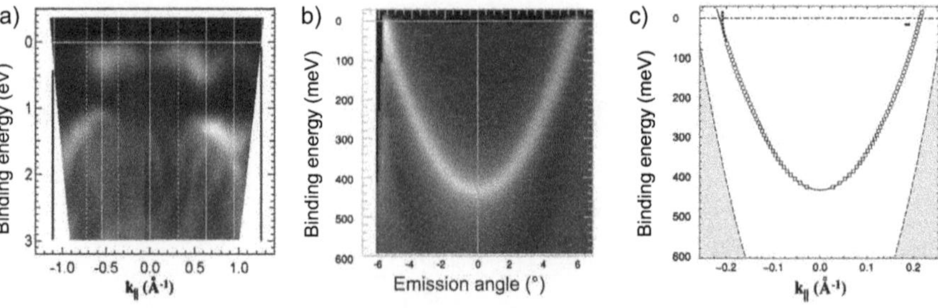

Figure 7.3 *(a) Bands of 1/3 ML of Sn/Si (111)-$\sqrt{3} \times \sqrt{3}$R30° (after [3]) Reprinted figure with permission from J. Lobo, A. Tejeda, A. Mugarza, and Michel EG, Phys. Rev. B 68, 235332 (2003). ©2003 by the American Physical Society. (b) Two-dimensional representation of the Cu(111) Shockley state dispersion (logarithmic scale) (after [2]). (c) Points obtained from the intensity maxima in (b) Figure reprinted with permission from F. Reinert, G. Nicolay, S. Schmidt, D. Ehm, and S. Hüfner, Phys. Rev. B 63, 115415 (2001). ©2001 by the American Physical Society.*

band structure for the $(\sqrt{3} \times \sqrt{3})$R30° reconstruction of the Sn/Si(111) interface (Fig. 7.3a). In this case, θ was varied by moving the analyzer with respect to the sample, while the azimuth angle ϕ is changed by rotating the sample around its normal. This procedure requires mobile analyzers. In the same figure, we represent the dispersion of the Cu(111) Shockley state which was obtained with a fixed two-dimensional detector (Fig. 7.3b,c). These fixed detectors can be bigger and have higher brightness, which reduces the acquisition time considerably and improves the angular and energy resolution[1]. Moreover, intensity depends on the measurement geometry and can be asymmetrical as observed on the two branches of Fig. 7.3b. In the parabolic dispersion of the surface state, the asymmetry between the right and left intensity reflects the effect of the incidence of light (polarization).

7.1.2 Energy and Momentum Distribution Curves

Energy Distribution Curves

Even if nowadays the two-dimensional representation is extensively widespread, many results are still presented in the form of "energy distribution curves" (EDC), where the photoemitted intensity is plotted according to the kinetic or the binding energy (Fig. 7.2). The photo-emitted intensity is usually expressed in arbitrary units as a function of the energy (for a given emission angle).

[1] With these detectors, however, care must be taken to compare the intensities of the different channels (especially those at the edges of the detector) and be sure of having a proper alignment.

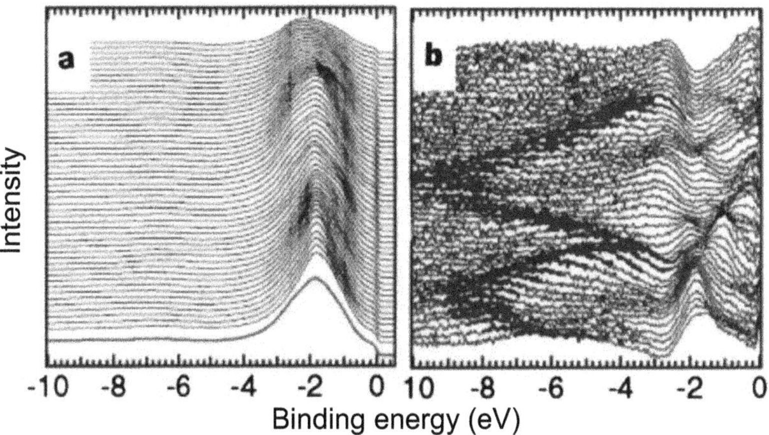

Figure 7.4 *Photoemission spectra for $Al_{71.8}Ni_{14.8}Co_{13.4}$ a) Series as a function of the polar angle (between 0° and 45°) b) Data from (a) normalized to the angle-integrated spectrum (from [4]). Adapted by permission from Springer Nature: Nature 406, 602 (2000), ©2000.*

The juxtaposition of EDC spectra as a function of the emission angle allows to visualize the dispersion of the states. Indeed, the peak positions in the EDC series give the dispersion relations ($E(k_{||})$) as shown in Fig. 7.3c obtained from the data of Fig.7.3b. The determination of a precise dispersion relation relies on the precise measurement of the peak position, so it is often necessary to fit the peaks. Peaks are usually described by Voigt functions, or simply by a Gaussian line shape with an adjusted $\sigma(E)$ width. The background of the EDCs, *i.e.*, the unstructured signal arising from the background, is not perfectly known theoretically. In principle, this background originates mainly from electrons that have undergone inelastic collisions. It can thus be described by a Shirley background [5], that is a curve which, at each energy E, is proportionate to the integral of the spectrum at lower binding energies. More generally, it is phenomenologically described with polynomials of slow variation.

A rather remarkable example of the background subtraction is encountered in quasicrystals. In these systems, the absence of periodicity reveals an infinite number of lattice parameters and reciprocal space vectors. The dispersion of $Al_{71.8}Ni_{14.8}Co_{13.4}$ without background subtraction (Fig. 7.4a) shows only broad structures of low dispersion. However, when the background is removed from the experimental data, a highly dispersive band appears between -11 and -4 eV (Fig. 7.4b).

Another representation that can enhance spectral signatures is the second derivative representation of the photoemission data, where $(-d^2I/dE^2)$ is reported. This representation highlights peaks and shoulders (Fig. 7.5). The statistical noise is also greatly increased, so this procedure requires experimental spectra with high statistics or smoothing of the original EDCs.

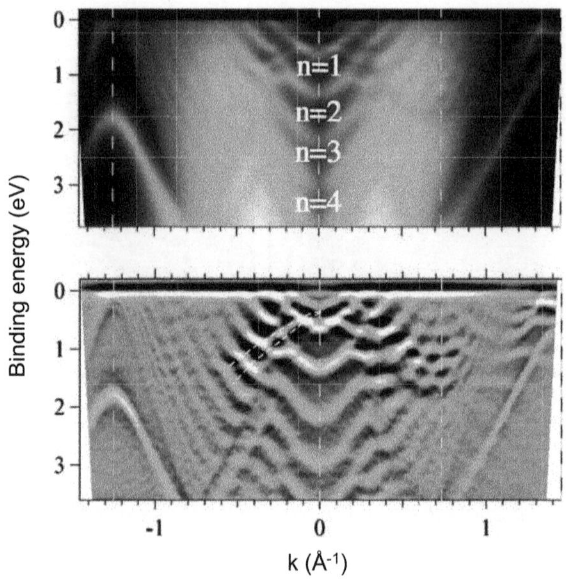

Figure 7.5 *Band structure of a surface alloy of Ag_2Bi on top of a Ag thin layer on Si. (a) Photoemission intensity. (b) Second derivative of the photoemission intensity, which allows to distinguish more clearly the details (according to [6]). Reprinted from M. Ogawa, P. M. Sheverdyaeva, P. Moras, D. Topwal, A. Harasawa, K. Kobayashi, C. Carbon and I. Matsuda, "Electronic structure study of ultrathin Ag (111) films modified by a Si (111) substrate and $\sqrt{3} \times \sqrt{3}$-Ag2Bi surface", J. Phys .: Condens. Matter 24, 115501 (2012).*

Momentum Distribution Curves

The same information contained in the EDCs can be obtained when performing sections of the two-dimensional representations as a function of the wave vector. These curves are called momentum distribution curves (MDC). They also allow us to obtain the dispersion relations by varying the energy of the considered cut. Figure 7.6 shows the two-dimensional representation, the EDCs and the MDCs near the Fermi surface of LSCO, a high Tc superconductor. The inelastic contribution to MDCs is not known from a theoretical point of view, but it can often be adjusted to a simple linear function. The peaks can then be fitted to a lorentzien function if the dependence with the wave vector normal to the Fermi surface is small [8, 9]. An experimental broadening may make it necessary to convolve the lorentzian by a gaussian.

The information contained in EDCs and MDCs is the same. Nevertheless, MDCs have the advantage of being constant-energy curves and, therefore, are unaffected by the deformation induced by the Fermi Dirac distribution. In this way it is possible to follow peaks even for energies above E_F, if the spectral weight and the signal-to-noise

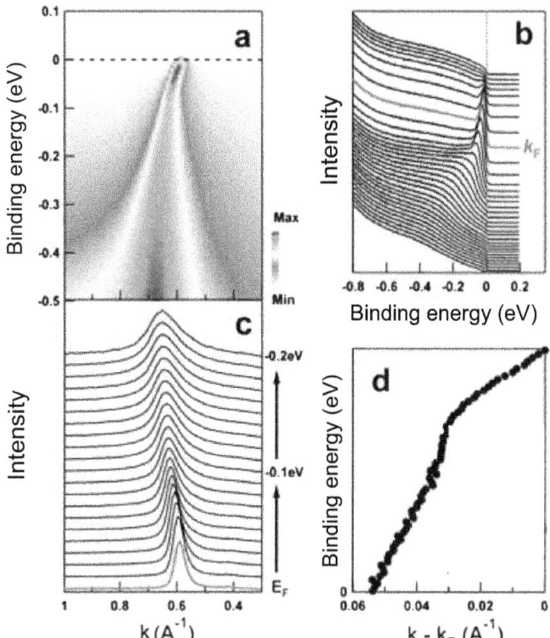

Figure 7.6 *Photoemission data for* $(La_{2-x}Sr_x)CuO_4$ *with* $x = 0.063$ *along the direction (0,0) -* (π, π) *at 20 K. (a) Raw data (b) EDCs (c) MDCs (d) Dispersion relation out of MDCs (from [7]). Reprinted from "Technical Reports: Electron-Phonon Coupling in High-Temperature Cuprate Superconductors as Revealed by Angle-Resolved Photoemission Spectroscopy" by X. J. Zhou, J. Hussain, Z.-X. Shen, Synchrotron Radiation News, Vol. 18: 3, pp 15-23 (2005), with permission of Taylor & Francis Ltd (www.tandfonline.com).*

ratio allow this. Matrix elements, on the other hand, affect MDCs more than EDCs, because the dependence of matrix elements on **k** is greater than that on energy [10].

7.1.3 Positive-negative energy symmetrization of the data

The symmetrization of the photoemission data is sometimes used to analyze spectral features near the Fermi level, such as the opening of band gaps. The measurement of a gap opening at the Fermi level is important for studies of metallicity or different electronic instabilities. However, the study of the Fermi level neighborhood is not easy, due to the Fermi-Dirac distribution that hides existing states when their population is small. One way to overcome this drawback is the normalization by the Fermi function, which will be explained in detail below. Another possibility is to symmetrize the photoemission spectra with respect to the Fermi level (Fig. 7.7). We

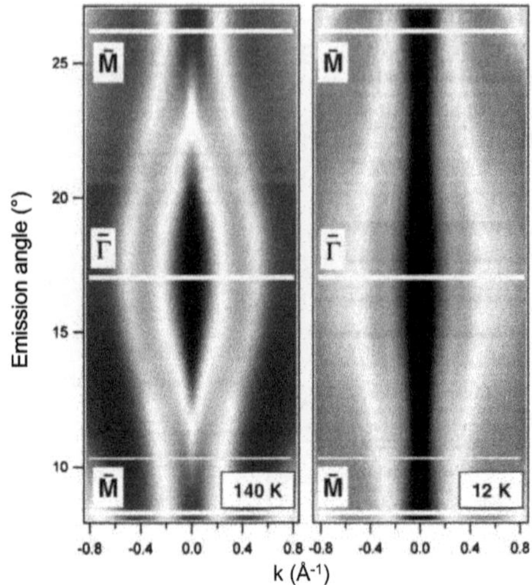

Figure 7.7 *Opening of a gap in the Mott transition of Sn/Ge (111) between 140 and 12 K (after [12]). Reprinted figure with permission from R. Cortés, A. Tejeda, J. Lobo, C. Didiot, B. Kierren, D. Malterre, E. G. Michel, and A. Mascaraque, Phys. Rev. Lett. 96, 126103 (2006). ©2006 by the American Physical Society.*

obtain the symmetric intensity I_{sym} according to

$$I_{sym} \propto I(k_F, E) + I(k_F, -E) \tag{7.5}$$

In this way, the Fermi Dirac function is eliminated. It is, therefore, possible to compare data at different temperatures without the influence of the Fermi function broadening, as for example on Sn/Ge (Fig. 7.7). The symmetrization allows to estimate the width of the gaps and its evolution as a function of the temperature, but this analysis only makes sense if the gap is symmetrical with respect to the Fermi level, since the symmetrization implies the electron-hole symmetry. That is why this method is widely used to study the superconductor gaps [11].

7.1.4 Normalizations to study states near the Fermi level

Each EDC measures the product of the spectral function by the Fermi distribution when the matrix elements vary only slightly. If it was possible to eliminate the effect of the Fermi-Dirac function, one could get access to the unoccupied states in an energy domain of the order of kT (corresponding to the thermally populated states). We will show how to achieve this goal.

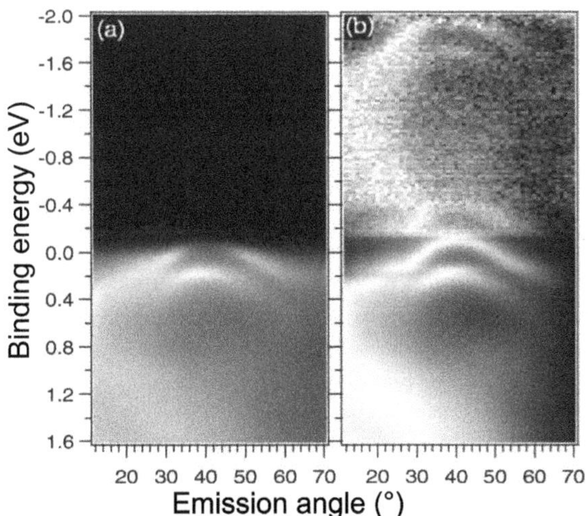

Figure 7.8 *(a) EDCs of Ni (111) without treatment, (b) normalization of each energy in (a) to have a maximum of angular contrast (according to [13]). Reprinted figure with permission from T. Greber, T. J. Kreutz, and J. Osterwalder, Phys. Rev. Lett. 79, 4465 (1997). ©1997 by the American Physical Society.*

One method for obtaining information on unoccupied states is to divide the experimental spectrum by the Fermi distribution. But since the Fermi function decays very fast with the energy above E_F, this procedure amplifies the statistical noise. It is, therefore, necessary to have very high quality experimental data with a very large signal-to-noise ratio and a much better energy resolution than the thermal broadening of the Fermi level ($4k_BT$) [14]. The experimental spectrum must be divided by the Fermi Dirac function further broadened due to the experimental resolution. The convolution of the Fermi Dirac distribution with a gaussian of FWHM ΔE_{exp} is approximately equivalent to an effective Fermi Dirac distribution [15] with

$$T_{eff} = \sqrt{T^2 + (\Delta E_{exp}/4k_B)^2} \tag{7.6}$$

This procedure gives information on the states above the Fermi level, in a limited energy domain, of the order of $4-5k_BT$. Beyond this limit, the experimental intensity is very low and the method does not extract information, as only amplified noise is obtained. This normalization procedure is very effective in highlighting narrow spectral patterns above E_F, as illustrated in the $CeCu_6$ Kondo Peak study [14]. Figure 7.8 shows another use of this technique to study the unoccupied bands of Ni (111).

It is instructive to analyze this procedure in detail to understand its limits. The experimental intensity corresponds to the spectral function $A(E, T)$ multiplied by the Fermi Dirac distribution $f(E, T)$ and convolved by the experimental function $g(E, \Delta E)$ (usually a Gaussian of FWHM ΔE is used):

$$I(E, T) = [A(E, T)f(E, T)] \otimes g(E, \Delta E) \tag{7.7}$$

To obtain spectral information above the Fermi level, the experimental intensity is divided by a Fermi function broadened by the experimental resolution:

$$\frac{[A(E, T)f(E, T)] \otimes g(E, \Delta E)}{f(E, T) \otimes g(E, \Delta E)} \tag{7.8}$$

However, since the convolution product is not distributive, we do not get $A(E, T) \otimes g(E, \Delta E)$. However, in the case of a a weak experimental broadening, $\Delta E \ll 4k_B T$, and we obtain a good approximation of $A \otimes g$. Interested readers can find all the details of this approach in the reference [16].

7.1.5 *Fermi surface representation*

The electrons obeying Fermi-Dirac statistics and therefore the Pauli exclusion principle must occupy different states. At zero temperature, states are occupied with a probability of 1 up to the Fermi level E_F and beyond this level, states are unoccupied. If a band is not completely filled, the system is metallic since occupied and unoccupied states are separated by an infinitesimal energy. As we saw above, the Fermi surface is defined as the intersection of the bands at E_F (Fig. 7.9a). It corresponds to the wave vectors $\vec{k}_F$ such that $E(\vec{k}_F) = E_F$. The Fermi surface is often presented by

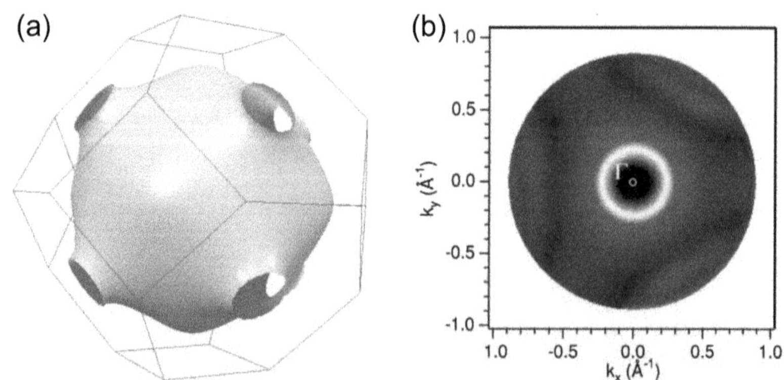

Figure 7.9 *(a) Fermi surface of bulk gold. The Fermi surface of the bulk does not exist along [111] directions, which allows the appearance of surface states (according to http://www.phys.ufl.edu/fermisurface/ http://www.phys.ufl.edu/fermisurface/) (b) Fermi surface of the surface state of Cu (111), which appears in the bulk gap of the direction [111]. After [17]). Figure reprinted with permission from F. Baumberger, T. Greber, and J. Osterwalder, Phys. Rev. B 64, 195411 (2001). ©2001 by the American Physical Society.*

2D sections of the Brillouin zone where the gray or color scale indicates the presence or absence of states at the Fermi level. The Fermi surface is a fundamental concept of condensed matter physics. Indeed, because of Pauli's principle, only electrons close to the Fermi level can be thermally excited (to thermally excite an electron of energy E, unoccupied states are required at $E + k_B T$). This is why the density of states at the Fermi level and also its topology play a fundamental role in most of the physical properties of solids (conductivity, magnetism, but also superconductivity or charge density waves). The experimental determination of the Fermi surface is therefore crucial for understanding a large number of properties.

The Fermi surface can be determined from the de Haas van Alphen oscillations. These oscillations are associated to the quantification of the electron energy in the presence of a magnetic field. The energy levels appear discretized according to the Landau levels. This method only applies for extremely well-ordered materials or with a small amount of defects and is limited to low temperatures. Other techniques for determining the Fermi surface exploit magneto-acoustic effect, Compton scattering, or positron annihilation. All these techniques are only adapted to the study of bulk materials and they provide rather indirect information on the Fermi surface and are difficult to apply in the case of complex systems. On the contrary, photoemission allows us to determine directly the band structure of the materials and therefore the Fermi surface, especially if the system is two- or one-dimensional, because the bands then depend only on $k_{||}$, which is unambiguously determined by photoemission.

We shall illustrate the study of the Fermi surface on noble metals. In fact, in these metals, the energy of the conduction band electrons is dominated by kinetic energy and consequently their distribution is quasi-isotropic in reciprocal space. The Fermi surface is thus formed of pseudospheres connected along the [111] directions by the "necks" (Fig. 7.9a). This shape allows to understand the gap opening in the [111] direction. Indeed, this topology of the Fermi surface with a gap localized in a particular direction of the Brillouin zone favors the appearance of surface states. This is indeed the case for the noble metals (Au, Ag, Cu) which present a surface state in the [111] direction called Shockley state, that can be seen in Fig. 7.9b for Cu(111).

k_F measurement

Determination of the Fermi surface requires the measurement of all the k_F vectors. It is expected to find a photoemission peak at the Fermi level for these values of the wave vector. However, the decrease of the spectral weight due to the Fermi function, the intrinsic or experimental broadening due to the experimental resolution or a significant dependence on the matrix elements can complicate this determination. In the following, we present the most commonly used procedures to obtain the Fermi surface.

Maximum intensity method

This method consists in determining the positions of the reciprocal space where the intensity at E_F becomes maximal [19–21] and which corresponds to the intersection of a band with the detection window placed at the Fermi level. This method is efficient and accurate when dispersions are important. In contrast, when the bands are narrow (low dispersion), we have spectral weight in a large range of wave vectors and the maximum intensity can be shifted with respect to the intersection of the band with the Fermi level (Fig. 7.10e).

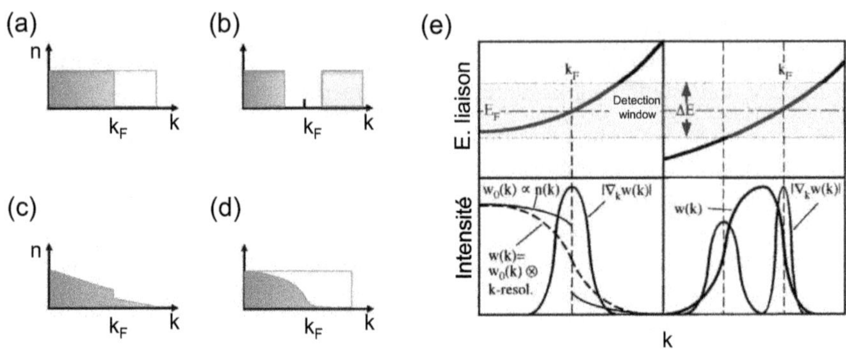

Figure 7.10 *(a)-(d): Density of states n(k) for different systems: a) metal b) band insulator c) Fermi liquid with electronic correlation d) correlated non-Fermi liquid system (e) Maximum intensity and maximum gradient methods. Dispersion of a system with a narrow band and with a wide band with respect to the detection window width (top). Representation of the measured intensity* $w(\mathbf{k})$ *as well as its derivative for systems with a narrow and a wide band (low) (after [18]) Figure reprinted with permission from Th. Straub, R. Claessen , P. Steiner, S. Hüfner, V. Eyert, K. Friemelt, and E. Bucher, Phys Rev. B 55, 13473 (1997). ©1997 by the American Physical Society.*

Maximum gradient method

The maximum gradient method is based on the variation of the density of states $n(k)$ [18, 22]. Fig. 7.10 shows $n(k)$ for different systems. A normal (weakly correlated) metal is characterized by a partially filled band and therefore there is a discontinuity in $n(k)$ for $k = k_F$, corresponding to the boundary between occupied and unoccupied states (Fig. 7.10a). For a band insulator, this discontinuity at k_F does not exist, since all the bands are either full or empty (Fig. 7.10b). The interactions in a Fermi liquid preserve the k_F discontinuity characteristic of an independent electron gas (Fig. 7.10c). However, the discontinuity decreases with the strength of the interactions as explained in Chapter 3. For systems that do not fall under the Fermi liquid model (non-Fermi liquids, Luttinger's liquid, ...), the discontinuity disappears but $n(k)$ always exhibits a fast variation at k_f (Fig. 7.10d).

Moreover, even in a Fermi liquid, the finite temperature and the experimental resolution suppress the discontinuity. Anyhow the integral:

$$n(\mathbf{k}) = \int_{-\infty}^{\infty} d\omega f(\omega) A(\mathbf{k}, \omega) \tag{7.9}$$

allows you to retrieve the $n(k)$ function from the experimental spectra. However, not all the possible energies are measured experimentally, thereby avoiding, for instance, the contribution of the secondary electrons. It is therefore necessary to integrate on a finite domain that contains the peak of photoemission. Figure 7.10 shows the principle of this method for narrow and wide band systems. The bands narrower than the detection window width give rise to a discontinuity of the experimental intensity $w(k)$. This discontinuity is responsible for the maximum of the $|\nabla_{\mathbf{k}} w(\mathbf{k})|$ gradient, which then coincides with k_F. In the case where the bands are wider than the detection window, $w(\mathbf{k})$ has a maximum and $|\nabla_{\mathbf{k}} w(\mathbf{k})|$ two maxima. The maximum of $|\nabla_{\mathbf{k}} w(\mathbf{k})|$ that is closest to the unoccupied states allows to accurately determine the Fermi surface. For the other maximum, its position in k depends on the ΔE width of the detection window, it is broader and less intense than the one at k_F [18].

7.1.6 Transport

From the study of the Fermi surface, it is possible to determine the two-dimensional conductivity of a system. The relationship between conductivity and Fermi surface is given by the Boltzmann equation which contains the velocity tensor. For a relaxation time independent of the wave vector [24]:

$$\sigma_{ij} = \frac{1}{2\pi^2}\frac{e^2\tau}{\hbar}\int_{SF}\frac{v_{ki}v_{kj}\,dk_F}{|v_k|} \tag{7.10}$$

with :

$$v_{ki} = \frac{1}{\hbar}\frac{\partial E}{\partial k_i}. \tag{7.11}$$

An isotropic surface has the same speed of carriers in all directions. At Fermi level, $|v_{ki}| = |v_{kj}| = |v_k| = v_F$ and the Boltzmann equation is:

$$\sigma = \frac{e^2\tau v_F}{4\pi^2\hbar}\int_{SF}\frac{dk_F}{|v_k|} \tag{7.12}$$

Furthermore, since the density of states at the Fermi level is:

$$D_{2D} = \frac{1}{2\pi^2}\int\frac{dk_F}{|\partial E/\partial k|} = \frac{1}{2\pi^2\hbar}\int\frac{dk_F}{|v_k|} \tag{7.13}$$

the conductivity can be written as:

$$\sigma = \frac{e^2}{2}(\tau v_F)v_F D_{2D} \tag{7.14}$$

with τ the carrier relaxation time, which leads to the Drude formula:

$$\sigma = \frac{ne^2\tau}{m^*} \tag{7.15}$$

As $n = k_F^2/2\pi$ for an almost free surface state, finally the conductivity is:

$$\sigma = \frac{k_F^2 e^2 \tau}{2\pi\, m^*} \tag{7.16}$$

To determine it, it is necessary to figure out τ, k_F et m^*. The quantities m^* and k_F are obtained in a standard way from the dispersions. τ is the characteristic time between two collisions:

$$\tau = \frac{\lambda}{v_F} \tag{7.17}$$

and the average free path λ can be obtained from the width of the peak at the Fermi level:

$$\lambda = \frac{1}{\Delta k} \tag{7.18}$$

7.2 Spectral signature analysis

7.2.1 *Electron-phonon coupling*

Electron-phonon coupling plays an important role on the physical properties of solids. It limits sometimes the electrical conductivity of solids but can also lead to singular phenomena such as superconductivity. In general, low-energy electronic excitations in a metal can be described by quasiparticles (an electron surrounded by a cloud of virtual excitations which, in the presence of electron-phonon coupling, correspond to virtual phonons). This cloud leads to an increase of the effective mass in $(1+\lambda)$, where λ is the strength of the electron-phonon coupling. Recently it has been shown that this coupling can manifest itself in photoemission spectra, in an energy range near the Fermi level of the order of the phonon energy (Debye energy). The spectral signature is the presence in the dispersion $E(k)$ of a slope change in the dispersion at the Debye energy (*kink*) and a narrowing of the band (effective mass increase) in this energy range (Fig. 7.11a). In consequence, electron-phonon coupling is observed in EDCs due to the splitting of the dispersion into two branches. There is a narrow peak associated with the mixing of electronic states with the collective mode and a larger peak, which follows the dispersion of the non-renormalized band (bare electrons). Figure 7.11b shows the two branches in manganese oxide $La_{1.2}Sr_{1.8}Mn_2O_7$ (LSMO). On one hand, a quasi-particle peak is observed below 50 meV and also a bump ("hump"), that disperses parabolically between 300 and 1800 meV. On the other hand, the dispersive peak becomes narrower at binding energies comparable to that of the phonon. Such a behavior appears because the system can no longer relax in a hole and a phonon and the average life of the final state increases.

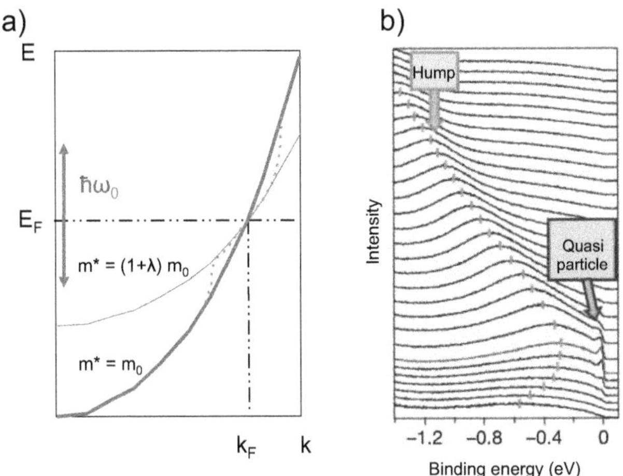

Figure 7.11 *(a) Diagram of the influence of the electron-phonon coupling on a band dispersion. At binding energies greater than the phonon energy $h\omega_0$, the intensity maxima follow the dispersion of the undressed electrons (non-renormalized effective mass m_0). For energies similar to the phonon energy, there is a renormalization of the band (narrow band). The effective mass increases, but at E_F both bands cross at the same point k_F. (b) LSMO band structure (after [23]). Adapted with permission from Springer Nature: Nature 438, 474 (2005), ©2005. http://www.nature.com/.*

The same phenomenon is observed for any coupling with other collective modes, such as magnons. For example, in Fe(110) (Fig. 7.12a-b), the maximum energy of the surface or bulk phonons is 30 meV. But the dispersion kink appears around 160 meV for the two surface states S_1 and S_2, which corresponds to the energy of an acoustic magnon. It is thus the signature of the electron-magnon coupling with widths controlled by $\Sigma_I \propto \omega^{3/2}$ with $\omega < \omega_0$ where ω_0 is the cutoff frequency of the acoustic magnon dispersion.

In the following, we will present the methods to quantitatively obtain information on the coupling between the electron and the collective modes.

From the Eliashberg fonction

All the relevant quantities for the electron-phonon coupling can be deduced from the Eliashberg function $\alpha^2 F(\omega; E, \vec{k})$, which is the phonon density weighted by the electron-phonon coupling. It describes the transition probability of a quasi-particle between the initial and final states defined by $(E, \vec{k})$ by exchanging a phonon of frequency ω. In Debye's simplified model (that does not take into account the details of phonon density), the Eliashberg fonction is [29]:

$$\alpha^2 F(\omega) = \begin{cases} \lambda(\omega/\omega_D)^2, \omega < \omega_D \\ 0, \omega \geqslant \omega_D \end{cases} \tag{7.19}$$

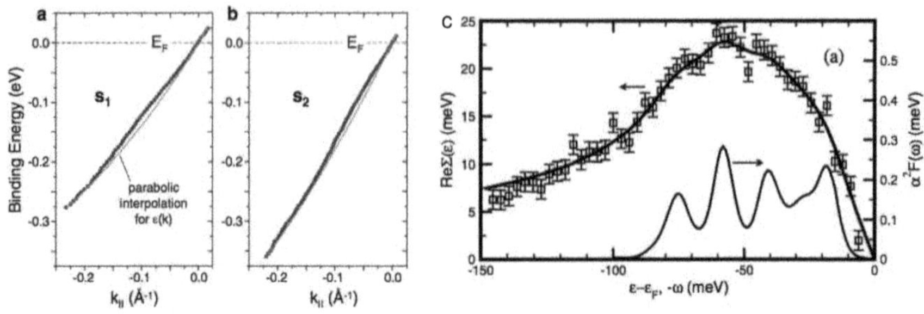

Figure 7.12 *Dispersion of the surface states of Fe(110) (a) S_1 and (b) S_2 (after [25]). Figure reprinted with permission from J. Schäfer, D. Schrupp, Eli Rotenberg, K. Rossnagel, H. Koh, P. Blaha, and R. Claessen, Phys. Rev. Lett. 92, 097205 (2004). ©2004 by the American Physical Society. (c) Be(10$\bar{1}$0). Real part of the self-energy, adjusted by the MEM method and Eliashberg function (after [26]). Figure reprinted with permission from Junren Shi, S.-J. Tang, Biao Wu, T. Sprunger, W. L. L. Yang, V. Brouet, X. J. Zhou, Z. Hussain, Z.-X. Shen, Zhang Zhenyu, and E. W. Plummer, Phys. Rev. Lett. 92, 186401 (2004). ©2004 by the American Physical Society.*

The relation between the real part of self-energy Σ_R and the Eliashberg function is:

$$\Sigma_R = \int_0^\infty d\omega \alpha^2 F(\omega) K(E/kT, \omega/KT) \qquad (7.20)$$

where

$$K(y, y') = \int_{-\infty}^\infty dx \frac{f(x-y)2y'}{x^2 - y'^2} \qquad (7.21)$$

and $f(x)$ is the Fermi distribution. Equation 7.20 allows us *a priori* to obtain the Eliashberg function from the photoemission data. The inversion of this expression, however, is mathematically unstable and the least squares method cannot be used. One option is to use the maximum entropy method (MEM) [30] and to impose that the Eliashberg function is positive (since it is a probability) and zero for energies above a given threshold and at the Fermi level. Figure 7.12c shows such an analysis for Be(10$\bar{1}$0), which finally allows to determine λ from:

$$\lambda = 2 \int_0^\infty \frac{d\omega}{\omega} \alpha^2 F(\omega). \qquad (7.22)$$

From double fitting

A less sophisticated method for obtaining λ is to determine the effective mass m renormalized by the electron-phonon coupling and the mass m^* without renormalization. These effective masses are obtained from the dispersions of the non-renormalized band (energies greater than $\hbar\omega_0$) and the renormalized band. Once known m and m^*, the electron-phonon coupling is simply the ratio:

$$\lambda = \frac{m^* - m_0}{m_0} \tag{7.23}$$

From Σ_R determination

Another method equivalent to the previous one is to obtain the real part of the spectral function Σ_R. Σ_R is obtained from the difference of the dispersion $E(\vec{k})$ with respect to the dispersion in the absence of the electron-phonon coupling $E_0(\vec{k})$ (a hypothesis on the shape of this dispersion is however necessary):

$$\Sigma_R = E(\vec{k}) - E_0(\vec{k}) \tag{7.24}$$

The electron-phonon coupling is:

$$\lambda = -\frac{d\Sigma_R}{dE}|_{E_F} \tag{7.25}$$

From temperature measurements

λ can also be determined from the study of the width variation of a dispersing peak as a function of the temperature. Indeed, the dispersing peak width is the inverse of the lifetime [29]:

$$\Gamma_{e-ph} = \pi\hbar \int_0^{\omega_{max}} \alpha^2 F(\omega')[1 - f(\omega - \omega') + 2n(\omega') + f(\omega + \omega')]d\omega' \tag{7.26}$$

with ω the frequency of the excited phonon, $\alpha^2 F(\omega)$ the Eliashberg function, $f(\omega)$ the Fermi distribution and $n(\omega)$ the Bose-Einstein distribution. At high temperature, *i.e.* $kT \gg \hbar\omega_{max}$ (with ω_{max} the maximum frequency of the phonon), the Bose-Einstein distribution is much larger than that of Fermi-Dirac:

$$n(\omega) \gg 1 > f(\omega) \tag{7.27}$$

and the associated width to the electron-phonon coupling is:

$$\Gamma_{e-ph} \simeq 2\pi\lambda k_B T \tag{7.28}$$

λ is then obtained from the width as a function of the temperature:

$$\lambda = \frac{1}{2\pi}\frac{d\Gamma}{d(k_B T)} \tag{7.29}$$

as shown for Sb(111) in Fig. 7.13.

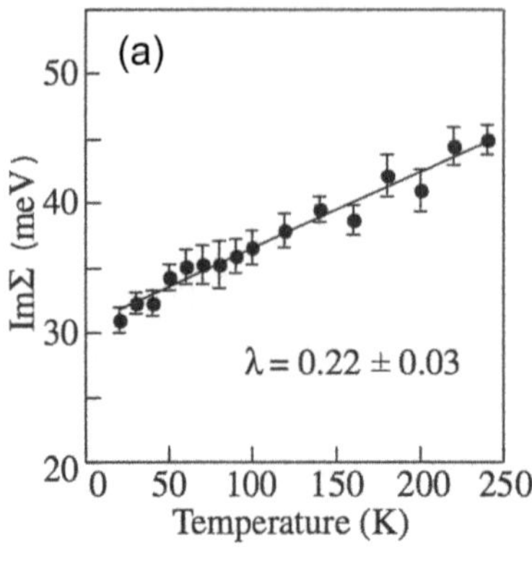

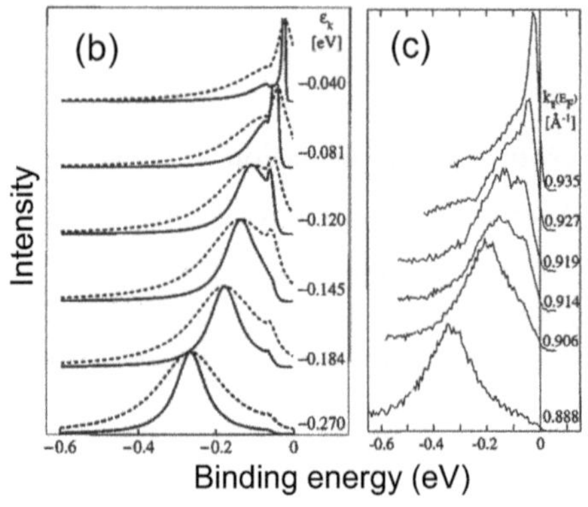

Figure 7.13 *(a) Temperature dependence of the imaginary part of the self-energy for Sb (111) [27]. Figure reprinted with permission from K. Sugawara, T. Sato, S. Souma, T. Takahashi, M. Arai, and T. Sasaki, Phys. Rev. Lett. 96, 046411 (2006). ©2006 by the American Physical Society. (b) Spectral function for Be(0001) in the Debye model for $\omega_D = 65$ meV and $\lambda = 0.65$. The continuous line corresponds to the electron-phonon contribution. The dashed line correspond to Σ_I plus a constant of 54 meV to simulate the impurity contribution. (c) Photoemission spectra corresponding to the simulations in (b) (after [28]). Figure reprinted with permission from S. LaShell, E. Jensen, and T. Balasubramanian, Phys. Rev. B 61, 2371 (2000). ©2000 by the American Physical Society.*

7.2.2 k perpendicular determination

The non-conservation of the perpendicular component of the wave vector in the photoemission process does not allow you *a priori* to determine the band structure for a three-dimensional compound.

However, there are methods to obtain the parallel and perpendicular components experimentally and thus to determine the band structure. One method consists of using surfaces of different orientations and is called the triangulation method [32]. It consists of measuring the same states on two single crystals cut along two different surfaces. For simplicity, the first surface is chosen to observe the desired band at normal emission. The second surface is chosen in a way that it allows you to measure the same initial state but the transition occurs for another polar angle, which is not know *a priori*. It is therefore necessary to perform a series of measurements varying θ_A in the correct azimuthal direction of the surface A. For a given angle θ_A, the energy of the transition will coincide on the surfaces A and B, allowing to identify the transition. Figure 7.14 shows how to obtain $k_\perp^A$ without hypothesis on the inner potential (*cf.* section 3.1.2). Since $k_\perp^A \sin\alpha_0 = k_{||}^B + k_{||}^A \cos\alpha_0$, $k_{||}$ and $k_\perp$ are:

$$k_\perp^A(\text{Å}^{-1}) = 0.512\sqrt{E_k(\text{eV})}\frac{\sin\theta_B + \sin\theta_A\cos\alpha_0}{\sin\alpha_0} \tag{7.30}$$

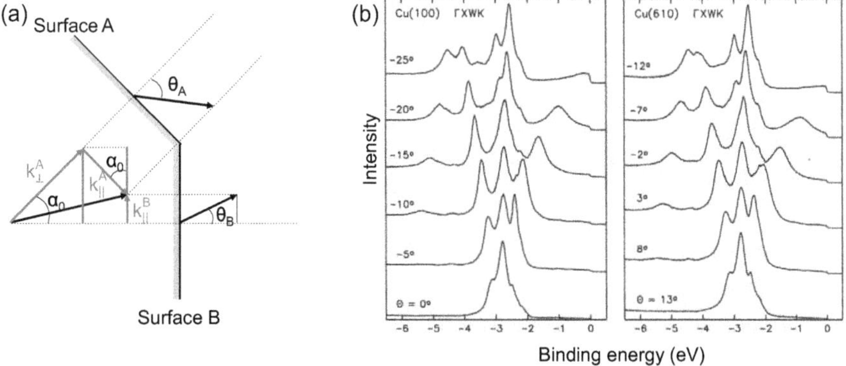

Figure 7.14 *(a) Geometry of the triangulation method to determine the two components (∥ et ⊥) of k. (b) Triangulation method with the surfaces of Cu(100) and Cu(610). The spectra correspond to the same point $\vec{k}$ of the bulk Brillouin zone and are measured at different angles for each surface (after [31]). Reprinted from Surf. Sci. vol. 400, R. Matzdorf and A. Goldmann, "High-resolution angle-resolved photoemission from Cu (610)", page 329, ©1998, with permission from Elsevier.*

$$k_{\|}^{A}(\text{Å}^{-1}) = 0.512\sqrt{E_k(\text{eV})}\sin\theta_A \tag{7.31}$$

where α_0 is the angle between surface normals, k is expressed in Å^{-1} and the energy in eV.

7.2.3 *Spin polarisation*

So far, we have not considered spin issues. In most systems, each state of wave vector k is at least twice degenerated for each orientation of the electron spin. This degeneracy can however be lifted in magnetic systems but also in non-centrosymmetric systems where spin-orbit coupling is important.

Spin-orbit coupling

The spin-orbit interaction is a relativistic effect that leads to the coupling of the spin degree of freedom with the lattice. This relativistic correction leads to the degeneracy of the bands similarly to the observed effect on atomic levels. For example, in semiconductors such as Si and at the Brillouin zone center, $6p$ degenerated bands ($\ell = 1$) (orbital and spin degeneration) are found when the spin-orbit interaction is ignored. The introduction of spin-orbit coupling leads to two distinct energy levels, like in the fine structure of the p states of the hydrogen atom. There are four-fold degenerated states associated with the total kinetic momentum $j = \ell + 1/2 = 3/2$, and two-fold degenerated states associated with the total kinetic moment $j = \ell - 1/2 = 1/2$. Let us recall the expression of the spin-orbit Hamiltonian:

$$H_{SO} = \frac{\hbar^2}{4m^2c^2}(\vec{\nabla} V \times \vec{p}) \cdot \vec{\sigma} \tag{7.32}$$

with V the crystalline potential and $\vec{\sigma}$ the spin operator. For a central potential, as in atomic physics, this expression is reduced to

$$H_{SO} = \lambda \vec{\ell} \cdot \vec{s} \tag{7.33}$$

where $\vec{\ell}$ is the orbital kinetic moment. Since $\vec{\ell} \cdot \vec{s} \propto j^2 - \ell^2 - s^2$, the eigenstates of H_{SO} are those of the total kinetic moment.

When one deviates from the Γ point in any direction, the degeneracy is raised but each band is at least doubly degenerated. This behavior is a consequence of two fundamental symmetries, viz., the inversion and the time reversal symmetries. To understand this, consider a Bloch state of + magnetization and of energy $\varepsilon(\vec{k}, +)$. The time reversal, which is a symmetry of the problem if no external magnetic field is applied, will transform this Bloch function into a Bloch function of the same energy but with opposite wave vector and opposite magnetization:

$$\varepsilon(\vec{k}, +) = \varepsilon(-\vec{k}, -) \tag{7.34}$$

Moreover, if the solid has inversion symmetry, it reverses the wave vector (polar vector) but not the spin (axial vector):

$$\varepsilon(\vec{k}, +) = \varepsilon(-\vec{k}, +) \tag{7.35}$$

The combination of these two symmetries leads to:

$$\varepsilon(\vec{k}, +) = \varepsilon(\vec{k}, -) \tag{7.36}$$

i.e. at the same point of the Brillouin zone, the two Bloch states of opposite magnetizations are degenerated.

Another effect of spin-orbit coupling has no equivalent in atomic physics and applies only if the solid does not have an inversion center. It leads to the separation of the opposite magnetization bands in wave vector except if $\vec{k} = -\vec{k}$, for instance, at the Brillouin zone center. Degeneracy lift is observed when the inversion symmetry is broken, for example, in non-centrosymmetric semiconductor materials of zinc-blende structure such as InGe, or even at the surface states of centrosymmetric crystals, where the surface breaks the inversion symmetry. This is the case, of Au(111) Shockley state. This surface state behaves like a nearly-free electron two-dimensional gas and the spin-orbit coupling is written (Rashba model):

$$H_{SO} = \frac{\mu_B}{2c^2}(\vec{v} \times \vec{E}) \cdot \vec{\sigma} \tag{7.37}$$

Since the potential varies only along z (the normal to the surface), equation 7.32 is written as:

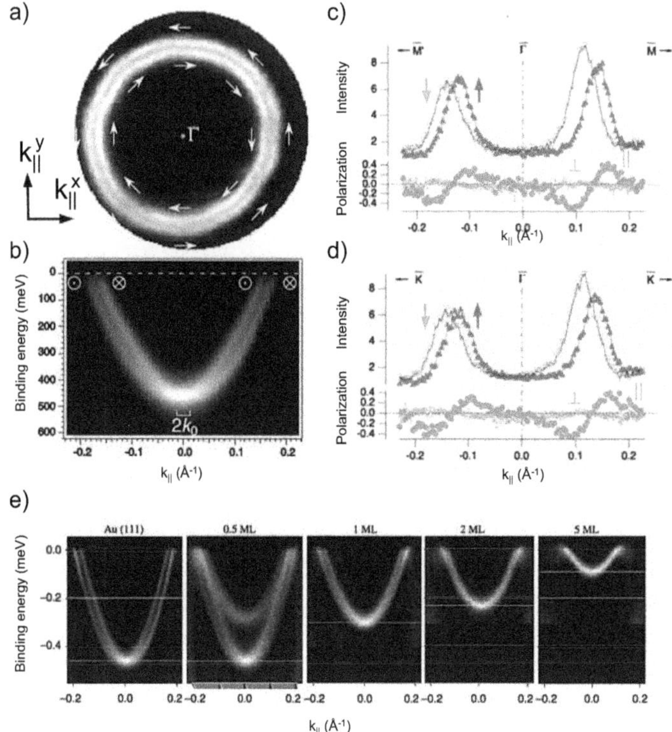

Figure 7.15 *(a)-(d): Au (111) a) Fermi surface integrated in spin b) Bands integrated in spin. The arrows indicate the direction of the spin according to a free-electron model. c) and d) MDCs at 170 meV resolved in spin (according to [33]). Reprinted figure with permission from M. Hoesch, M. Muntwiler, V. N. Petrov, M. Hengsberger, L. Patthey, M. Shi, M. Falub, T. Greber, and J. Osterwalder, Phys. Rev. B 69, 241401 (R) (2004). ©2004 by the American Physical Society. (e) Ag (Au) films deposited at room temperature (after [34]). Reprinted from D. Malterre, B. Kierren, Y. Fagot-Revurat, S. Pons, A. Tejeda, C. Didiot, H. Cercellier and A. Bendounan, "ARPES and STS investigation of Shockley states in thin metallic films and periodic nanostructures", New. J. Phys. 9, 391 (2007). ©IOP Publishing & Deutsche Physikalische Gesellschaft. CC BY-NC-SA.*

$$H_{SO} = \frac{\mu_B}{2c^2}(\vec{v} \times \vec{E}) \cdot \vec{\sigma} = \alpha_R(\vec{e}_z \times \vec{k}_{||}) \cdot \vec{\sigma} \tag{7.38}$$

with $\vec{v}$ the electron speed and $\vec{E}$ the electric field oriented perpendicular to the crystal surface ($\vec{e}_z$). We obtain two parabolic dispersions shifted by $\Delta k_{S.O.}$ proportionally to the spin-orbit coupling:

$$E_{\pm(\vec{k}_{||})} = E_0 + \frac{\hbar^2(\vec{k}_{||} \pm \Delta\vec{k}_{S.O.}/2)^2}{2m^*} \tag{7.39}$$

The Fermi surface corresponds to two concentric circles associated to the two opposite directions of the spin (Fig. 7.15a-d). The splitting of these two bands can be perfectly tailored when creating silver films on gold (Fig. 7.15e).

7.2.4 *Final-state and matrix element effects*

Final state effects

The photoemitted intensity depends on the matrix element controlling the transition from the initial to the final state. For the transition to occur, it is necessary that the photon energy allows the system to reach a final state. It is therefore possible, for some photon energies, that the *existing* initial state is not observed simply because the optical transition cannot take place. Figure 7.16 shows that this effect is more important at low photon energy, because final states are more likely to be accessible when $h\nu$ is large, especially due to the broadening ΔE of the states because the lifetime decreases when energy increases. The photoelectron intensity is therefore not directly related to the density of states as illustrated in 7.16b. This figure shows, for the d band of Cu(111), that the experimental intensity is well described by a simulation with a one-step model that takes into account final state effects. However, the experimental intensity disagrees with the spectral density of the initial states.

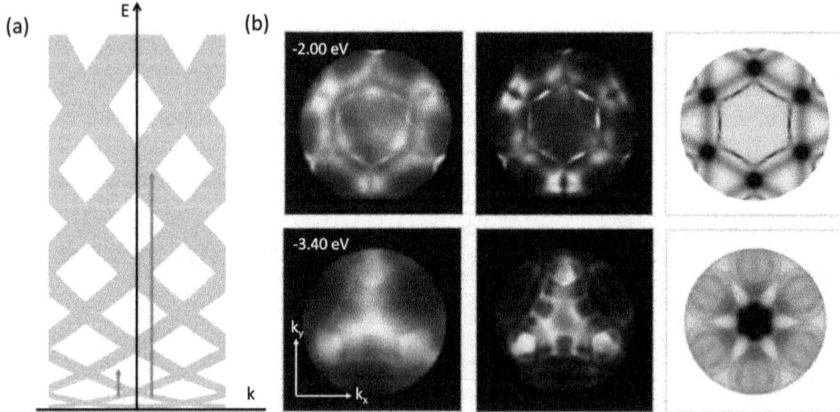

Figure 7.16 *Final state effects. (a) Final-state effects are more important at low energy because it is more difficult to find an accessible final state. (b) Isoenergetic sections of Cu(111) d bands. Comparison between experiment (left), one-step photoemission calculations (center) and the spectral density of initial states (after [35]). Reprinted from A. Winkelmann, C. Tusche, A. A. Ünal, M. Ellguth, J. Henk and J. Kirschner, "Analysis of the electronic structure of copper via two-dimensional photoelectron momentum distribution patterns", New. J. Phys. 14, 043009 (2012). ©IOP Publishing & Deutsche Physikalische Gesellschaft. CC BY-NC-SA.*

Another example of final-state effects are the Umklapp processes, *i.e.*, the appearance of replicas of the bands because of a periodicity. Fig. 7.17 shows the Fermi surface of Ag-Au/Si(111) (the most intense circles at left and right). Replicas of lower intensity appear on this map of zero binding energy. These replicas are connected by translations of a reciprocal vector of the $\sqrt{21} \times \sqrt{21}$ reconstruction and its equivalent domain rotated by 21.8° (*cf.* the diagram of the translation of the Fermi surface by the reciprocal lattice vectors in the figure).

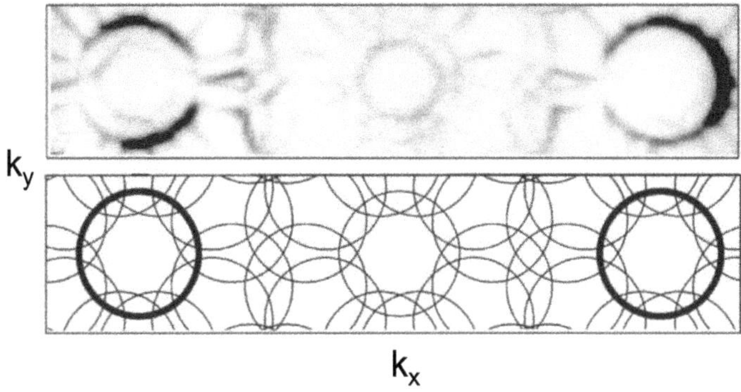

Figure 7.17 *(Top) Experimental Fermi surface of the $\sqrt{21} \times \sqrt{21}$ reconstruction of Ag-Au on Si (111). Two electron pockets are observed. (Bottom) Replicas of the Fermi surface by the reciprocal lattice vectors of the $\sqrt{21} \times \sqrt{21}$ reconstruction (after [36]). Figure reprinted with permission from J. N. Crain, K. N. Altmann, C. Bromberger, and F. J. Himpsel, Phys. Rev. B 66, 205302 (2002). ©2002 by the American Physical Society.*

Matrix element effects

Matrix elements can be exploited to obtain information on the system. For example, if two states have a different symmetry with respect to a mirror plane, matrix elements can identify them. In this way, it is possible to distinguish σ from π orbitals in self-organized molecules on surfaces. For a molecule arranged along the high symmetry directions of the substrate and with a main axis normal to the surface, the σ states do not give intensity if $\vec{A}$ is horizontal. The other polarization cancels the contribution of π orbitals to the measured intensity. An analogous situation may appear in surface reconstructions. For example, the deposition of Cs on Si (001) gives rise to two domains of dimers (reconstructions 2×1 and 1×2). For each dimer π and π^* orbitals are filled. Figure 7.18 shows PEEM with linear polarization. The photoemission signal changes dramatically as a function of the angle between the dimers and the polarization, because the states are symmetrical with respect to the plane perpendicular to the dimer. The matrix elements are zero when $\vec{A}$ is perpendicular to this mirror plane.

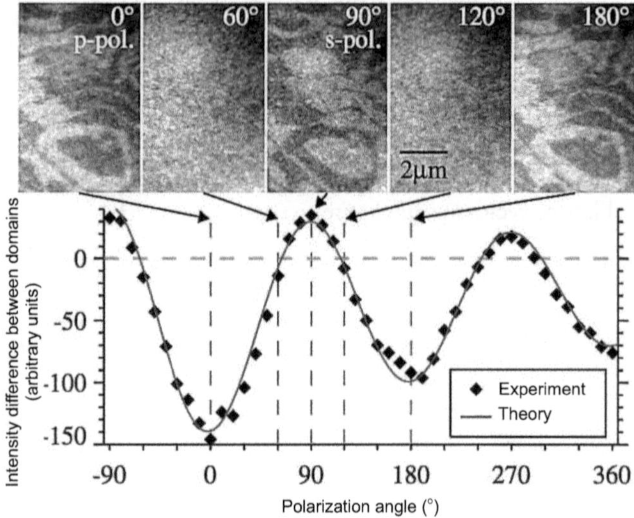

Figure 7.18 *Contrast polarization dependence of the domains 2 × 1 and 1 × 2 depending on the angle between the dimers and the polarization. The continuous line is the comparison with the theory, superimposed on an exponential decay due to the Cs desorption over time. [37]. Figure reprinted with permission from D. Thien, P. Kury, M. Horn-von Hoegen, F.-J. Meyer zu Heringdorf, J. van Heys, M. Lindenblatt, and E. Pehlke, Phys. Rev. Lett. 99, 196102 (2007). ©2007 by the American Physical Society.*

7.2.5 Other applications

Low dimensional systems

A low dimensional system has at least one of the three spatial dimensions of a reduced size, which often leads to different physical properties of those of three-dimensional systems. Electronic instabilities may appear, leading to phase transitions. Low dimensional systems are widespread. Some three-dimensional compounds, whose electronic properties have a very important anisotropy (for instance those with very weakly coupled planes or presenting atomic chains) have a phenomenology of low dimensional systems. Also surfaces are intrinsically two-dimensional systems, sometimes with localized electronic states (surface states).

The absence or a very small dispersion in $k_\perp$, is a good criterion for defining a low dimensional system (2D). Figure 7.19a shows, on one hand, the dispersion parallel to the surface of one to four layers of graphene. All these systems disperse significantly in the surface plane. On the other hand, the dispersion in the perpendicular direction is very different. This dispersion is completely negligible for a single layer of graphene corroborating the two-dimensional character of its electronic structure. On the other hand, from 2 layers there is a relation between $k_\perp$ and $k_\parallel$, indicating a coupling between planes.

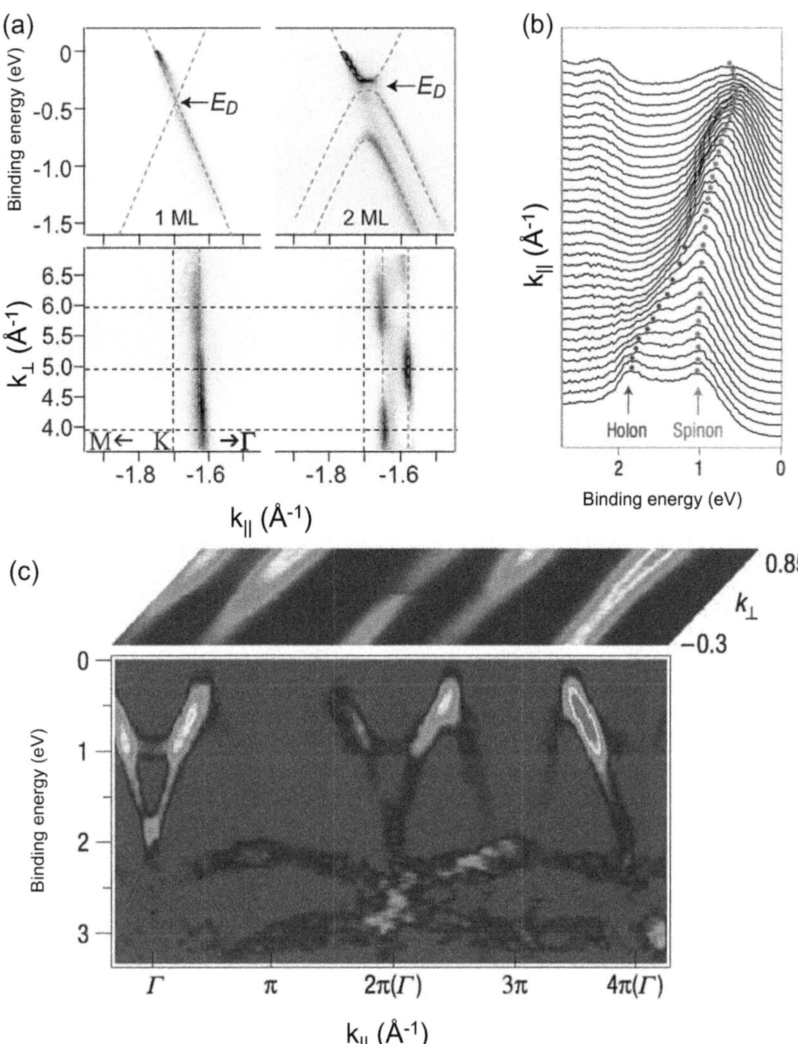

Figure 7.19 *(a) Graphene on SiC(0001). π and π^* bands for 1 to 2 monolayers of graphene. (Top) Energy dispersion versus $k_{||}$. (Bottom) Energy dispersion versus $k_\perp$ (after [38]). Figure reprinted with permission from T. Ohta, A. Bostwick, J. L. McChesney, T. Seyller, K. Horn, and E. Rotenberg, Phys. Rev. Lett. 98, 206802 (2007). ©2007 by the American Physical Society. (b) EDCs on $SrCuO_2$ illustrating the two branches associated with spin-charge separation (c) Top: distribution as a function of the wave vector for $E \simeq 0.5$ eV. Bottom: band structure for $k_\perp = 0.7Å^{-1}$ (after [39]). Adapted by permission from Macmillan Publishers Ltd: Nature Phys. 2, 397 (2006), ©2006.*

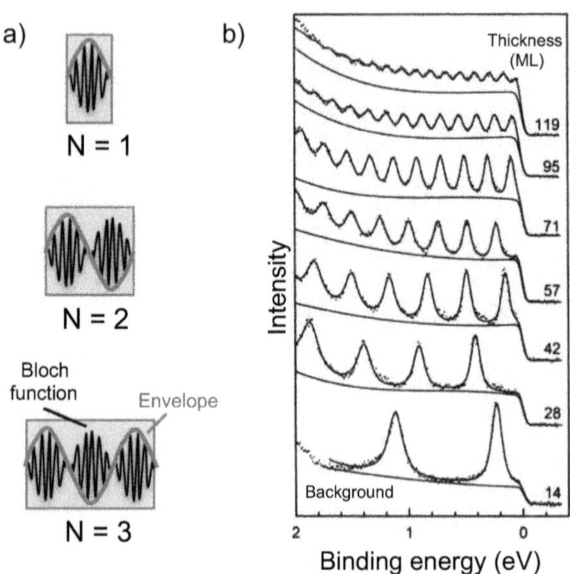

Figure 7.20 *(a) Quantum well diagram in an interferometer model. Bloch waves oscillate rapidly and are modulated by an envelope that must satisfy the boundary conditions. (b) Ag/Fe(100) quantum wells observed by photoemission at normal emission. This system allows to have very uniform films between one and a hundred monolayers (according to [40]). Reprinted from J. Electron. Spectrosc. Rel. Phenom. vol. 114-116, T. Miller, J. J. Paggel, D.-A. Luh, T. C. Chiang, "Quantum-well photoemission spectroscopy of atomically-uniform films", page 513, ©2001, with permission from Elsevier.*

After observing the low dimensionality of a system, some strange properties can be studied. For instance, in one dimension, the Fermi liquid model no longer applies, which means that the low energy excitations can no longer be described by quasiparticles. The one-dimensional Luttinger liquid model leads to exotic electronic properties: the low-energy excitations are collective excitations characterized by a charge-spin separation. The photoemission spectrum should present two structures associated to the dispersion of the spinon (spin excitation) and the holon (charge excitation), each one with a characteristic behavior. This case is illustrated by the photoemission spectra of the quasi-one-dimensional $SrCuO_2$ compound (Fig. 7.19bc) which show the dispersion of 2 bands attributed to holons and spinons.

Another effect of the low dimensionality is electronic confinement which leads to the appearance of the discrete states characteristic of a quantum well. Figure 7.20 shows how new quantum well states appear for silver films that grow layer by layer on Fe (001). The confinement induces there a discretization of the wave vector and thus a discretization of the electronic states. The more confined the particle, the greater is the separation in energy among the states. The figure shows that as the film thickness increases, the separation between quantized states decreases and tends towards a continuum for thick films.

Finally, consider the surface states that are confined perpendicularly to the surface. They are naturally trapped in a well consisting of the surface potential of one side and the gap of the bulk band structure of the other. These two-dimensional states are very sensitive to the surface potential and can be modified by the morphology of the surface, such as in vicinal surfaces. These stepped surfaces induce a periodic potential in the plane that can further confine surface states in the surface plane, in the direction perpendicular to the steps. Indeed, by modeling the potential of the steps with a Kronig-Penney model [42, 43], the dispersion is:

$$E(k) = \frac{\hbar^2}{2m^* L^2}[\cos^{-1}(|T| \cos kL) - \phi]^2 \tag{7.40}$$

where $T = |T|e^{i\phi}$ is the transmission coefficient across the barriers, that depends on the energy and on the phase shift ϕ of the transmitted wave. The transmission coefficient and the potential barrier $U_0 a$ are related by:

$$T = \frac{q}{q + iq_0} \tag{7.41}$$

with $q = \sqrt{2m^* L^2}/\hbar$ and $q_0 = m^* U_0 a/\hbar$. The energy variation of the band minimum is:

$$\Delta E = \frac{\hbar^2}{2m^* L^2}[\cos^{-1}(|T|) - \phi]^2 \tag{7.42}$$

Figure 7.21 shows the parabolic dispersion in several vicinal surfaces of Cu. The previous expression can be used to obtain the potential barrier of the steps. For the (332) steps, the barrier is 0.9 eV Å and for the (221) steps it is 1.35 eV Å.

Another procedure for obtaining the periodic potential in the plane is to use the gaps that appear in the band structure at the Brillouin zone edge. The potential can be decomposed into a Fourier series, each component will mainly affect one of the gaps of the band structure. For a potential with a single Fourier component V_G, the unique gap has an amplitude of $2|V_G|$. When multiple gaps exist, the principle is the same but with more complex expressions. Figure 7.22 shows several gaps in the dispersion of the Au(788) surface state reinforced by a self-organized network of Co islands. From the comparison of theory and experiment, the potential can finally be determined [44].

Phase transitions

Photoemission is also a very powerful tool for studying phase transitions that change the structure of bands. Particularly interesting examples are superconducting transitions or metal-insulator transitions (charge or spin density waves, Mott transitions...). Figure 7.23 shows, for example, the study of the order parameter of a transition. This is the phase transition in Cr(110) between a metallic state and a spin density wave

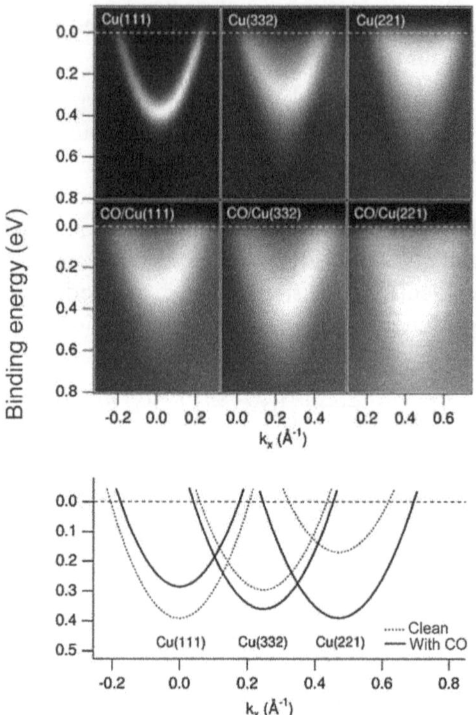

Figure 7.21 *Photoemission spectrum of Cu(111) Shockley state and two vicinal surfaces, for bare surfaces and upon adsorption of 0.2 MC of CO (after [41]). Figure reprinted with permission from F. Baumberger, T. Greber, B. Delley, and J. Osterwalder, Phys. Rev. Lett. 88, 237601 (2002). ©2002 by the American Physical Society.*

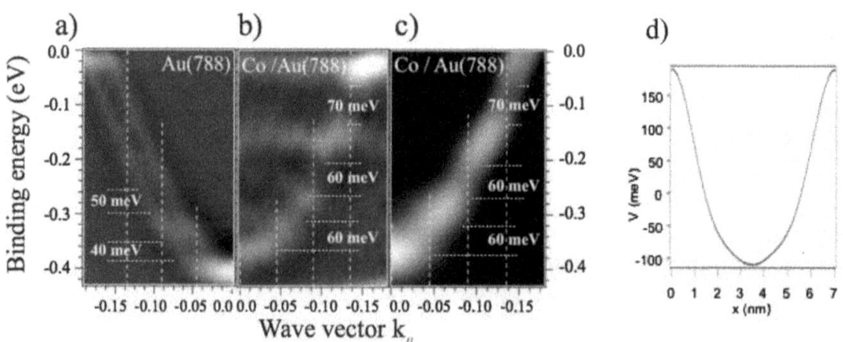

Figure 7.22 *Second derivative of the bands of (a) Au (788) and (b) Co/Au(788) substrate. c) Simulated spectra for Co/Au (788) (d) Surface potential deduced from (c) [44].*

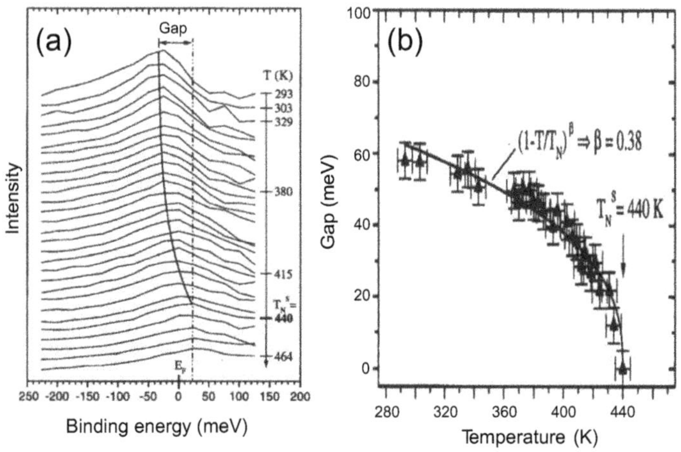

Figure 7.23 *Cr(110). a) Spectra as a function of temperature. b) Gap versus temperature obtained from the fit of the spectra in a) [45] Figure reprinted with permission from J. Schäfer, Eli Rotenberg, G. Meigs, SD Kevan, P. Blaha, and S. Hüfner, Phys. Rev. Lett. 83, 2069 (1999). ©1999 by the American Physical Society.*

state. Photoemission shows the gradual opening of a gap when the temperature decreases. The gap has a temperature dependence according to the law $(1 - T/T_N)^\beta$ with $\beta = 0.38 \pm 0.05$ and a transition temperature characteristic of the surface of $T_N = T_N^S = 440 \pm 10$ K.

Bibliography

[1] J. Koralek *et al.*, Phys. Rev. Lett. **96**, 017005 (2006).
[2] F. Reinert *et al.*, Phys. Rev. B **63**, 115415 (2001).
[3] J. Lobo, A. Tejeda, A. Mugarza, and E. Michel, Phys. Rev. B **68**, 235332 (2003).
[4] E. Rotenberg, W. Theis, K. Horn, and P. Gille, Nature **406**, 602 (2000).
[5] D. A. Shirley, Phys. Rev. B **5**, 4709 (1972).
[6] M. Ogawa *et al.*, Journal of Physics: Condensed Matter **24**, 115501 (2012).
[7] X. J. Zhou, Z. Hussain, and Z.-X. Shen, Synchrotron Radiation News **18**, 15 (2005).
[8] N. Armitage *et al.*, Phys. Rev. B **68**, 064517 (2003).
[9] M. Hasan *et al.*, Phys. Rev. Lett. **92**, 246402 (2004).
[10] K. Rossnagel *et al.*, Phys. Rev. B **63**, 125104 (2001).
[11] A. Damascelli, Z. Hussain, and Z.-X. Shen, Rev. Mod. Phys. **75**, 473 (2003).
[12] R. Cortes *et al.*, Phys. Rev. Lett. **96**, 126103 (2006).
[13] T. Greber, T. Kreutz, and J. Osterwalder, Phys. Rev. Lett. **79**, 4465 (1997).
[14] D. Ehm *et al.*, Phys. Rev. B **76**, 045117 (2007).
[15] T. K. T. Greber, P. Aebi, and J. Osterwalder, Phys. Rev. B **58**, 1300 (1998).
[16] D. Ehm *et al.*, Physica B **312-313**, 663 (2002).

[17] F. Baumberger, T. Greber, and J. Osterwalder, Phys. Rev. B **64**, 195411 (2001).
[18] T. Straub *et al.*, Phys. Rev. B **55**, 13473 (1997).
[19] E. Rotenberg and S. Kevan, Phys. Rev. Lett. **80**, 2905 (1998).
[20] G. G. et al., Phys. Rev. B **55**, R13353 (1997).
[21] P. Aebi *et al.*, Phys. Rev. Lett. **72**, 2757 (1994).
[22] F. Ronning *et al.*, Science **282**, 2067 (1998).
[23] N. Mannella *et al.*, Nature **438**, 474 (2005).
[24] I. Matsuda *et al.*, Phys. Rev. B **71**, 235315 (2005).
[25] J. Schäfer *et al.*, Phys. Rev. Lett. **92**, 097205 (2004).
[26] J. Shi *et al.*, Phys. Rev. Lett. **92**, 186401 (2004).
[27] K. Sugawara *et al.*, Phys. Rev. Lett. **96**, 046411 (2006).
[28] S. LaShell, E. Jensen, and T. Balasubramanian, Phys. Rev. B. **61**, 2371 (2000).
[29] B. McDougall, T. Balasubramanian, and E. Jensen, Phys. Rev. B **51**, 13891 (1995).
[30] J. E. Gubernatis, M. Jarrell, R. N. Silver, and D. S. Sivia, Phys. Rev. B **44**, 6011 (1991).
[31] R. Matzdorf, Surf. Sci. **400**, 329 (1998).
[32] E. Kane, Phys. Rev. Lett. **12**, 97 (1964).
[33] M. Hoesch *et al.*, Phys. Rev. B **69**, 241401(R) (2004).
[34] D. Malterre *et al.*, New J. Phys. **9**, 391 (2007).
[35] A. Winkelmann *et al.*, New J. Phys. **14**, 043009 (2012).
[36] J. Crain, K. Altmann, C. Bromberger, and F. Himpsel, Phys. Rev. B **66**, 205302 (2002).
[37] D. Thien, P. Kury, M. H. von Hoegen, and F.-J. M. zu Heringdorf, Phys. Rev. Lett. **99**, 196102 (2007).
[38] T. Ohta *et al.*, Physical Review Letters **98**, 206802 (2007).
[39] B. Kim *et al.*, Nature Phys. **2**, 397 (2006).
[40] T. Miller, J. Paggel, D.-A. Luh, and T.-C. Chiang, Journal of Electron Spectroscopy and Related Phenomena **114-116**, 513 (2001).
[41] F. Baumberger, T. Greber, B. Delley, and J. Osterwalder, Phys. Rev. Lett. **88**, 237601 (2002).
[42] L. Davis, M. Everson, R. Jaklevic, and W. Shen, Phys. Rev. B **43**, 3821 (1991).
[43] O. Sanchez *et al.*, Phys. Rev. B **52**, 7894 (1995).
[44] C. Didiot *et al.*, Phys. Rev. B **76**, 081404(R) (2007).
[45] J. Schäfer *et al.*, Phys. Rev. Lett. **83**, 2069 (1999).

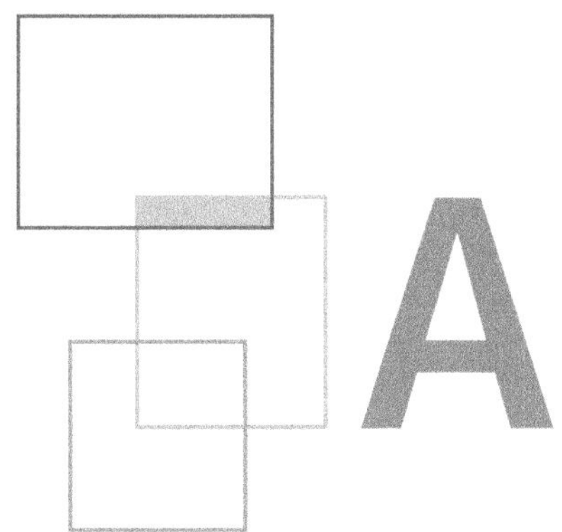

Photoemission with a detection slit parallel to the polar rotation

We can now study the determination of the wave vector from the emission angle when using a two-dimensional detector with a detection slit parallel to the polar rotation. The measured angles are the angle α on a two-dimensional detector placed at an angle θ_0 with respect to normal emission (Fig. A.1a). The outgoing electron has a total wave vector equal to $k_0 = 0.512\sqrt{E_k}$ Å^{-1}. We have to convert $(\alpha, \theta_0, \phi_0)$ angles to the polar coordinates (θ, ϕ) in the $OK_xK_yK_z$ coordinate system of the sample, in order to determine all the components of $\vec{k}$, i.e., the two $k_{||}$ components and the $k_\perp$ component.

To obtain the coordinate transformation, we shall consider the axes $OK_x''K_y''K_z''$ where the $OK_y''K_z''$ plane is defined by O and by the detector slit, which is projected on the direction OK_y'' (Fig. A.1 b). We see that:

$$k_x'' = 0 \tag{A.1}$$

$$k_y'' = k_0 \sin\alpha \tag{A.2}$$

$$k_z'' = k_0 \cos\alpha \tag{A.3}$$

We can now consider a slightly more general coordinate system $OK_x'K_y'K_z'$ whose axis OK_z' coincides with the OK_z axis of the sample coordinates (and which is therefore shifted by an angle θ_0 from the previous axis OK_z''), but with $OK_y'' = OK_y'$. So we have:

$$k_x'' = k_0 \sin(\theta - \theta_0) \tag{A.4}$$
$$k_z'' = k_0 \cos(\theta - \theta_0) \tag{A.5}$$

If we use trigonometric equations, we obtain:

$$k_x'' = k_0(\sin\theta \cos\theta_0 - \cos\theta \sin\theta_0) \tag{A.6}$$
$$k_z'' = k_0(\cos\theta \cos\theta_0 + \sin\theta \sin\theta_0) \tag{A.7}$$

and substituting in the Equations A.5, we have:

$$k_x' = k_z'' \sin\theta_0 \tag{A.8}$$
$$k_z' = k_z'' \cos\theta_0 \tag{A.9}$$

Finally, we obtain the $OK_xK_yK_z$ coordinate system of the sample by a ϕ_0 rotation of $OK_x'K_y'K_z'$ around the axis $OK_z = OK_z'$, being ϕ_0 the angle defining the center of the detector slit.

$$k_x' = k_0 \cos(\phi - \phi_0) \tag{A.10}$$
$$k_y' = k_0 \sin(\phi - \phi_0) \tag{A.11}$$

$$k_x' = k_0(\cos\phi \cos\phi_0 + \sin\phi \sin\phi_0) \tag{A.12}$$
$$k_y' = k_0(\sin\phi \cos\phi_0 - \cos\phi \sin\phi_0) \tag{A.13}$$

therefore

$$k_x' = k_x \cos\phi_0 + k_y \sin\phi_0 \tag{A.14}$$
$$k_y' = k_y \cos\phi_0 - k_x \sin\phi_0 \tag{A.15}$$

and, for the inverse transformation, we obtain:

$$k_x = k_x' \cos\phi_0 - k_y' \sin\phi_0 \tag{A.16}$$
$$k_y = k_y' \cos\phi_0 + k_x' \sin\phi_0 \tag{A.17}$$
$$k_z = k_z' \tag{A.18}$$

and finally the measured quantities are related with the (θ, ϕ) in the system of reference of the sample by applying the relations A.16-A.18.

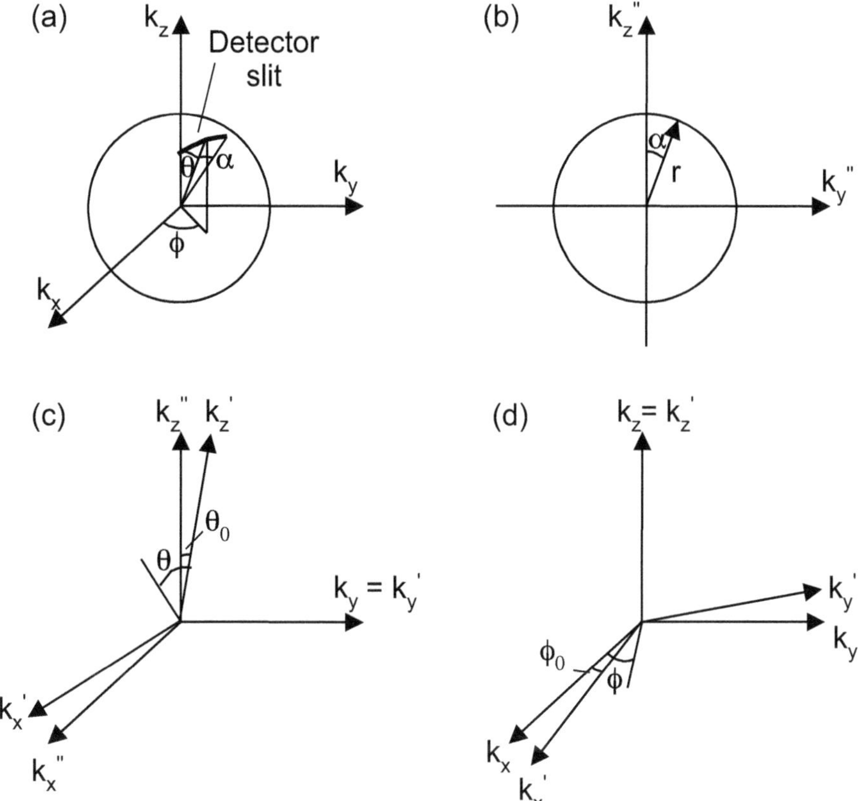

Figure A.1 *(a) Emission angle of the electron in the $OK_xK_yK_z$ coordinates of the sample (b) $OK_x''K_y''K_z''$ coordinate system where $OK_y''K_z''$ is defined by the intersection O of the three rotation axis and the detector slit (c) $OK_x'K_y'K_z'$ coordinate system whose OK_z' axis coincides with the OK_z axis of the sample coordinates (and which is thus at an angle θ_0 from the previous OK_z'' axis), but with $OK_y'' = OK_y'$ (d) $OK_xK_yK_z$ is obtained by rotation of $OK_x'K_y'K_z'$ around the $OK_z = OK_z'$ axis. The rotation angle is ϕ_0 defining the center of the detector slit.*

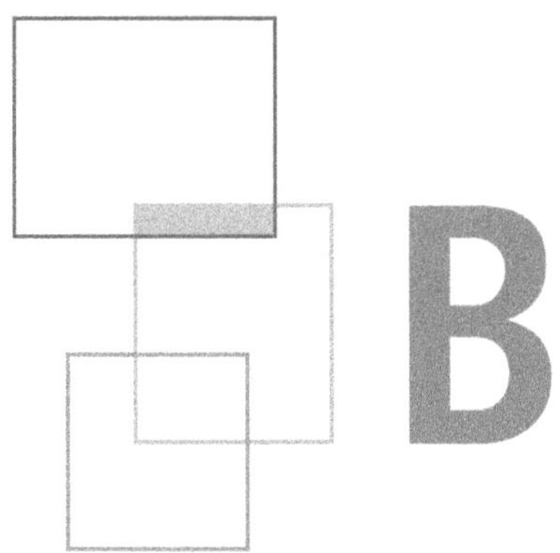

Quality of a core level fit

We have previously introduced the principles in order to deconvolute a core level. The deconvolution procedure consists in determining the number of components and their shape. Very often the number of components is largely known. It is important not to introduce an excessive number of components without physical meaning.

Once the spectrum is fitted, a criterion is needed to quantify the agreement between the experiment and the model. One possibility is the quality factor defined by [1]:

$$\chi^2(\mathbf{p}) = \sum_{i=1}^{N} \frac{[I_{mod}(i, \mathbf{p}) - I_{exp}(i)]^2}{\sigma(i)^2} \tag{B.1}$$

where $\mathbf{p}$ are the free parameters, $\sigma(i) = \sqrt{I_{exp}(i)}$ is the statistical error and the numerator corresponds to the squared residuals $R(i) = I_{mod}(i, \mathbf{p}) - I_{exp}(i)$, with I_{mod} the model function and I_{exp} the measured data. The adjustment consists of finding the minimum of $\chi^2(\mathbf{p})$. It is clear that a model with a larger number of adjustable parameters leads to a better fit. To optimize the number of parameters, it is useful to define a χ^2 normalization according to:

$$\chi^{2*}(\mathbf{p}) = \frac{\chi^2(\mathbf{p})}{N - P} \tag{B.2}$$

where P is the number of independent parameters in the model function I_{mod} and N is the number of points in the spectrum. This χ^{2*} should tend to 1 if the noise is statistical.

Once a minimum of χ^2 or of χ^{2*} is reached, it is useful to study the R residuals or the residuals normalized to σ to observe if they are distributed statistically. Figure B.1 shows an example of such a representation. This is the core level fit of Na(110) with a surface-related doublet and a second bulk-related doublet. The residuals only show statistical noise.

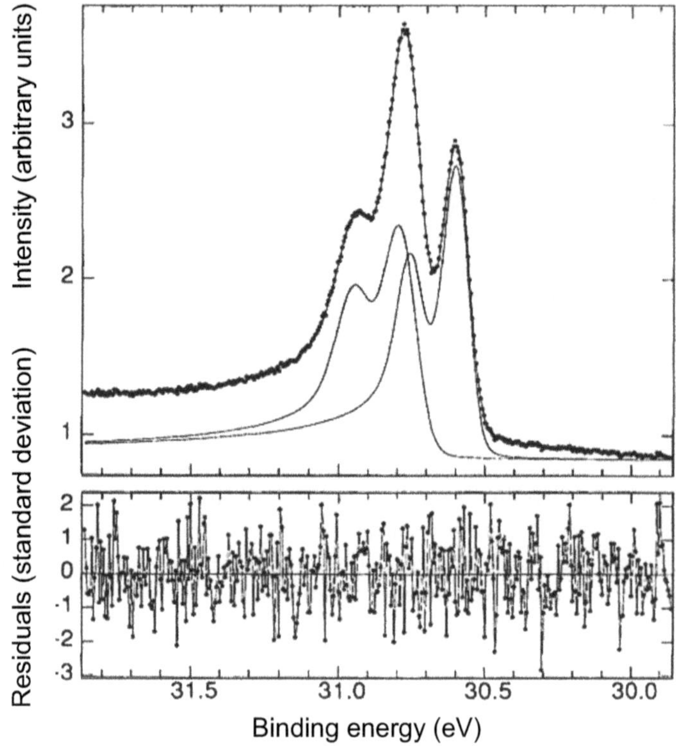

Figure B.1 *2p core level of Na(110) at 80 K (after [2]). Reprinted figure with permission from D. M. Riffe, G. K. Wertheim, and P. H. Citrin, Phys. Rev. Lett. 67, 116 (1991). ©1991 by the American Physical Society.*

The previous residual representation is intuitive and allows to find the regions where the agreement is poorer. Moreover, correlations in residuals can be quantified through the Abbe number n_A:

$$n_A = \frac{1}{2}\frac{\sum_{i=1}^{N-1}[R(i+1) - R(i)]^2}{\sum_{i=1}^{N}[R(i)]^2} \tag{B.3}$$

If $n_A = 0$, residuals are correlated and the model is inappropriate; for $n_A = 2$, residuals are anti-correlated and for $n_A = 1$, residuals are statistically distributed, indicating a good fit.

Once the correct fit is achieved, the errors of the different independent parameters must be estimated. If the set of parameters **p** minimizes χ^2, the function can be approximated by a parabola close to the minimum. Then, by varying the parameter p_i by Δp_i intervals, χ^2 reaches its initial value $\chi^2(p_i)$ plus 1 [1]:

$$\chi^2(p_i + \Delta p_i) = 1 + \chi^2(p_i) \tag{B.4}$$

On the other hand, the Fourier expansion of the left-hand side of B.4 is:

$$\chi^2(p_i + \Delta p_i) = \chi^2(p_i) + \frac{\partial^2 \chi^2}{\partial p_i^2} |_{p0} (\Delta p_i)^2 \tag{B.5}$$

since the first derivative is zero at the minimum. From the two previous relations, we obtain as the independent parameters:

$$\Delta p_i = \left(\frac{2}{\frac{\partial^2 \chi^2}{\partial p_i^2}} \right)^{1/2} \tag{B.6}$$

which allows to determine the error of each parameter. However, if the parameters are coupled, it is necessary to determine the errors from the matrix:

$$\frac{\partial^2 \chi^2}{\partial p_i \partial p_j} \tag{B.7}$$

Bibliography

[1] R. Hesse, T. Chassé, P. Streubel, and R. Szargan, Surf. Interface Anal. **36**, 1373 (2004).
[2] D. Riffe, G. Wertheim, and P. Citrin, Phys. Rev. Lett. **67**, 116 (1991).

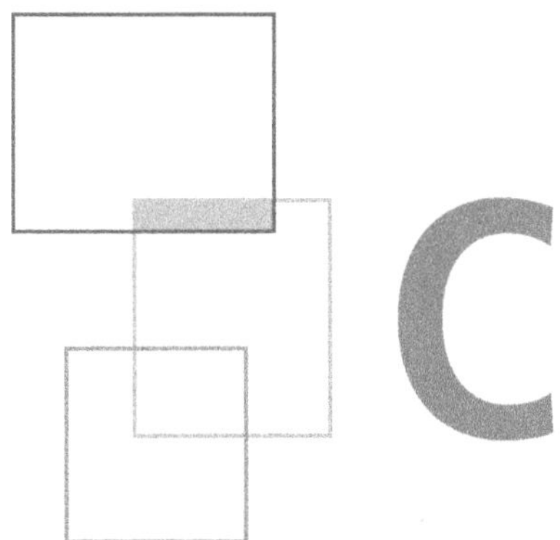

Fermi level determination

The experimental determination of the Fermi level (E_F) of a metal in photoemission is trivial. E_F corresponds here to the threshold observed at high kinetic energy and associated with the last occupied states. In the case of the Shockley state of gold (111), this surface state is metallic and photoemission allows us to observe the occupied part of it. Figure C.1 (a) shows a gray-scale intensity map where the x-axis corresponds to the emission angle and the y-axis to the kinetic energy of the photoelectrons. The separation between the occupied and unoccupied states, easily identifiable on such a figure, allows qualitatively to observe the kinetic energy associated with the Fermi level E_F. This representation is not the most appropriate for precisely determining the position of E_F. Indeed, this position in a gray-scale representation depends on the contrast or on the color scale. For a quantitative analysis, a serie of spectra is represented, each spectrum corresponding to a detection angle (Fig. C.1 (b)). These spectra allow us to follow with precision the dispersion of the state towards the Fermi energy. The red spectrum corresponds to the Fermi wave vector (k_F). E_F is precisely determined from the inflection point of the spectral weight between the occupied states and the empty states. This point can be obtained by adjustment of the spectrum at k_F or the integral of all the spectra measured by the Fermi-Dirac function (Fig. C.1(c)).

The situation is far less clear in a semiconductor, where the Fermi level is in the gap, its exact position depending on the doping of the sample. There is therefore no threshold at the measured image to determine the Fermi level. On the other hand, in semiconductors, the Fermi level of the surface and the bulk are not necessarily separated from the maximum of the valence band by the same energy. In the

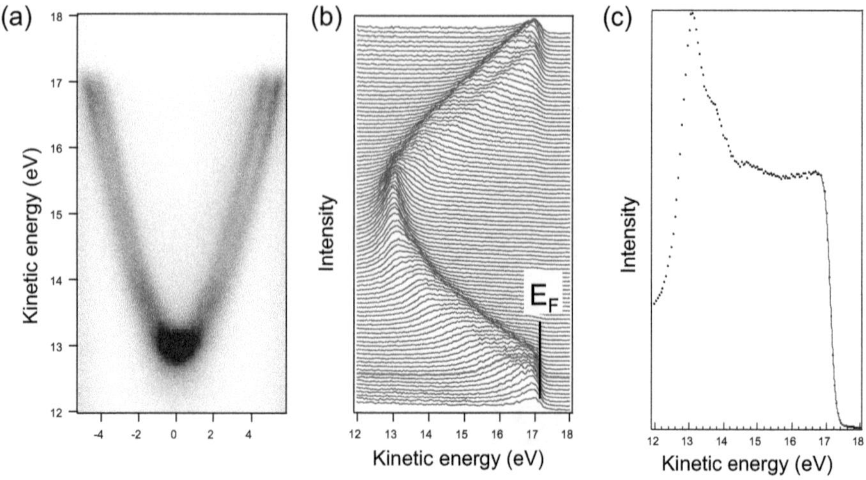

Figure C.1 *Fermi level of a metal. (a) Au (111) band structure. Intensity map as a function of the emission angle and the kinetic energy. (b) Set of spectra corresponding to the previous image. The highlighted spectrum shows the boundary between the occupied and unoccupied states and determines the Fermi level E_F. It corresponds to the Fermi k_F wave vector. (c) Adjustment of the threshold corresponding to the integral of all the spectra in (b).*

bulk, the Fermi level depends only on temperature and doping. On the surface, the Fermi level is often fixed by the metallic surface states. Indeed, if the surface band is metallic, the Fermi level of the surface is the boundary between the occupied and unoccupied surface states (we speak then of *pinning* of the Fermi level). Therefore, the Fermi level position at the surface is determined by the metallic surface state. The valence or conduction bands on the surface and in bulk will therefore be at different energies. Figure C.2 illustrates how the position of the maximum of the valence band and the minimum of the conduction band evolve as a function of the depth in the material. The bands are thus curved when approaching the surface, with a curvature that depends on p or n dopings. However, it is not this surface property of the semiconductors that complicates the experimental determination of E_F by photoemission but rather the effect of surface photovoltage [2, 3]. This effect can change the amplitude of the band bending depending on the photon flux and the temperature, as if there was an applied voltage between the bulk and the surface. The actual position of E_F, *i.e.*, in the absence of photovoltage, cannot therefore be determined directly from a photoemission measurement. What is the physical mechanism of this photovoltage effect? In a semiconductor with a given band bending studied by photoemission, holes are created in the band and these holes always accumulate in the least bonded region of the valence band (Fig. C.3). This results in the total or partial compensation of the initial band bending. The photovoltage introduced in this way depends on the photon flux and the temperature (because of the screening dynamics). The boundary between occupied and unoccupied states of a metallic

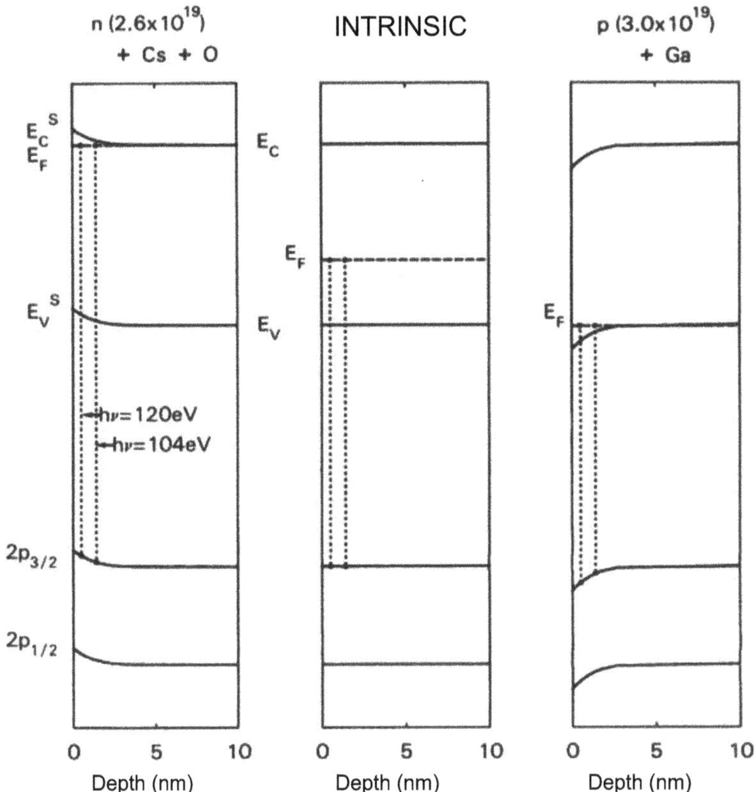

Figure C.2 *Evolution of the maximum of the surface valence band* E_V^S *and the minimum of the conduction band* E_C^S *as a function of the depth. The energies of the core levels also vary with depth. The Fermi level* E_F *of bulk and surface are at the same kinetic energy. The direction of the band bending depends on the doping type, n or p. Intrinsic samples have no band bending (according to [1]). Reprinted figure with permission from F. J. Himpsel, G. Hollinger, and R. A. Pollak, Phys. Rev. B, 28, 7014 (1983). ©1983 by the American Physical Society.*

surface state does not give here the actual position of the surface Fermi level in the absence of photon flux.

Determining the Fermi level of a semiconductor is therefore complex. The Fermi level of a semiconductor can a priori be obtained from the Fermi level of a metal in electrical contact with the semiconductor. In the absence of surface photovoltage, as bulk and surface are in equilibrium, E_F can be determined from a metal in electrical contact with the bulk, in order to determine the surface Fermi level. Similarly, the measurement of E_F of a metal evaporated on a semiconductor or on a system with a surface metallic band gives the Fermi level at the surface, which coincides with

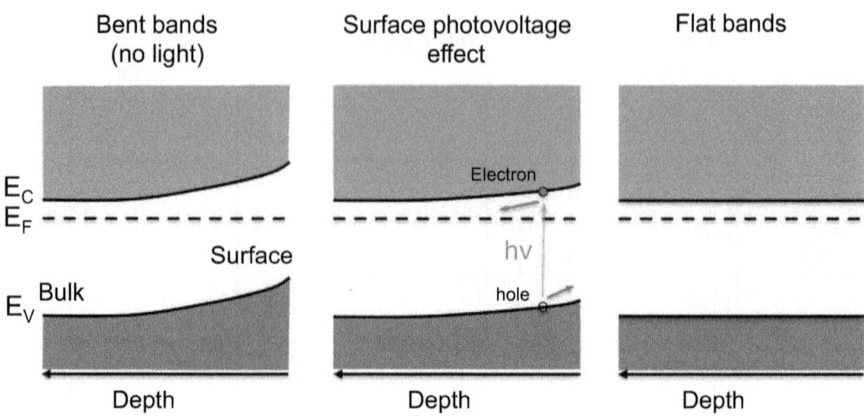

Figure C.3 *Position of the maximum of the valence band and the minimum of the conduction band as a function of the depth (n-doped semiconductor). (Left) semiconductor with band bending in the absence of illumination. (In the center) a high photon flux for a given temperature causes holes to accumulate near the surface. (On the right) a very high photon flux leads to a situation of flat bands, where the surface and the bulk are in equilibrium. For a p type semiconductor, the band bending is inverted. The movement of electrons and holes thus changes direction with respect to the n-type semiconductor.*

the bulk Fermi level in the absence of surface photovoltage. However, to use these methods, the effect of photovoltage must first be eliminated.

A first method consists of evaporating a metal or studying a metallic surface state on *n* and *p* doped substrates and analyzing the kinetic energy of the apparent Fermi level, since the displacement due to the photovoltage must be reversed with *n* or *p* dopings. Other possibilities are the use of a photon flux and a temperature which minimize the photovoltage effect or, on the contrary, maximize it so that the initially bended bands become flat and thus reach a controlled situation. The evolution towards flat bands can be easily verified by analyzing the evolution of the kinetic energy of the valence band (or of a core level, which evolves rigidly with respect to the valence band as a function of the photovoltage). When the position of the core levels no longer changes, the bands have reached the flat situation and the Fermi levels of surface and bulk are at the same energy.

Bibliography

[1] F. Himpsel, G. Hollinger, and R. Pollak, Phys. Rev. B **28**, 7014 (1983).
[2] J. Demuth, W. Thompson, N. DiNardo, and R. Imbihl, Phys. Rev. Lett. **56**, 1408 (1986).
[3] M. Alonso, R. Cimino, and K. Horn, Phys. Rev. Lett. **64**, 1947 (1990).

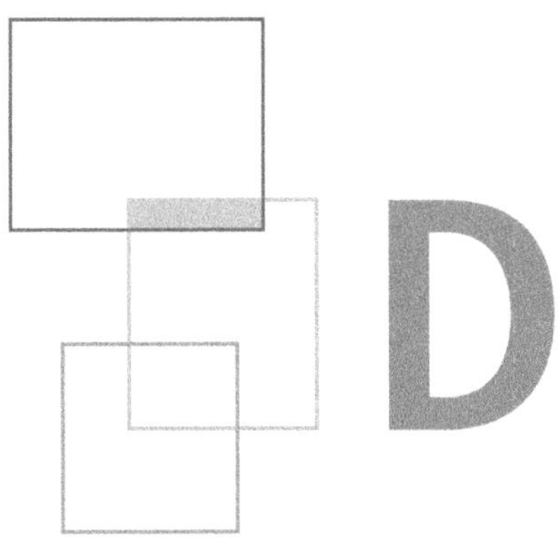

Acronyms

AES	*Auger Electron Diffraction*
AES	*Auger Electron Spectroscopy*
ARUPS	*Angle Resolved Ultraviolet Photoemission Spectroscopy*
ARPES	*Angle Resolved Photoemission Spectroscopy*
BIS	*Bremsstrahlung Isochromat Spectroscopy*
CCD	*Charge-Coupled Device*
CDAD	*Circular Dichroism in the Angular Distribution*
CHA	*Concentrical Hemispherical Analyser*
CIS	*Constant Intitial State*
CMA	*Cylindrical Mirror Analyser*
CRR	*Constant Retarding Ratio*
EDC	*Energie Distribution Curves*
ERC	*Electron Cyclotron Resonance*
ESCA	*Electron Spectroscopy for Chemical Analysis*
EXAFS	*Extended X-ray Absorption Fine Structure*
FAT	*Fixed Analyser Transmission*
FEL	*Free Electron Laser*
FWHM	*Full-Width at Half-Maximum*
HHG	*High Harmonic Generation*
IMFP	*Inelastic Mean Free Path*
IPES	*Inverse Photoemission Spectroscopy*
KDP	*Potassium dihydrogen phosphate*

LEED	*Low Energy Electron Diffraction*
MDC	*Momentum Distribution Curves*
MEM	*Maximal Entropy Method*
OPA	*Optical Parametric Amplifier*
PED	*PhotoElectron diffraction*
PEEM	*PhotoEmission Electron Microscopy*
PES	*Photoemission Electron Spectroscopy*
PhD	*Photoelectron Diffraction*
RA	*Regenerative Amplifier*
SASE	*Self-Amplified Sponteneuous Emission*
SEXAFS	*Surface Extended X-ray Absorption Fine Structure*
TOF	*Time-of-Flight*
UHV	*Ultra High Vacuum*
UPS	*Ultraviolet Photoemission Spectroscopy*
UV	*UltraViolet*
XPD	*X-ray Photoelectron Diffraction*
XPEEM	*X-ray PhotoEmission Electron Microscopy*
XPS	*X-ray Photoemission Spectroscopy*
XR	*X-rays*
2PPE	*Two Photon PhotoEmission*

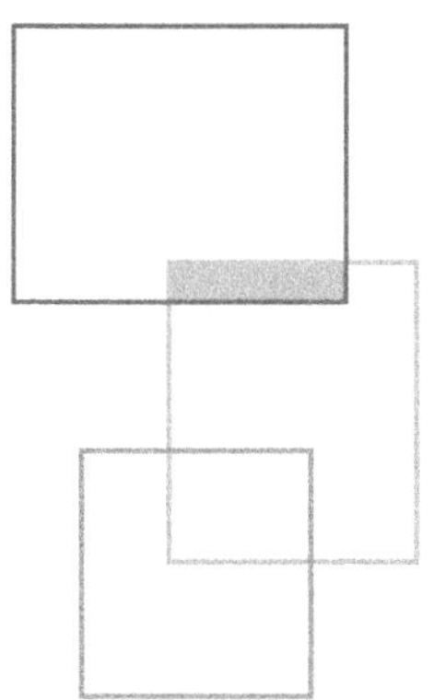

Index

A

B

C

D

E

F

G

H

I

K

L

M

N

O

P

Q

R

S

T

U

V

W

X

Z

www.ingramcontent.com/pod-product-compliance
Ingram Content Group UK Ltd.
Pitfield, Milton Keynes, MK11 3LW, UK
UKHW022050260726
13993UKWH00001B/33

9 782759 820658